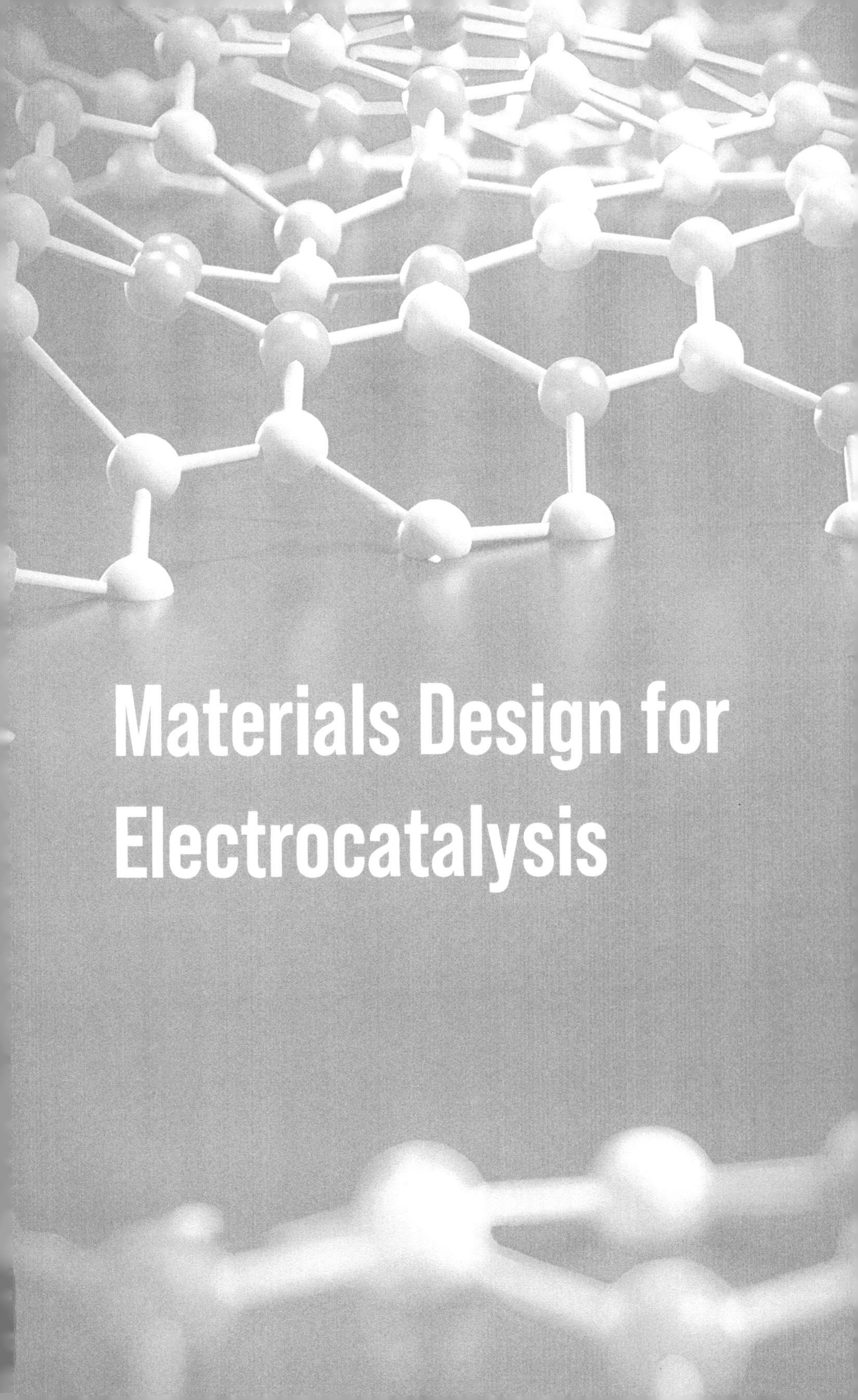

Materials Design for Electrocatalysis

Materials Design for Electrocatalysis

edited by
Sagarika Bhattacharya
Biswarup Chakraborty

JENNY STANFORD
PUBLISHING

Published by

Jenny Stanford Publishing Pte. Ltd.
101 Thomson Road
#06-01, United Square
Singapore 307591

Email: editorial@jennystanford.com
Web: www.jennystanford.com

British Library Cataloguing-in-Publication Data
A catalogue record for this book is available from the British Library.

Materials Design for Electrocatalysis

ISBN 978-981-5352-32-0 (Hardcover)
ISBN 978-1-042-02558-9 (eBook)

DOI: 10.1201/9781042025589

Dedicated

to our beloved parents

Contents

Preface

Although electrochemistry is a well-established field, with roots dating back to the 1790s, it has become an integral part of both fundamental and applied research in chemistry, physics, materials science, and industrial innovation. Modern applications of electrochemistry include electrocatalysis for sustainable development, such as water splitting for hydrogen production, CO_2 reduction to carbonaceous fuels, and N_2 or NO_x reduction to ammonia. Electrocatalytic and electrochemical processes are also being explored for the removal of microplastics and organic and inorganic contaminants from wastewater. Moreover, significant progress has been made in the development of fuel cells and Li-ion and/or various metal batteries. In this book, we aim to showcase the development of electrocatalytic materials for the diverse electrochemical studies stated above. Activity, stability, and selectivity in electrocatalysis are the key challenges. Physicochemical characteristics and product selectivity during electrocatalysis depend on several factors, including the synthesis process and structure-function relationships. However, sustainability, cost-effectiveness, and eco-friendly materials are the central challenges in materials development for electrocatalysis. In this book, we have attempted to introduce the readers to the diverse categories of materials discovered in the past few decades for the development of electrodes. There are nine chapters in this book summarizing the crucial steps in designing electrocatalysts, including synthetic techniques for different electrocatalytic materials, and microscopic and spectroscopic characterization techniques commonly employed to understand the surface and bulk structure of the materials or electrodes. We have included a separate chapter to explain the details of the fabrication of materials on the electrode surface. In two chapters, we have given glimpses of some specific electrocatalytic reactions. We have segmented these two chapters according to their mode of operations, like cathodic reduction reactions, hydrogen evolution reaction, CO_2 reduction, nitrogen/nitrate reductions, and organic reductions. In another analogous chapter, anodic reactions like water oxidation and organic biomass oxidation are summarized.

There are several invaluable books available in the market covering the fundamentals of electrochemistry. However, in this book, our target is to deliver intuitive and practical knowledge about electrocatalysis, emphasizing "how to design a catalyst or how to perform electrocatalysis in the laboratory." This book will be an ideal reference for first-year PhD students or beginners in any branch of chemistry who have an interest in the applications of materials chemistry for electrochemical energy conversion. This book can also be an ideal reference book for the master's-level students of universities or colleges who are keen to learn different domains of research in material chemistry or catalysis. This book can also provide guidance to early-career teachers of universities, research institutions, and national laboratories to teach the aligned courses to postgraduate students or research scholars. We have also covered up-to-date literature in different chapters, which will be very helpful to researchers to follow the recent discoveries in materials science and electrocatalysis.

We immensely thank all the authors who have generously contributed their subject knowledge in the form of chapters. All the corresponding authors have several years of research and teaching experience in top-tier teaching or research institutes in India, such as the Indian Institute of Technology, BITS Pilani, Amity, and Chanakya University. They are active researchers in the area of materials chemistry and electrocatalysis. We would appreciate readers providing their valuable feedback and suggestions.

Biswarup Chakraborty
Sagarika Bhattacharya
June 2026

Chapter 1

Classifications of Electrocatalytic Materials

Mrinal Kanti Adak,[a] Dolly Singh,[b] and Sagarika Bhattacharya[b]
[a]*Department of Chemistry, Indian Institute of Technology Delhi, Hauz Khas, New Delhi 110016, India*
[b]*Department of Chemistry, National Institute of Technology Raipur, Chhattisgarh 492010, India*
mrinal@iitd.ac.in

1.1 Historical Landscape of Electrochemistry

Electrochemistry is a significant branch in the current age for versatile applications, especially in the field of green energy. The beneficial practical application makes the techniques widely interesting to the scientific community. The development of the science of electrochemistry has a historical background, and some milestones are showcased in the following way [1–3]:

Materials Design for Electrocatalysis
Edited by Sagarika Bhattacharya and Biswarup Chakraborty

ISBN 978-981-5352-32-0 (Hardcover), 978-1-042-02558-9 (eBook)
www.jennystanford.com

DOI: 10.1201/9781042025589-1

1570: William Gilbert uses the term "electricus" for static electricity.
1793: First battery by Alessandro Volta.
1800: Nicolson used Volta's battery for water splitting.
1806: Sir Humphry Davy proposed the electrical theory of chemical affinity.
1835: Michael Faraday defined the terms of electrochemistry.
1839: W. R. Grove defined the first practical fuel cell.
1890: Calomel electrode by Wilhelm Ostwald.
1893: Discovery of the electron by J. J. Thompson.
1920: Sabatier principle for OER.
1939: Electrochemical Society of Japan.
1949: International Society of Electrochemistry.
1983: Binary alloy for ORR.
2002: Alloy of Pt with transition metals.
2014: Defect engineering for OER.
2016: Proposed LOM process in theory.
2017: Pt single-atom catalyst.
2019: Single atom alloy for OER, HER.

The ongoing era is facing challenges of energy scarcity and environmental degradation due to rapid urbanization, exhaustion of fossil fuels, and unrestrained industrialization. Therefore, competent green renewable energy technologies are urgent for environmental sustainability and clean energy production. Electrocatalytic techniques show promising potential in clean energy production and storage.

The performance of electrocatalytic materials relies on several factors, including activity, selectivity, durability, and cost-effectiveness. In this regard, controlling factors like surface active sites, electrolyte solution nature, and electrode potential are the vital points to achieve the target. The sluggish reaction kinetics hinder the path to achieve the goal of ultimate reactivity, wherein restricted mass transfer, complex electron transfer, and adsorption or desorption of active species are considered key factors [4]. So, the regulation of electrons at the active sites of the electrocatalyst's surface is crucial. In such cases, the nature of electrocatalytic materials is the most important factor in electrocatalytic factor in electrolysis. The selection of materials for the specific catalytic reaction is based on their physical, chemical, mechanical, and durability properties. The choice of materials should also be low-cost, environmentally sustainable, and recyclable in nature for commercial applications.

This chapter discusses the types of heterogeneous materials used for electrocatalytic applications. The factors related to these materials for efficient electrocatalytic activity are discussed to find out the key findings of the materials' activity.

1.2 Applications of Electrocatalysis

Electrocatalysts and their applications are dependent on the complex pathways of electrocatalytic reactions. The adsorption of the reactants, intermediate formation, desorption of the products from the catalyst surface, and their relative interaction play a vital role in generating the ultimate products. In the pursuit of renewable, clean energy sources through metal-air batteries and fuel cells, redox reactions like oxygen evolution reaction (OER), hydrogen evolution reaction (HER), and oxygen reduction reaction (ORR) play a crucial role [5]. The other part of electrocatalytic applications is reducing greenhouse gases and generating valuable feedstock. In this regard, electrochemical CO_2 reduction (CO_2R) to valuable products is an advanced solution for CO_2 utilization to close the carbon cycle [6]. The electrooxidation of organics to value-added products through low-energy consumption is another advantage of this technology [7].

Oxygen evolution reaction. In a water splitting electrolyzer, the application of a certain external potential leads to the OER at the anode, and oxygen gas is produced. The four-electron oxidation is associated with it and makes it more complex with a higher overpotential than the cathodic reaction. The OER pathway depends on the pH of the electrolytes, and a general reaction can be expressed as follows [8]:

In acidic or neutral conditions:

$$2H_2O \longrightarrow 4H^+ + 4e^- + O_2 \tag{1.1}$$

In alkaline conditions:

$$4OH^- \longrightarrow 2H_2O + 4e^- + O_2 \tag{1.2}$$

In acidic electrolyte medium, H_2O adsorption on the catalyst surface is crucial to boost the OER performance. In an alkaline medium, hydroxyl ion adsorption is the first step, followed by different intermediate formations. The free energy for each elemental state in OER is close to 1.23 V [8].

Hydrogen evolution reaction. HER takes place at the cathode of the electrolyzer, and water is split into H_2 from the surface of the electrocatalyst under a certain external potential. The water splitting reaction pathway depends on the pH of the electrolytes as follows:

In acidic conditions:

$$2H^+ + 2e^- \longrightarrow H_2 \tag{1.3}$$

In neutral or alkaline conditions:

$$2H_2O + 2e^- \longrightarrow H_2 + 2OH^- \tag{1.4}$$

In the HER process, three reaction steps are considered for the generation of H_2, such as the Volmer step, Tafel step, and Heyrovsky step. The Volmer step is responsible for the hydrogen adsorption (H_{ads}) on the catalyst surface for both acidic ($H^+ + e^- \longrightarrow H_{ads}$) and alkaline/neutral conditions ($H_2O + e^- \longrightarrow H_{ads} + OH^-$). The Heyrovsky ($H_{ads} + H_{ads} \longrightarrow H_2$/$H_{ads} + H_2O + e^- \longrightarrow H_2 + OH^-$) and Tafel step ($H_{ads} + H_{ads} \longrightarrow H_2$) lead to the final step for the H_2 generation. So, the HER proceeds through the Volmer–Heyrovsky, Volmer–Tafel, or Volmer–Heyrovsky/Tafel pathway. The free energy for an optimal HER catalyst should follow an almost zero value for H_{ads} and H_2 desorption [8].

Oxygen reduction reaction. It is a cathodic half-reaction in fuel cells or metal-air batteries that follows sluggish dynamics with high overpotential. The ORR follows four- and two-electron transfer process, where the electron ORR is considered suitable for fuel cells and metal-ion batteries. In other cases, two electron ORR pathway is important in wind or solar energy applications for hydrogen peroxide production [9]. In acidic and alkaline media, the ORR process follows as [10]:

In acidic conditions:

$$O_2 + 4H^+ + 4e^- \longrightarrow 2H_2O \tag{1.5}$$

$$O_2 + 2H^+ + 2e^- \longrightarrow H_2O_2 \tag{1.6}$$

$$H_2O_2 + 2H^+ + 2e^- \longrightarrow 2H_2O \tag{1.7}$$

In alkaline conditions:

$$O_2 + 2H_2O + 4e^- \longrightarrow 4OH^- \quad (1.8)$$

$$O_2 + H_2O + 2e^- \rightarrow HO_2^- + OH^- \quad (1.9)$$

$$H_2O + HO_2^- + 2e^- \rightarrow 3OH^- \quad (1.10)$$

Electrocatalytic CO_2 reduction reaction (CO_2RR). In CO_2, carbon is present in its highest oxidation state of +4. So, the conversion of CO_2 to valuable products proceeds through the transfer of electrons to carbon, which is renamed as CO_2 reduction or CO_2 hydrogenation [6, 11]. In this regard, ECO_2R is more advantageous than other hydrogenation processes in terms of simple processes, ambient conditions, and quick production. The ECO_2R process occurs at the cathodic chamber and combines with electrochemical water splitting, which follows the general reaction as [6]:

$$xCO_2 + nH^+ + ne^- \longrightarrow \text{product} + yH_2O \quad (1.11)$$

In the ECO_2R process, the OER takes place at the anodic chamber, and water is considered the only renewable and flexible origin of electrons and protons. In this regard, a suitable electrocatalyst is highly demanded to activate the inert CO_2 molecule for high-value products.

Electrooxidation of organics. Electrooxidation reactions (EORs) are important owing to their energy-saving products, which are highly beneficial than the sluggish OER. The presence of low theoretical voltages for organic EORs like urea oxidation reaction (0.37 V), hydrazine oxidation reaction (−0.33 V), ammonia oxidation reaction (0.06 V), and glucose oxidation reaction (0.05 V) makes them promising as an alternative to OER [7]. But the multi-electron process and slow kinetics hinder the EORs from gaining the expected outcome. During the electrocatalytic process, hydroxyl ions are adsorbed on the anode surface, which also transform into crystal lattice oxygen as follows [12]:

$$MO_x + H_2O \rightarrow MO_x[\cdot OH] + H^+ + e^- \quad (1.12)$$

$$MO_x[\cdot OH] \rightarrow MO_{x+1} + H^+ + e^- \quad (1.13)$$

These two forms of oxygen molecule interact with the catalyst surface adsorbed organic molecule through both direct and/or indirect oxidation processes.

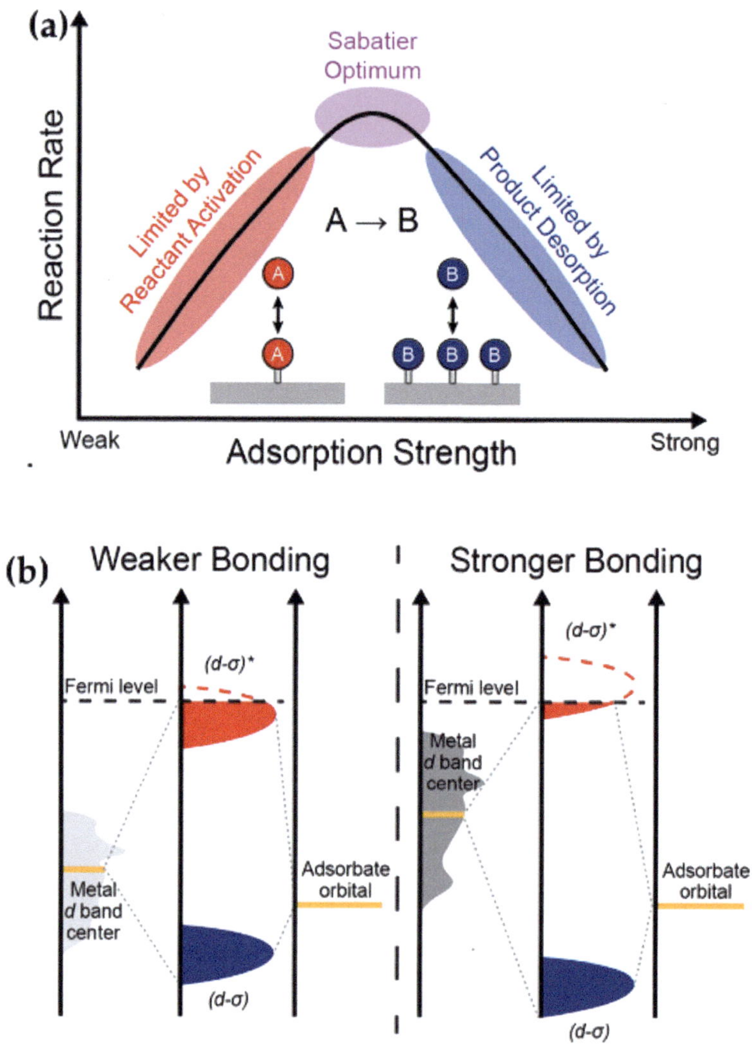

Figure 1.1 A schematic presentation about the (a) Sabatier principle and (b) the role of the Fermi level and d-band for bonding formation (reprinted from [13]).

1.3 Characteristics of Electrocatalytic Materials

Materials can be used for electrocatalytic applications, but not all materials are electrocatalytically active. So, to behave as a good electrocatalyst, materials should have good electrocatalytic reaction-related activity. In this respect, good adsorption and interaction characteristics with the substrate, products, solvent, and electrolyte ions at the materials' surface interfaces are crucial [13]. The ability of a material as an electrocatalyst can be explained through "d-band theory" and "Sabatier principle" (Fig. 1.1).

The surface adsorption activity of the electrocatalyst can be explained by gas-phase adsorption related to d-band theory. The Fermi level position of the electrocatalyst materials (d-band position) and the valence orbital position of the adsorbate for the bonding and antibonding hybrid orbitals formation decide the fate of the electrocatalyst's activity [14, 15]. According to the Sabatier principle, an ideal catalyst will bind to reactants at an intermediate strength. The principle is based on thermodynamics, where the Brønsted (Bell)–Evans–Polanyi (BEP) relationship suggests that the free activation energy is directly proportional to the reaction free energy [16]. In such cases, the substrate's nature, the catalyst's interaction, the cell setup, the nature of the electrolyte, ambient conditions, and the applied potential are also important.

1.4 Classification of Materials

The selection of materials as electrocatalysts is an important task in developing clean energy technology for green sustainability. The electrocatalytic reaction involves a solid-phase electrode, a liquid-phase electrolyte, and a gas-phase diffusion, where the multistep electron and proton transfer process takes place through several intermediate steps. Such complex reactions hinder the choice or design of perfect materials for suitable activity. In an effective and fundamental path, materials can be tuned through atomic arrangements, where strong intrinsic activity with flexible control may influence the materials for expected parameters.

1.4.1 Noble Metal–Based Materials

Noble metal–based materials are considered benchmark catalysts in electrochemical reactions. Platinum group metals (PGMs) are examples of noble materials, including Pt, Ru, Rh, Ir, and Pd. In the electrocatalytic investigation, Ru and Ir are used as commercial OER catalysts, and Pt acts as an HER catalysts, owing to its high current density and fast kinetics [17].

However, benchmark noble metal catalysts should be replaced by earth-abundant, lower-cost materials owing to higher price, low abundance, dissolution, and agglomeration of the noble metals, which leads to their restriction in large-scale applications. So, the replacement of noble metals with earth-abundant materials is very significant and urgent to avoid energy-related complications. Considering the activity of the noble materials, several steps are taken to enhance the stability of the materials under harsh conditions. In such cases, the reduction of the amount of noble metals with nano-structure design, single atom catalyst, alloy formation, modification with support, and heteroatom loading shows constructive effort for commercial applications. The above modifications may vary the structural co-ordination of the PGMs which affects the d-band structure and Fermi level positions for better electrocatalytic activity.

1.4.2 Non-Noble Metal Catalysts

Noble materials–based electrocatalysts are excellent for electrocatalytic applications but have a high cost, non-abundance, and lower stability making a barrier for large-scale delivery processes. Therefore, the earth's abundance of non-noble metals is considered and preferred for the futuristic goal of green energy and electrochemical conversion. The researchers also study carbon-based, intermetallic materials, nonmetals, and their heterostructures to find suitable electrocatalysts. The types of non-noble materials and their source of activity in terms of structural prospectus for electrocatalytic applications are discussed below.

I. Alloys materials:

Alloys are single-phase materials where different metal atoms/nonmetals form a solid solution with stable phase and atoms placed in the random lattice sites. They can be demonstrated as a mixture of two or more metals or intermetallic compounds as demonstrated in Fig. 1.2 [18].

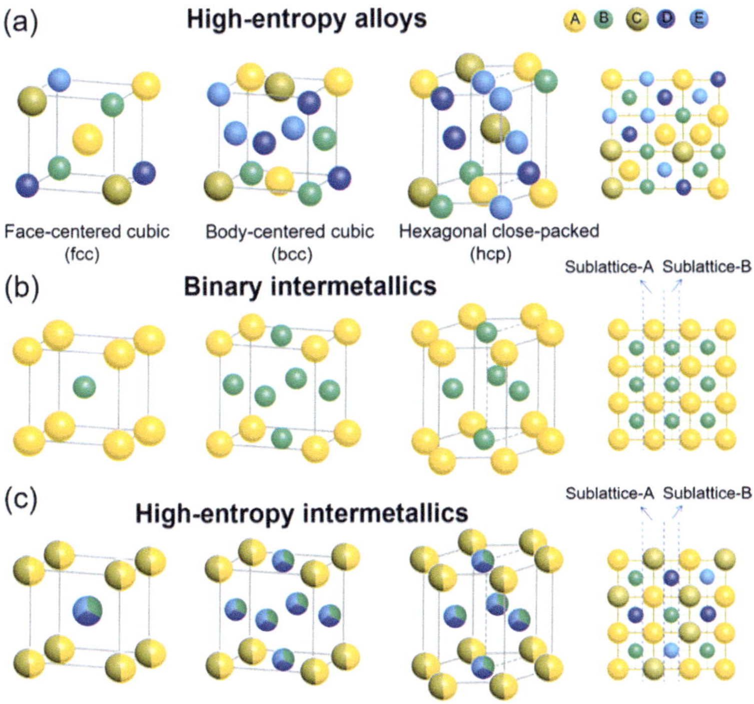

Figure 1.2 Schematic presentation of crystal structure and lattice structure models for (a) high-entropy alloys, (b) binary intermetallics, and (c) high-entropy intermetallics (reprinted from [18]).

For electrocatalytic reactions, the electrocatalytic activity of nickel is enhanced by alloy formation with Mo, Fe, Co, and Cu metals. In the hydrogen oxidation reaction (HOR), Ni surface acts as an active site for hydrogen adsorption, where Mo influences the electric properties of Ni by reducing its binding energy for better HOR activity [19]. However, the dissolution of Mo during the electrocatalytic process in the electrocatalytic medium is a crucial problem for the stability of the catalyst [20–22]. For the NiCu alloy, the maximum HOR activity is observed for 40% Cu incorporation in nickel, owing to the role of copper in improving the intrinsic and antioxidation properties of Ni [23]. For nitrate reduction, Wang et al.

choose Ni, Nb, and Ti metals for the formation of single atom alloys with Cu from 3d, 4d, and 5d transition metals based on their stability, nitrate ion adsorption, selectivity, and activity [24]. Ni/Cu(111) shows efficient results in terms of both activity and selectivity for nitrate to ammonia conversion than other single atom alloys owing to the upgrading adsorption ability of the N-contain intermediates in the presence of Ni. In a comparative study among Ni/C, Ni_3Co_1/C, Ni_3Fe_1/C, and Ni_3Cu_1/C, the Fe-based catalyst displayed higher -OH adsorption than others in an alkaline medium and detected in exchange current density of HOR activity. Elena et al. reported a dissolution-precipitation mechanism for the above cases which inactive the catalysts for HOR at 0.7 V_{RHE} potential due to oxidation of Ni to Ni^{2+} [25]. Hexagonal close-packed NiFe nanoparticles are decorated on the nitrogen-doped carbon shell and the related OER activity shows 226 mV overpotential at 10 mA cm^{-2} current density. An N-doped carbon shell's presence influences NiFe's electronic properties and shows better activity than bare NiFe [26]. A trimetallic NiCoSn alloy shows better stability than commercial Pt/C catalysts during oxygen reduction and methanol oxidation reactions (MORs), owing to the synergistic effects of Ni, Co, and Sn metals that enhance large catalytic surfaces, active sites, and electron transport facilities. The poison tolerance activity of the NiCoSn is also more advanced than that of the benchmark catalyst, indicating the suitability of the alloy [27]. Nanoporous NiMnFeMo exhibits ultrahigh catalytic activity as both anode and cathode materials during electrocatalytic water splitting. The alloy compound delivers a 1000 mA cm^{-2} current density at 290 and 570 mV overpotential for HER and OER, respectively. Ni acts as a reaction center in the alloy, where other metals control electronic properties and enhance hydrogen adsorption capability [28]. High valent Mo and W doped and thermal annealing on Co_2FeAl alloy show superior HER activity under alkaline conditions. $Co_2FeAlMo_{1.5}$ annealed at 750°C, and the resultant electrocatalyst shows a 71 mV overpotential at a 10 mA cm^{-2} current density. The doping of high valence metal ions influences the electronic and crystal structure of the alloy, where annealing leads to improvement in crystallinity and morphological characteristics, which is reflected in the electrocatalytic performance [29]. Hasannaeimi et al. utilized an amorphous $Ni_{80}P_{20}$ alloy for electrocatalytic MOR. The strong d-orbital contribution by Ni at the

Fermi level and charge distribution at sites by high electronegative P account for high MOR activity [30].

Along with the binary or tertiary alloys, there are high-entropy alloys (HEAs), where five or more metals take part in the alloy formation. Cu, Ni, Mo, W, and Co metals are used to form CuNiMoWCo HEAs on the Cu foam surface. The HER activity indicates a 22 mV overpotential at 10 mA cm^{-2}, which shows superior activity than Pt/C. The presence of different metal d-orbitals causes orbital overlap and electronic interaction, which influence the adsorption capabilities of surface species [31]. The OER activity of nitrogen-doped FeCoNiCrN alloy shows a 269.7 mV overpotential at 10 mA cm^{-2} but is unable to act as an active cathode. The involvement of nitrogen toward the FeCoNiCr lattice causes lattice distortion that generates tensile strain. The improved catalytic performance is generated from the synergistic effect of different metals, their oxidation state, and electronegativity [32]. In another study, hydrothermally synthesized S-doped NiCoFeCu nanoparticles show OER activity under alkaline conditions with a 261 mV overpotential (10 mA cm^{-2}). The sulfurization process leads to shifting of the valence states of constituent metals that indicate distortion in the crystal structure, as reported for the nitrogen-doped alloy, and activates the materials for better electrocatalytic activity. Such types of electron density distribution under sulfurization form metal oxyhydroxide states and act as active sites [33]. HEA can be formulated as 12-component (MnNiCuCoVFeMoPdPtAuRuIr), 13-component (AlAgAuCoCuFeIrMoNiPdPtRhRu), 14-component (AlAgAuCoCuFeIrMoNiPdPtRhRuTi), or 16-component (MnNiCuCoVFeMoCrPdPtAuRuIrAgRhOs) alloys, where arc melting and melt spinning techniques are used for the targeted products [34].

II. Intermetallic materials:

Intermetallic materials are formed from two metal elements or from one metal with another metalloid through integer ratios [35]. But the main difference with the solid-state alloy is that component metals demonstrate substantial different characteristics which comprise individual crystal structure and high-ordered atomic arrangements [36]. The general formula of the intermetallic materials is represented as MnNn, where M is the active metal, N is the counterpart, and m, n are integers. Intermetallic materials show different electronic and crystal structure activity than their constituent elements, leading to new structural and catalytic activity. Long-range ordered NiZn nanoparticles shows intrinsic OER activity by the synergistic role of Ni and Zn at the atomic level. The presence of Zn in the Ni influences its electronic structure and acts as a catalytic promoter. Simultaneously, the 2p and valence shell of Ni is shifted toward Zn upon intermetallic formation, leading to a slightly positive nature of the Ni center [37]. Ma et al. analyzed 21 intermetallic compounds based on Sb and Bi (Bi based: Ag_3Bi, Au_2Bi, CuBi, PtBi, PdBi, $IrBi_2$, $RhBi_2$, $NiBi_3$, $CoBi_3$; Sb based: $AuSb_2$, Ag_3Sb, Cu_2Sb, PdSb, PtSb, FeSb, $IrSb_2$, RhSb, NiSb, $OsSb_2$, RuSb, CoSb) to check their novelty in the electrocatalytic nitrate reduction (NO_3RR) [38]. Among them, CoSb is chosen owing to the equilibrium among the intermediates related to NO_3RR for the minimization of energy barrier. The outstanding activity of hexagonal CoSb is reflected as 96.3% Faradaic efficiency, 89.1% NH_3 selectivity, stable nature, and 11.88 mW cm^{-2} power density in Zn-NO_3^- battery. Yang et al. synthesized Brewer intermetallic Mo_6Co_7, W_6Co_7, and W_6Fe_7 for electrocatalytic HER and OER activity under alkaline conditions [39]. The intermetallic follows the activity order as follows $Mo_6Co_7 > W_6Co_7 > W_6Fe_7$ in both cases. In case of individual metal atoms, the activity follows as Co > Fe > Mo > W. The dissolution of Mo in the electrolyte through the formation of molybdate ion and the formation of cobalt hydroxide and oxyhydroxide species at the electrocatalyst surface leads to the generation of electrocatalytic active nature. In a comparative HER analysis between Co_7Mo_6 and Fe_7Mo_6, the first one show higher activity owing to stronger electronic interaction between Co and Mo that modifies the optimum adsorption/desorption of the reaction intermediates [40]. Nickel antimonide intermetallic electrocatalyst acts as bifunctional electrocatalyst for overall water splitting under alkaline conditions [41]. The observed overpotential for HER is ~119 mV at 10 mA current density, where for HER ~200 mV overpotential is marked at 50 mA. Kelvin probe force microscopy (KPFM) mapping analysis is used to characterize catalytic activity. At high negative potential, the work function of the catalyst increased that indicates a weak M-H bond and helps HER process. Co_3W and WNi_4 intermetallic compounds are combined on nickel skeleton to form Co_3W-WNi_4/Ni heterostructure for electrocatalytic HER and

HOR under alkaline conditions [42]. The presence of electroactive multisite boosts the HER process and a low overpotential ~55 mV is required to reach 100 mA cm^{-2} current density. The heterostructure formation helps for the ordered structure as Co-W-Ni alloy that leads to the generation of Co-Ni-Co and Co-W-Co multisite for H* and OH* adsorption for the electrocatalytic process. 3D transition metal-based high entropy intermetallic FeCoNiAlTi (dealloying for 15h) presents 88.2 mV overpotential at 10 mA cm^{-2} current density with 40.1 mV dec^{-1} Tafel slope for HER [43]. The partial density of states of the d-orbitals is shifted from the Fermi level to more antibonding electronic states that weaken the hydrogen adsorption and enable the electrocatalytic HER process.

However, the shortcomings in alloy formation are their dissolution in harsh conditions or at high overpotentials. The other related problems include particle agglomeration, catalyst stability, and phase stability. In such cases, involving a suitable substrate and its interaction with the alloy particles is very important to avoid agglomeration problems [44].

III. Metal oxides (MO_2):

The tremendous use of non-noble MO_2s as electrocatalysts is reported in the literature owing to their structural flexibility, compositional diversity, low cost, earth abundance, easy synthesis, and environmental sustainability.

(a) Simple transition metal oxides (TMOs)

Due to their simple synthesis procedure, earth-abundant 3d TMOs are randomly used for electrocatalytic study. In such cases, TiO_2, ZnO, NiO, Fe_2O_3, CuO, MnO_2, Sc_2O_3, CoO, ZrO_2, WO_3, MoO_3, V_2O_5 metal oxides are highlighted by the researchers. TiO_2 exhibits three crystalline forms—anatase, rutile, and brookite—where the first one mostly shows photocatalytic efficiency, and the other two are related to electrocatalytic activity. Low-temperature oxide synthesis, easy nanostructure oxide synthesis with high surface area, conductive nature, strong promoting effect, hypo d-electron character, and durable characteristics make TiO_2 beneficial for electrocatalytic suitability [45].

The choice of ZnO as an electrocatalyst is based on its stability, high surface area, and electrode-electrolyte interactions. NiO mainly shows electrocatalytic OER activity, where Ni(II) is converted to Ni(III) under alkaline conditions. Fe_2O_3 is a highly abundant material in the earth's crust and shows an important role in terms of economic, stability, and catalytic prospects. The active redox characteristics of Fe_2O_3 enable its electrocatalytic activity, which plays a significant role in electrocatalytic applications [46].

CuO is the most optimal TMOs for the generation of hydrogen electrocatalytically [35]. Ghosh et al. reported higher electrocatalytic efficiency of chiral CuO than the achiral one, owing to the formation of a spin-parallel intermediate that occurs in chiral where it becomes forbidden for latter, leading to the production of H_2O_2 as a byproduct (Fig. 1.3) [47].

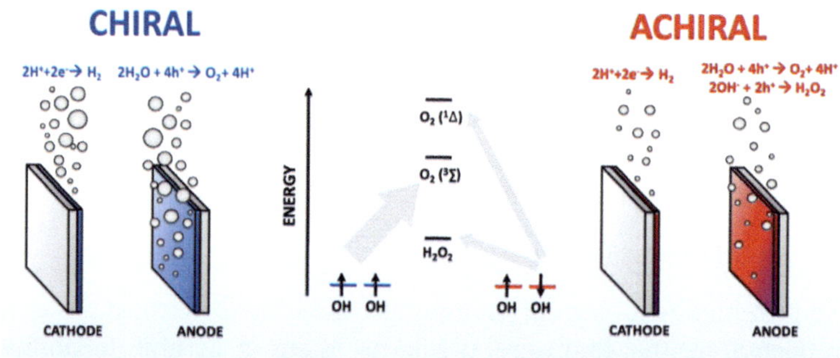

Figure 1.3 Formation of H_2O_2 as a byproduct due to the presence of achiral CuO molecules (reprinted from [47]).

(b) Perovskite

The high impact and powerful nature of perovskite oxides in materials science arises owing to their ordered atomic arrangement and crystalline nature. Perovskite oxides are represented in the ABO_3 chemical formula where A and B correspond to alkaline/rare-earth and transition metal cations, respectively. A general structure of perovskite is given in Fig. 1.4, which shows that "A" atom corresponds to the 12-fold oxygen coordination, and "B" follows the 6-fold oxygen coordination.

Nickel-based perovskite materials show importance in the electrocatalytic investigation not only from an economic view but also from structural factors, including mixed oxidation states, oxygen vacancies, and layered nickelate structures [49].

Hydrothermally synthesized $GdFeO_3$ perovskite materials act as bifunctional electrocatalysts for OER and ORR reactions [50]. In the OER study, a porous $GdFeO_3$ nanostructure provides diffusion channels for electrolyte penetration and exhibits a 340 mV overpotential noted at 10 mA cm^{-2} current density (η_{10}). Nanorod-shaped $LaNiO_3$ is studied for electrocatalytic purposes, where the influence of Ni^{2+} and Ni^{3+} is separately examined to identify the Ni-O bond role [51]. Ni^{3+} follows strong hybridization between Ni 3d and O 2p with enhanced electronegativity, and ionization energy than its lower oxidation state (Ni^{2+}), resulting in low activity for Ni^{3+} contain materials. In terms of e_g orbital filling at a low spin state, Ni^{2+} contains one more number of electrons in e_g orbital than Ni^{3+} that are related to observed OER activity [51]. The holes are introduced in the electrocatalytic structure for high performance and for this purpose, divalent (Sr^{2+}) ion is doped in the place of trivalent (La^{3+}) [52]. It helps double-exchange phenomena in the electrocatalysts for an efficient conduction pathway from O 2p to surface active site. Cobalt phosphate-modified $La_{0.7}Sr_{0.3}MnO_3$ electrocatalyst demonstrates good electrocatalytic OER activity with η_{10} = 220 mV. Oxygen vacancy creation is introduced in the OER active $LaCoO_3$ catalysts through the addition of Sr^{2+} at the "A" site and Ar plasma treatment ($La_{1-x}Sr_xCoO_{3-\delta}$) [53]. Such type of oxygen vacancy upgrade the OER activity through the increase of active sites and enrichment of intrinsic activity. In $La_{0.6}Ca_{0.4}CoO_3$, the "B" site modification by Mn leads to the tightening of M-OH bond and reduce electrocatalytic activity. The greater number of d-electron occupancy causes the presence of electrons at the antibonding orbitals that weakens the M-OH bond and helps in better activity. In a comparative study between Co and Mn, the first one present in +3 state where the later one shows mainly +4 oxidation state and leads to reduction in catalytic activeness [54]. Fe- and Co-based double perovskites Nd_2FeCoO_6, Eu_2FeCoO_6, Gd_2FeCoO_6, and Dy_2FeCoO_6 are studied to check their electrocatalytic activity [55]. Among them, Eu_2FeCoO_6 shows efficient electrocatalytic oxygen (η_{10} = 220 mV) and hydrogen evolution (η_{10} = 238 mV) activity in the presence of a 1 M NaOH electrolyte. The presence of large oxygen vacancies boosts the catalytic efficiency for Eu-contain materials than others, where di-μ-oxo-bridged of Co-Co, Fe-Co, or low coordinated Fe act as active sites. ORR-active double perovskite type $BaLnMn_2O_5$ (Ln = Y, Gd, Nd, and La) materials show predominant performance for larger radius of La, Nd than the smaller Gd, Y [56].

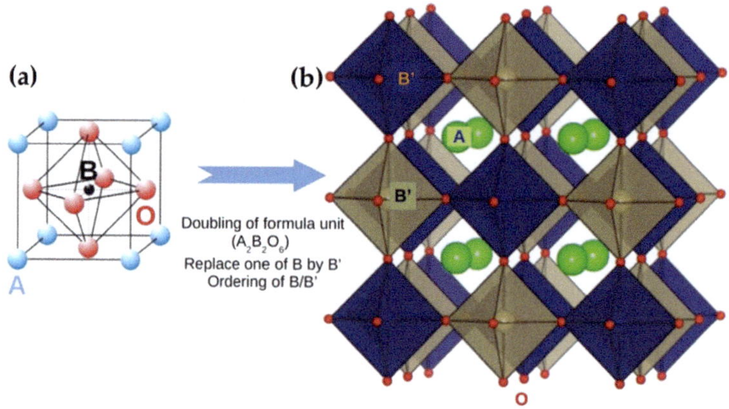

Figure 1.4 Schematic presentation of (a) perovskite and (b) double perovskite structure (reprinted from [48]).

(c) Spinel

Spinel is represented as the AB_2O_4 formula, where A represents a divalent cation in the tetrahedral site and B is a trivalent cation in the octahedral site. The presence of different metal ions with different electronic configurations, valence, oxidation state, radius, and spin state leads to a variety in electronic and chemical characteristics. The transformation of cations in the different sites of spinel is also noticed, and based on this, two types of spinels are identified, namely, normal spinel and inverse spinel. In a normal spinel, A^{2+} occupies the tetrahedral site and both B^{3+} occupy the octahedral site ($[A^{2+}]_{Td}[B^{3+}B^{3+}]_{Oct}O_4$), whereas in inverse spinel, A^{2+} and one of B^{3+} occupy in the tetrahedral site and the other B^{3+} occupies the octahedral site ($[A^{2+}B^{3+}]_{Td}[B^{3+}]_{Oct}O_4$) (Fig. 1.5).

In electrocatalytic applications, Fe-based p-type semiconductor $NiFe_2O_4$ inverse spinel is used vastly in diverse fields. OER study of $NiFe_2O_4$ under alkaline conditions leads to surface reconstruction that generates metal oxyhydroxide and multivalent metal ions on the catalytic surface that improve the reaction activity [59]. Phosphorus induced into the nanosheet of $NiFe_2O_4$ leads to higher electrocatalytic OER performance (η_{10} = 264 mV) than the undoped spinel. The involvement of P in the catalytic lattice helps in the lowering of the energy band gap and shifts the O 2p band center to the Fermi level, boosting the lattice contribution in electrocatalytic reaction through lattice oxygen oxidation mechanism (LOM) [60]. Tungsten ion doping in $NiCo_2O_4$ favors better electrocatalytic OER activity than $NiCo_2O_4$ alone [61]. The presence of W^{6+} initiates electron transfer facility from Co to W that modifies the valence state and e_g orbital filling of cobalt, which helps the generation of metal oxyhydroxide during the electrocatalytic reaction and improves intrinsic activity. According to the electronic orbital distribution diagram, the Ni^{2+} has a $3d^8$ valence state with $t_{2g}{}^6\ e_g{}^2$ state that creates electron–electron repulsion in t_{2g} orbital and affects the oxide ion bridge for indirect electron transfer from O to Co/W (Fig. 1.6).

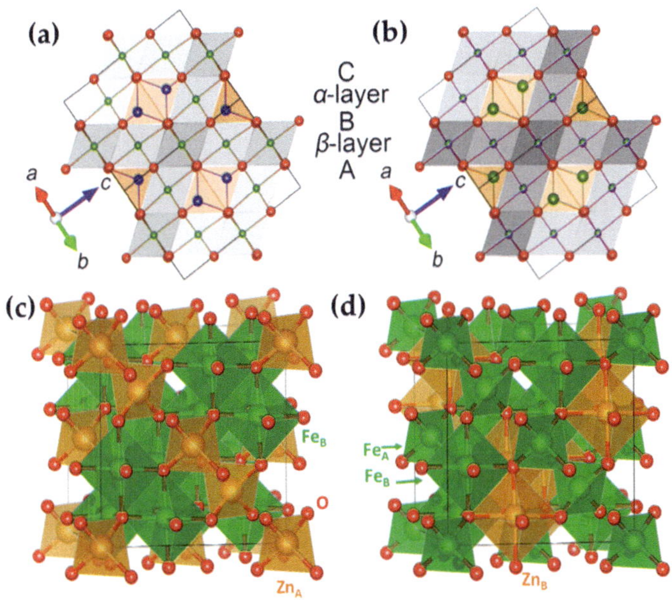

Figure 1.5 Crystal structure of (a) normal and (b) inverse spinel (reprinted from [57]). Crystal structure of zinc ferrite as (c) normal and (d) inverse spinel (reprinted from [58]).

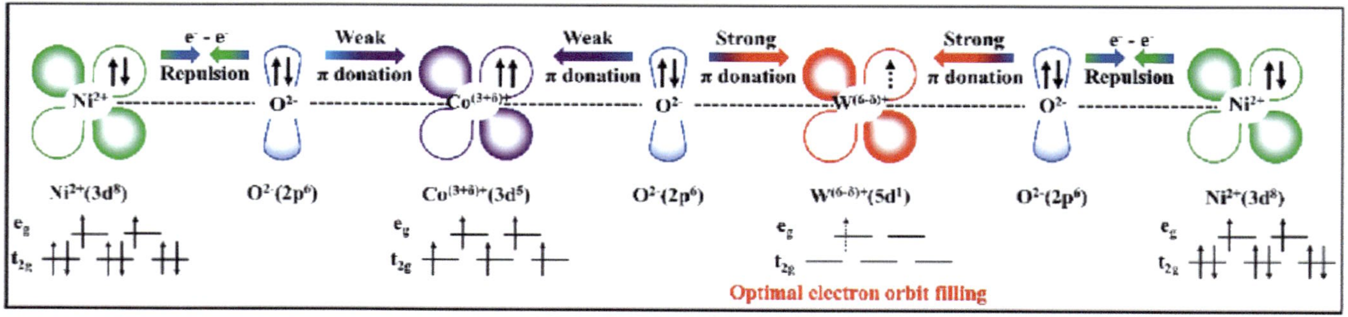

Figure 1.6 A schematic presentation about the charge distribution through electronic orbital for W-doped $NiCo_2O_4$ (reprinted from [61]).

The $Co^{(3+\delta)+}$ contains a t_{2g}^3 e_g^2 electronic configuration, as W captures its valence shell electron that favors the weak pi bond formation between O 2p and Co 3d. In the case of $W^{(6-\delta)+}$, the e_g^1 electronic configuration leads to a strong pi electron contribution to the oxide ion bridge. In this pathway, the e- transfer from Co to W improves intrinsic properties that are reflected in the interfacial OER activity. Copper doping at the "A" site of $NiCo_2O_4$ leads to the partial substitution of Ni^{2+}/Ni^{3+} by Cu^{+}/Cu^{2+} that enables new active sites with lowering of the electronic band gap [62]. The presence of Cu as a dopant in the lattice structure improves its conductive nature with the generation of oxygen vacancies, which are beneficial for better hydrogen adsorption and related HER activity. Hierarchical porous $Ni_{0.6}Fe_{2.4}O_4$@NC synthesized from NiFe-based metal-organic framework (MOF) exhibited both ORR and OER electrocatalytic activity and durability under alkaline conditions [63]. Here, the presence of N-doped graphitic carbon protects the spinel structures and reduces aggregation probability. In the lattice structure, defect formation through the Fe^{3+} and Ni^{2+} vacancies leads to the modification of surface-active sites with better charge transport and ion exchange that boosts the adsorption of intermediate for electrocatalytic reactions. These cations defect sites are responsible for the oxygen adsorption to oxygen anion and hydration for ORR. The impact of temperature on $ZnCo_2O_4$ spinel causes lattice distortion, which influences the electrocatalytic OER performance [64]. The high calcination temperature reduces the magnetic moment of the spinel, which diminishes the electrocatalytic OER activity. The high-spin Co^{3+} enables the formation of a spin channel between Co-O-Co with better electron transportation and an effect on OER activity.

(d) Oxyhydroxides

Metal oxyhydroxides are composed of metal cations, which is coordinated by oxide and hydroxide groups. In the structural characteristics, the oxyhydroxides are different from their oxides and hydroxides species. Transition metal-based oxyhydroxides draw significant attention in electrocatalytic investigation owing to their high performance, durable nature, low cost, and active surface sites. The formation of metal oxyhydroxide is done through the oxidation and deprotonation of metal hydroxide, oxide, and metal at higher potentials. The phase and crystallinity of a specific oxyhydroxide material are important factors in their electrocatalytic activity. As an example, FeO(OH) exists in four phases, like α-, β-, γ-, and δ-FeO(OH) [65, 66] (Fig. 1.7).

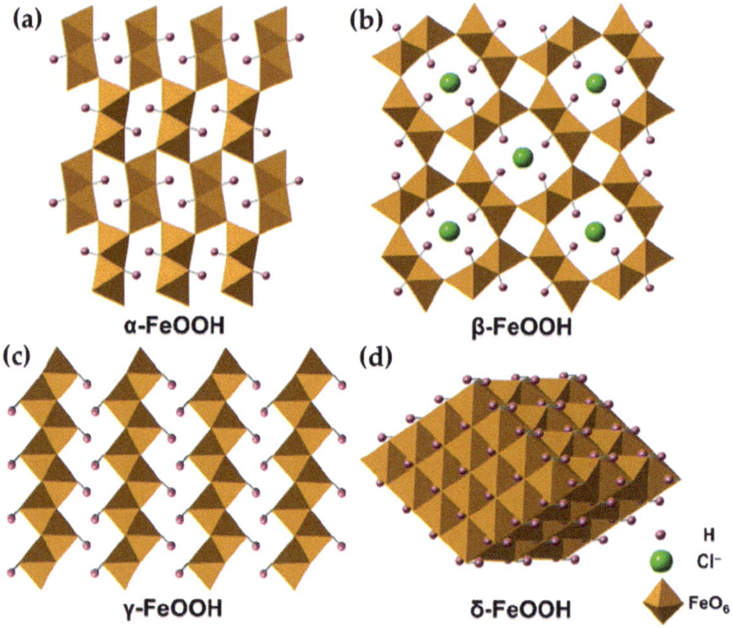

Figure 1.7 Polyhedral crystal structure of (a) α-FeO(OH), (b) β-FeO(OH), (c) γ-FeO(OH), and (d) δ-FeO(OH) (reprinted from [65]).

The OER activity under 1.0 M KOH demonstrates better activity for FeNiOOH (252 at 10 mA cm^{-2}) than the benchmark IrO_2 catalyst. The long-term chronoamperometry test of FeNiOOH indicates surface reconstruction with the reduction of terminal oxygen and enhancement of terminal hydroxyl groups. In this regard, cobalt-based oxyhydroxides are noticeable for their strong metal-oxygen bond strength and durability compared to Ni-based materials [67]. Enman et al. used different substrates like Au, Pt, Pd, C, and Cu to check the efficient activity of FeOOH [68]. The highest OER activity is observed for the Au substrate, owing to better charge transport, adsorption strength by substrate, and stable local atomic structure of $[FeO_6]$. The electron-withdrawing effect of Au leads to the enhancement of the metal oxidation state that favors the increased adsorption of iron oxyhydroxide on the gold surface.

Among hierarchical CoOOH and MnOOH, the last one is inactive to OER and CoOOH acts as OER active materials (η_{10} = 320 mV) with a higher durable nature under alkaline conditions [69]. The presence of high valent Cr in the CoOOH lattice boosts the electrocatalytic OER efficiency of unmodified CoOOH [70]. The Fermi level electronic state of Cr is more suitable for the intermediate *OOH formation than the Co site and is reflected in the operando in situ spectroscopic analysis. Higher density of states of Cr 3d electrons leads to the edge of the Cr-site being more beneficial than the pure edge Co sites for oxygen generation. Encapsulation of cobalt-iron oxyhydroxide into a carbon shell leads to better OER activity (η_{10} = 254 mV) and long-term stability [71]. The presence of a core shell acts as a conducting framework for charge transfer from the oxyhydroxide materials and performs better than the non-encapsulated material. The formation of an oxyhydroxide layer on nickel foam is reported by Bao et al., where sulfur-doped FeNiOOH, CoNiOOH, NiNiOOH, and CuNiOOH are examined for their electrocatalytic OER and HER activity in alkaline medium [72]. Trimetallic oxyhydroxide CoFeCuOOH shows 170 mV overpotential (10 mA cm^{-2}), which is better than its bimetallic CoFeOOH components [73]. The ^{18}O isotope labeling experiment indicates the active participation of lattice oxygen for CoFeCuOOH for oxygen evolution, which is totally absent for related bimetallic oxyhydroxide. Metal oxyhydroxides (M = Fe, Ti, Nb, Sn, Co, Ni, Cu, Ag, and Au) supported FeOxHy are used for electrocatalytic OER to study the synergistic influence of M-OOH [74]. Mondal et al. investigated the influence of Fe, Ni, and Mn incorporation in Ni-(oxy) hydroxide during alkaline OER activity and alcohol dehydrogenation [75]. The presence of Mn in Ni-(oxy)hydroxide improves the alcohol dehydrogenation, where Fe prefers OER. In the first case, Ni(O)OH acts as the active species with its facile formation, whereas in the second case of Fe incorporation leads to anodically shifted Ni(O)OH formation, resulting in poor alcohol oxidation. Such type of variable reactivity for different transition metals suggests the modulation of Ni(O)OH structure and catalytic performance. The Fe-, Ti-, Nb-, and Sn-related oxyhydroxides have poor performance compared to the other five. The application of in situ Raman spectroscopy indicates the generation of negatively charged superoxide radicals or peroxyl groups for active metals, which are totally absent for Ti, Sn, and Nb. In such cases, a strong synergistic interaction between

FeO_xH_y and M-OOH is considered, which prefers the formation of active oxygen species. In another study, the 100 ppb Fe^{3+}, Co^{2+}, Mn^{2+}, and Cu^{2+} incorporation in the 1 M KOH electrolyte during OER study of NF-NiOOH leads to the variation of electrocatalytic activity [76]. The high activity is observed for iron-containing systems, where Fe largely extends to the catalytic surface and actively participates in the electrocatalytic reactions. This observation is diluted for other cations due to lower penetration and lower tendency to participate in the electrocatalytic reaction. The cation exchange between Ni^{2+} and Fe^{3+}/Co^{3+} in the NiO_6 octahedron leads to the formation of hydrotalcite-like structure with slight lattice distortions, but others faced strong distortions and disfavored the catalytic efficiency. S-FeNiOOH shows higher electrocatalytic performance than the others, owing to the smaller density of states at the Fermi level, facilitating easy electrical conduction for the catalytic reaction. The synergistic effect of S and NiFe on the oxyhydroxide layer enhances intermediate formation and reduces the energy barrier for the targeted products. rGO- and cobalt-nickel-based Prussian blue analogs (Prussian blue abbreviated as PB) derived rGO/CoNiPBd-OOH demonstrated superior electrocatalytic OER performance [77]. The nanosized structure, high surface area, and synergistic effect between rGO and PB lead to good electrocatalytic performance.

(e) Oxyhalides

Metal oxyhalides are considered a group of compounds that contain metal-centered oxyhalide groups, which can be generalized as $[MO_mX_n]$ (M = metal, X = halogens). In the formula, M is coordinated by both the oxide and halide ions rather than single types [78]. The introduction of heteroanionic materials can tune the electrocatalytic properties of the materials. The presence of different anions creates differences in electronegativity, which will be beneficial for the new electronic modifications of the investigated catalysts. $Mn_8O_{10}Cl_3$ and $FeMn_7O_{10}Cl_3$ are used for acidic electrocatalytic OER studies where Mn-Cl plays an important role in the material's stability and activity [79]. The presence of a chloride or oxychloride unit directs the high efficiency in the OER activity where the oxide part cares about materials stability. The electrocatalytic reaction oxidizes $Mn^{2+/3+}$ to Mn^{4+} and this transformation takes speedily in the presence of Fe. As a consequence, the Mn^{3+} with elongated Mn-Cl prefers the OER, whereas a high valence retarded the reaction kinetics. Layered perovskite piezeocatalyst Bi_4TaO_8Cl performed 1.5 mmol g^{-1} h^{-1} hydrogen production rate with simultaneous H_2O_2 generation [80]. The piezoelectric characteristics lessened its surface charges through capacitive current and faradic current, where the first one is related to the reduction of redox reaction probability and the second one is related to the transformation of charge through the solution and the catalyst's surface, encouraging the redox reaction. Stress creation on the Bi_4TaO_8Cl leads to the deformation of nanoplates and favors hydrogen production (Fig. 1.8).

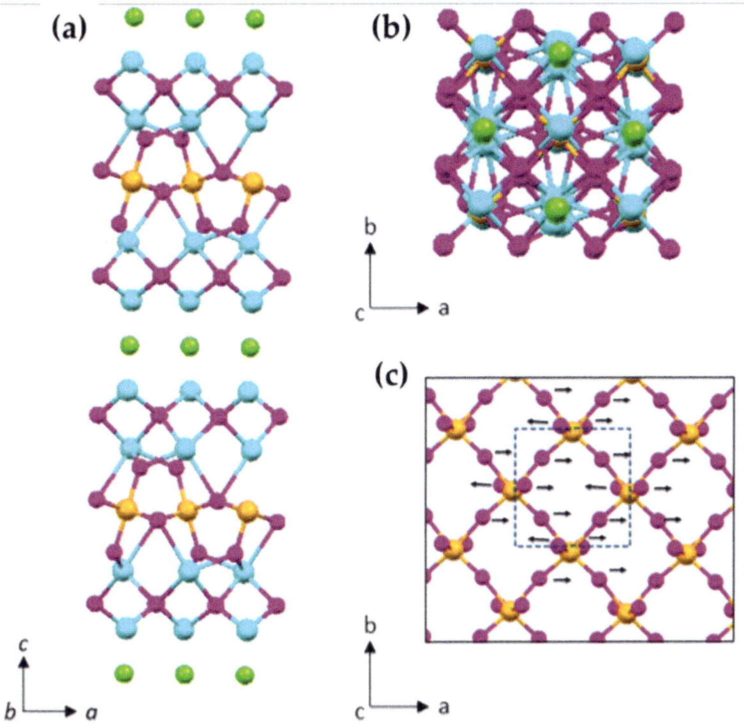

Figure 1.8 Crystal structure of Bi_4TaO_8Cl through (a) ac plane and (b) ab plane. (c) Displacement of oxygen atoms from the lattice under pressure (reprinted from [80]).

In the strontium cobalt oxyhalides, the presence of fluoride (Sr_2CoO_3F) influences catalytic activity much more than the other halides [81]. The strong electronegativity of fluorides leads to the creation of oxygen vacancies at the cobalt site and boosts the ionic conductivity for better electrocatalytic activity. The CV analysis suggests a higher electrochemical redox current for Co^{2+}/Co^{3+} for Sr_2CoO_3F than other halide analogs. Such performance is an indication of a higher electroactive surface for fluoride-containing oxyhalides. Doping of Fe in the Co site of Sr_2CoO_3Cl leads to higher electrocatalytic performance for both OER and ORR [82]. The electronic interaction between Co–Fe and corresponding charge transfer energies influences the rate-determining step of OER. In a comparative electrocatalytic analysis among bismuth oxyhalides (BiOCl, BiOBr, and BiOI), the iodide species show potential performance for diclofenac detection in terms of their activity, conductivity, and active surface area [83]. Good repeatable performance and stability of the BiOI indicate the suitability of the materials for diclofenac detection in river water and human urine. Fe doping at the Bi site of BiOCl leads to a higher tetragonal nature that influences local Bi-O electronic configurations and improves electrocatalytic OER activity [84]. A 270 mV overpotential is recorded for Fe-doped BiOCl, lower than undoped material owing to the higher conductivity, greater polarization, and high ECSA that boosts the number of active sites for electrocatalytic reactions. In this regard, the layered structure of BiOCl is very important, where $[Bi_2O_2]^{2+}$ is sandwiched between the $[Cl_2]^{2-}$ plates [85]. Such types of structural orientations facilitate electronic polarization and are beneficial for surface charge separations. The repeating of $[Bi_2O_2]^{2+}$ layers along c-directions is responsible for the anisotropy nature with ferroelectric characteristics, which are the adsorption of polar species on the catalytic surfaces. Involvement of Fe impurity in BiOCl leads to the higher catalytic impact in the OER performance and a 270 mV overpotential is recorded at 10 mA cm^{-2}. Lower electrochemical activity and ECSA of BiOCl are improved by the iron doping with improved conductivity, polarization, and active sites. Bi_3O_4Br electrocatalysts show better electrocatalytic HER activity (η_{50} = 662 mV) under alkaline conditions than its pristine one (η_{50} = 736 mV) [86]. The presence of Ni improves ECSA, mass activity, and the photocurrent response of the material, which leads to efficient charge transfer efficiency. Rare earth metals containing metal oxyhalides, and high entropy materials are not used randomly for the electrocatalytic purpose, which may have a good output in the electrocatalyst fields [87].

(f) Wolframite type oxides

The scheelite structure of ABO_4 is highly interesting to the scientific community for electrocatalytic applications. A lot of cations in the periodic table are reported in this structural type, where "A" is a low valent cation (viz., alkali, alkaline earth, Co, Ni, Fe, Mn, Cu, etc.) and "B" is a high valent cation (W, Mo, Os, etc.). $CaWO_4$ is the first structure of Scheelite and is known as the ore of tungstate. In our previous study, the tungstate and molybdate of iron and nickel ($FeWO_4$, $FeMoO_4$, $NiWO_4$, $NiMoO_4$) are used for the electrocatalytic OER under alkaline conditions (Fig. 1.9).

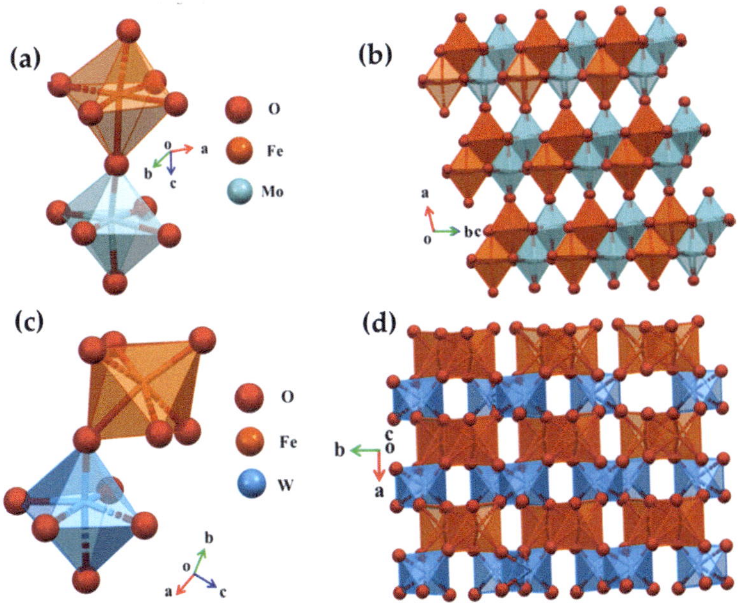

Figure 1.9 Polyhedral presentation of (a,b) $FeMoO_4$ and (c,d) $FeWO_4$ crystal structure (reprinted from [21]).

The tungstate-based oxides demonstrate durable characteristics for long-term analysis, whereas for molybdenum, an in situ metal oxyhydroxide active phase formation is noticed through the leaching of molybdate ions from the catalytic surface [21, 22, 88, 89]. The in situ-formed metal oxyhydroxide shows higher electrocatalytic activity than the pristine one, and therefore metal molybdates are more electrocatalytically active than tungstate. Investigation from density functional theory (DFT) explained that $FeWO_4$ is energetically more favorable than $FeMoO_4$, which suggests the weak bond arrangements in the latter case [21]. As a consequence, the molybdate ions are easily leached out from the electrode with the formation of electrocatalytic active iron oxyhydroxide and show more efficient performance than pre-catalyst materials. The pre-catalysts $NiMoO_4$ and $FeMoO_4$ are both transformed to the active oxyhydroxide species during electrocatalytic OER activity. In a comparative view, $FeO(OH)_{ED}$ shows 244 mV overpotential at 10 mA cm^{-2}, where it is observed as 238 mV for $NiO(OH)_{ED}$ at 20 mA cm^{-2}. The leaching of Mo from the lattice site is the crucial factor for the OER active phase formation. In this regard, $FeMoO_4$ loses 51% of Mo from the catalyst surface in 8 h, whereas $NiMoO_4$ takes 6 h to leach 58% Mo, and the related activity is reflected in the higher OER performance for $NiO(OH)_{ED}$. Rare earth metals (Ce, Pr, and Nd)-based vanadates are used for the electrocatalytic sensing of metronidazole drugs [90]. Among them, $CeVO_4$ exhibits better sensing activity with an extensive linear range of 0.02–75 μM, up to 4.5 nM detection, high stability, and selectivity characteristics. For theobromine (TBR) detection, the composite of $BaWO_4$/carbon nanofiber shows efficient results where synergistic interactions, π–π stacking, and hydrogen bonding play an important role for catalytic sensing [91]. The heterocyclic ring of TBR is influenced by the electrocatalyst's electrostatic and π–π interaction that improves the sensing performance of the composite material. The resulting value highlighted 5.2 nM low detecting limit within a wide potential range of 0.01 to 1726 μM.

IV. Metal chalcogenides/phosphides/nitrides:

Transition metal chalcogenides show more efficient characteristics in the electrocatalytic activity than the MO_2s owing to their high electrical conductivity. The chemical stability, robust nature, corrosion resistance, and tunable electronic structure fascinate them with interest in electrocatalytic application.

The CoTe@CoS heterojunction electrocatalyst showed OER activity under alkaline conditions where leached TeO_3^{2-} and SO_3^{2-} in the electrolyte transformed to TeO_4^{2-} and SO_4^{2-}, respectively, and reabsorbed on the cobalt oxyhydroxide electrode surface [92]. The adsorption of chalcogenide ions reduces the free energy, which suggests a positive direction for OER response. As a result, the density of states on the Fermi energy level improved by surface adsorption through the enhancement of electron conductivity and better charge transfer. Among the nickel chalcogenides, the researchers strongly recommend nickel sulfides as electrocatalysts due to their robustness in alkaline and acidic conditions [93]. Pumera et al. used different layered structure materials, suggesting higher HER activity for the Co, Ni, and Bi-based materials but better for Co in the OER study [94]. The $BiPS_4$ shows the lowest onset potential −0.40 V vs. RHE, but the structural degradation is noticed with the leaching of P in acidic medium. However, good structural stability is observed for $CoPS_3$ than the others. In such categories, p-block metal-based chalcogenides display interesting features for electrocatalytic water splitting. The combination of chalcogenides with Group 13 (GaS, InSe); Group 14 (GeSe, SnSe, $SnSe_2$, SnS, SnS_2); and Group 15 (Bi_2Se_3, Bi_2Te_3, Sb_2S_3, Sb_2Te_3) are beneficial for HER [95]. The placement of double chalcogenides in a single metal catalyst is functioned for better electrocatalytic activity. In a comparative electrocatalytic activity for HER and OER between $FeNiSe_2$ and FeNiSSe, the latter shows better efficiency in both cases (η = 91.2 and 213 mV at 10 mA cm^{-2}, respectively). The higher BET surface area, pore area, and pore volume support higher electrocatalytic activity for FeNiSSe materials. The weak hydrogen adsorption for sulphides also reflected in the resultant catalytic activity than the selenide-based materials [96]. In the CO_2RR, chalcogenides play an active role for lattice covalency, which is beneficial for improvement in d-electron density [97]. These consequences show better CO adsorption with improved selectivity for C2 products, owing to lower CO surface concentration.

In water electrolysis, metal phosphides are fascinating due to tunable modulation of metal structure by phosphorus with high conductivity and earth abundance. Strong covalent bond formation between metal and phosphorus in metal phosphides provides its robust characteristics during long-term electrocatalytic study. In this regard, Ni_xP_y, Co_xP_y, Mo_xP_y, W_xP_y, and Fe_xP_y are frequently studied for water splitting [98]. Introduction of a second metal in binary metal phosphides compositions leads to ternary compositions as $M_xM_yP_z$, which is highly beneficial in the electrocatalytic activity than the parent one. In such cases, the compositions like $Co_{2-x}Rh_xP$ [99],

$Ni_{2-x}Ru_xP$ [100], $Fe_{0.5}Co_{0.5}P$ [101], and $Ni_{1.6}M_{0.4}P$ [102] show efficient results in the electrocatalytic study. However, some challenges are still pending with it, including single-phase formation, scalable synthesis process, commercial applications, etc.

Transition metal nitrides (LTMN) are efficient for the energy electrolysis, owing to their unique electronic configuration related to noble metals [103]. In this regard, Ti_2N, V_2N, Cr_2N, Mn_4N, Fe_4N, Co_3N, Ni_3N, ZrN, Nb_4N_3, Mo_2N, Hf_3N_2, Ta_3N_5, W_2N, etc., show interesting electrocatalytic activity in the OER, HER, and nitrogen electroreduction. The incorporation of nitrogen in the interstitial sites of the metal atom modulates the electronic structure of the parent atom. The interstitial nitrogen may oxidize or leach during the electrocatalytic study and reasonable for the in situ electrocatalytic-active materials [104].

V. Polyoxometalates

Polyoxometalates (POMs) are a class of redox-active nanosized clusters, where different MO_2s are linked together by oxygen throughout the three-dimensional network. Modification of cobalt-based POM by MoS_2 and Pd nanoparticles leads to the formation of hierarchical heterostructures in the form of fibers [105]. In electrocatalytic water splitting, modified POMs exhibit better OER and HER activity, whereas MoS_2 and Pd influence the catalytic activity through the enrichment of active sites and synergistic effect. In an alkaline medium, the choice of materials exhibits high enough than the OER-related benchmark catalysts (Ir/C or IrO_2/C). The stable nature of the materials during 8 h chronoamperometry analysis suggests its suitability for long-term practical applications. Copper-based POMs formulated as $[\{Cu^{II}(2,2'\text{-bpy})(H_2O)_2\}][\{Co^{II}W^{VI}{}_{12}O_{40}\}\{Cu^{II}(2,2'\text{-bpy})(H_2O)\}\{Cu^{II}(2,2'\text{-bpy})\}]\cdot 2H_2O$ demonstrated high HER activity at near-neutral medium (pH = 4.8) [106]. The HER process follows a 520 mV overpotential at 1 mA cm^{-2} current density and proposes a proton-coupled electron transfer (PCET) mechanism. The high overpotential of the Cu-based POM materials arises from metal hydride formation at the rate-determining step with slower charge transfer kinetics. Das et al. reported $[Li(H_2O)_4][\{CuI(H_2O)_{1.5}\}\{Cu^{II}(H_2O)_3\}_2\{W^{VI}{}_{12}O_{36}(OH)_6\}]\cdot N_2\cdot H_2S\cdot 3H_2O$ POMs with bifunctional electrocatalytic behavior for neutral water splitting [107]. The HER and OER processes follow 443 mV and 418 mV overpotential at

1 mA cm^{-2} with a Faradaic efficiency of 84 and 80%, respectively. Differential pulse voltammetry (DPV) study within the potential region of +0.2 V to −0.1 V indicates above POMs use five electron transfers in the overall electrocatalytic water reduction. The tungstate centers surfer reduction behavior during electrocatalytic reduction. Barrel-shaped POMs like $[\{H_3O\}_4\{Na_6(H_2O)_{22}\}][\{CuI(H_2O)_3\}_2\{CuII(H_2O)\}_3\{B\text{-}\alpha\text{-}Bi^{III}W^{VI}{}_9O_{33}\}_2]\cdot 7H_2O$ and $Li_4[\{NH_4\}_2\{H_3O\}_3\{Li(H_2O)_5\}][\{Cu^{II}(SH)\}\ \{(Cu^{II}CuI_{1.5})(B\text{-}\alpha\text{-}Bi^{III}W^{VI}{}_9O_{33})\}_2]\cdot 9H_2O$ (abbreviated as NaCu-POM and LiCu-POM, respectively) are applied for electrocatalytic water reduction [108]. The NaCu-POM and LiCu-POM are used as electrocatalysts for HER where W(VI) centers act as active sites rather than the copper centers. The presence of $Cu^{I/II}$ creates intra-cluster cooperative interactions with tungstate centers that are beneficial for the hydrogen formation from water at neutral conditions.

1.4.3 Nonmetals as Electrocatalysts

In electrocatalytic applications, carbon-based materials show good characteristics and are demandable owing to their abundance, diverse morphology, conductive nature, and easy modifications by heteroatoms or transition metals. In some cases, carbon electrodes show comparable activity like noble metals and therefore they are promising alternatives of expensive electrocatalysts under harsh conditions. Their ability to form different-dimensional structures with variable porosity enhances electrocatalytic efficiency. Such functionality is advantageous for forming heterostructures with transition elements for better conductivity, adsorption, and surface sites for chemical reactions [109].

Carbon-based materials from biomass origin show potential and versatile activity in physiochemical reactions. The production of biomass may be questionable from the sustainable view either from burning carbohydrate-based materials (from plants or animals) or from biological waste. In such cases, pyrolysis, hydrothermal, molten salt, and ionothermal carbonization are vastly used by the researchers. In the electrocatalytic field, these materials show high impact value owing to their developed surface area, intense electrical conductivity, low cost, adjustable pore size distribution, stability, and environment-friendly nature [110]. These carbon-rich materials are used mostly as auxiliaries to advance other material's activity in terms of active site distribution, prevent agglomeration, increase surface area, higher stability, and cost-effectiveness. Such type of alternation through composite or doping formation improves electron distribution characteristics, band-gap alternation, and adsorption–desorption characteristics that act as vital points in the subsequent electrochemical viewpoint.

The introduction of biomolecules, amino acids, proteins, or carbohydrate-based molecules can enhance the electrocatalytic activity and stability of carbon materials. Reported studies show that the incorporation of L-cysteine [111], tyramine [112], L-lysine [113], etc., with reduced graphene oxides are effective as a metal-free system for the electrocatalytic reaction process. In such cases, the involvement of heteroatoms provides better functional activity for the resultant materials through an improved electron transfer conductive way at the catalytic surfaces. The presence of heteroatoms in different atomic positions enhances charge distribution capability and acts as active sites for electrocatalytic reactions. The size and electronegativity differences between carbon and heteroatom influence the electronic conductivity and form defects in the structural skeleton [114]. Defect formation in carbon-based materials is an important strategy to enhance catalytic efficiency in the electrocatalytic path. In the ORR, such types of materials are highly appraisable, but the identification of definite carbon atoms is a challenging task [115], owing to the deactivation of active sites, corrosion at carbon backbone, or limitations of related characterizations [104].

1.4.4 Layered Transition Metal Oxides

One important class of functional materials is TMOs. Among their many benefits are their vast range of catalytic potential, composition diversity, simplicity of synthesis, environmental friendliness, significant flexibility, and crystal structure and functionality. Their semiconducting nature makes them naturally poor in catalysis. Compared to their bulk counterparts, TMOs can perform better catalytically when reduced to a 2D structure with a larger lateral dimension [116].

Commonly utilized transition metals for the synthesis of layered oxides include Ti, Mn, Mo, Nb, Ta, and W. These oxides can take on several structural shapes, such as honeycomb arrangements, layered α-$NaFeO_2$, and layered perovskites. Alkaline or alkaline earth cations with large radius are intercalated in the interlayer channels TMO_6 octahedral host layers, where TM stands for transition metal elements. In a honeycomb layered structure, for instance, the TMO_6 octahedra are joined by edge-sharing to create a configuration resembling a honeycomb. But in a layered perovskite structure, a perovskite layer is created by corner-sharing TMO_6 octahedra.

Well-defined layered oxides have also been created at high temperatures and pressures using solid-state synthetic methods. This synthetic technique results in the catalyst with poor catalytic efficacy owing to their large micrometer size and lower surface area. Layered MO_2s have generated a lot of attention in electrocatalytic applications until novel 2D oxide nanomaterials are produced using liquid-phase exfoliation processes.

Alkali/alkaline earth metal ions and transition MO_2 layers form layered TMOs, and edge-sharing MO_6 octahedral units comprise the MO_2 layers. In the MO_2 layer, oxygen as a donor center is attached to the metal ions, creating prismatic, tetrahedral, or octahedral structures with the layered arrangements. In a hexagonal lattice, the oxygen atoms may occupy three distinct locations. For instance, the O_3-type structure of lithium cobalt oxide, or $LiCoO_2$, the lithium ion, and the transition metal layers adopted octahedral oxygen (O) coordination in the stacking unit.

To improve the electrochemical and interfacial characteristics of LTMO during charging and discharging procedures, it is crucial to comprehend its electronic structure. In the molecular orbital diagram, the oxygen 2p and metal (M) orbitals hybridize in the smallest unit of octahedral MO_6. The oxygen 2p orbitals are completely occupied by electrons of the d-block elements. The valence states mostly occupied by t_{2g} and e_g of Mn(I) metal orbitals due to the higher energy of metal orbitals with respect to oxygen 2p orbitals. Metals' outermost t_{2g} and e_g orbitals suitable foe redox reactions are linked with intercalated Li^+. This idea of molecular orbitals is particularly pertinent to band theory when studying bulk crystals of LTMO composed of edge-sharing MO_6 units.

Mn-based oxides are one of the promising materials for conducting electrocatalytic applications, categorically OER reactions, owing to their variable oxidation states, unpredictable bonding patterns, and thermodynamic stability with low toxicity. Hence, manganese is an active element in O_2-evolution reactions, as evidenced by the presence of μ-oxo-bridged tetrameric Mn_4Ca clusters responsible for carrying out photosynthesis in plants. Several investigations have been conducted to replicate the biological enzymes constituted with manganese oxides. with different crystallographic and morphological features, e.g., 2D layered δ-MnO_2, where MnO_6 units are organized in an edge-sharing fashion [117].

Bulk-layered MO_2s frequently have low catalytic efficiency because of things like restricted active sites, delayed mass transfer, and poor electrical conductivity. Catalytic performance could easily be enhanced by exfoliating layered TMOs into mono or multilayer nanosheets of micrometer or sub-micrometer thicknesses by the successful exfoliation of graphene. The exfoliated materials exhibit unique properties on comparison to their bulk due to the increased surface polarization. Figure 1.10 represents the insertions of the TMOs into the 2D multilayer nanosheets with an increase in surface polarizations.

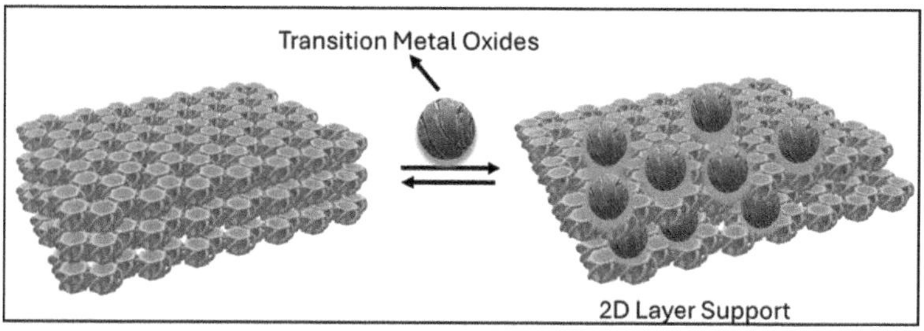

Figure 1.10 Schematic diagram synthesis of layered TMOs.

1.4.5 LTMO as an Electrocatalyst for Oxygen Evolution

In aqueous electrolyte solutions, the OER is known to be the opposite of green fuel processes, including nitrate reduction, carbon dioxide reduction, and HER. The widespread application of OER has been facilitated by the availability of straightforward and affordable anodic techniques that may be combined with green fuel in aqueous solutions. However, the slow kinetics of the OER frequently limits the total kinetics of green-fuel electrochemistry. For example, a two-electron transfer reduction during the water-splitting process produces one H_2 molecule, which is necessary to produce green hydrogen.

In contrast, the formation of a single O_2 molecule in the OER process requires a four-electron transfer. The OER's overall efficiency is constrained by these several electron-transfer mechanisms. Electrocatalysts are therefore required to enhance the OER kinetics. Despite being well-known to be effective OER catalysts, RuO_x and IrO_x are too expensive to be used widely. As an alternative, several TMOs have been designed and used as OER catalysts, such as perovskite structures, spinel, and LTMO (viz., Li_xCoO_2 and Na_xCoO_2 ($0 < x < 1$)) [118]. These reasonably priced catalysts can be customized by using surface reconstruction, element doping, and alkali metal ion extraction. [119] Although we focus on the LTMO catalyst's synthetic methods in this section, the fundamental mechanics and descriptors of the OER are discussed first and apply to all TMO catalysts in the OER.

1.4.6 Layered Metal Non-oxide Compounds

Various layered metal non-oxide materials, for example, transition metal sulfides (LTMS), transition metal carbides (LTMC), LTMN, and MXenes, comprise the layered metal non-oxide family. These four layer-structured, transition-non-oxide metal-based materials are discussed in this section. The petroleum and chemical industries used LTMS in the 1920s [120]. LTMS are commonly denoted by the formula MS_2, where M comprises elements such as Mo, Re, Ti, Ta, W, and Nb. These materials were utilized in hydrodesulfurization and the hydrogenation of aromatics, ketones, and olefins. Two-dimensional alternating S-Mo-S sheets make up the prototype LTMS, molybdenum disulfide (MoS_2). The Mo atom forms covalent bonds with three S atoms in the top layer and three S atoms in the bottom layer of each monolayer [121]. The material is appropriate for lubricating and sliding applications due to the weak van der Waals forces that exist between these S-Mo-S sheets. For many years, MoS_2 has also been used as a catalyst for hydrogen evolution processes (HER) because of its broad bandgap and semiconducting characteristics. In nanoparticulate MoS_2 with different perimeter diameters, research conducted by Thomas F. Jaramillo's group showed a direct relationship between catalytic activity and the number of edge sites.

This revelation has inspired intense research into increasing the percentage of MoS_2 exposed edge sites. Two primary therapeutic approaches are thus established: (i) activating inactive in-plane sites and (ii) exposing all possible edges. A prevalent and effective technique of converting an in-plane site into an active one is doping with a heteroatom. Doping with a heteroatom is a popular and effective method for turning in-plane sites into active ones. For example, doping with Ag atoms was discovered to be productive in modifying the electronic structures of the MoS_2 basal plane. Theoretical calculations also supported the activation of in-plane S sites between and near Ag atoms, which improved catalytic activity [122].

Besides, another dopant, a single fluorine atom, can also modify the chemical environments of MoS_2 with an increase of its electronegativity. The fluorine-doped MoS_2 electrode was fabricated using a plasma etching approach. Doping of fluorine activated the basal plane of MoS_2, leading to a five-fold increase in activity compared to clean margins. Similarly, Co ions were shown to interact with edge S atoms in a transition metal doping research, forming a hexagonal shape and changing the intrinsic activity of MoS_2 [123]. Another potential tactic to activate interfacial sulphur (S) sites is to interact with Ru nanoparticles in an inert basal plane. Think about creating pores, exfoliating into monolayers, and nanosizing to optimize edge exposure. The edge site of the MoS_2 was exposed by Jaramillo and coworkers with a remarkable enhancement of the HER. Highly ordered double gyroid MoS_2 thin films were fabricated by using template-assisted electrodeposition [124].

Chemical vapor deposition techniques (CVD) were implemented to grow MoS_2 thin films on a substrate with regulated layers by Yang and coworkers. With the increase in the number of deposited layers, the catalytic activity increased [125]. Later, Yu and his group investigated and proposed that layer-dependent electrocatalysis originates from electron hopping in the vertical direction, rather than the number of edge sites as previously thought [126]. Research on bulk MoS_2 has improved its performance, structure, and catalytic active site. These in-depth studies have improved the performance, structure, and catalytic activity of MoS_2 by proving it a suitable material for electrocatalysis, free of noble metals.

1.4.7 Multicomponent Layered Materials

Layered structures were also noted for the transition metal phosphorus sulfides (MPS_x), selenides, and nitrides. Mainly, Mn, Cr, Ga, Sn, Cd, Mn, Fe, Co, and Ni were utilized for the preparation of layered transition metal MPS_x. Mostly, the crystal structures were characteristic of monoclinic space group C2/m and contained D_3h symmetry. Typically, these metals form ionic and divalent bonds with P_2S_6 clusters, resulting in the chemical formula $M_2^{2+(}P_2S_6)^{4-}$ [127]. Van der Waals forces drive the interlayer interactions in this sandwich-like structure of S-M-S layers, which is reminiscent of MoS_2. These materials are extensively employed in magnetism and spintronic applications because of their enormous band gaps and specific ferromagnetic characteristics.

However, $NiPS_3$, $CoPS_3$, and $BiPS_4$ were prior investigated as one of the active materials for hydrogen evolution (HER), oxygen evolution (OER), and oxygen reduction (ORR), although with low and insufficient catalytic activity. In addition, Zhang and his group developed an amorphous lithium embedded into the palladium phosphosulfide nanodots, with a large number of vacancies due to lithiating the inert multilayer of $Pd_3P_2S_8$, and an improvement in its electrochemical activation was observed. The $Pd_3P_2S_8$ was reported to be a remarkable and stable catalyst for the HER with a starting potential of −52.0 mV against RHE [128]. The catalytic activity can be altered by structural modulations. Like this, Wang et al. employed carbon doping to activate the basal planes of $NiPS_3$, which enhanced hydrogen adsorption on surface sites and controlled the filled state of the valence band [129]. These investigations demonstrate how 2D nanomaterials' electrical structure and surface can be altered to increase catalytic activity. Although the research on metal MPS_x as an electrocatalyst is still in its infancy, thus paving a path for the future.

According to a recent study, layered materials are effective at identifying active locations and establishing structure-activity relationships. A major problem with stacked electrocatalysts is their structural stability. On the one hand, 2D layers with large lateral dimensions tend to agglomerate due to their high surface energy, which diminishes their peculiar 2D structural characteristics. However, 2D materials might undergo structural changes during catalysis. Hence, it is anticipated that the increasing structural stability of 2D nanomaterials would receive more attention. Recently, 2D materials have gained the attention of scientists as a potential catalyst for HER, OER, and ORR. On the achievement of rational design and ultrathin 2D nanosheets, plenty of opportunities exist for expanding their applications to electrochemical reduction of carbon dioxide (CO_2RR) and nitrogen reduction reaction (NRR).

1.4.8 Layer-Structured Electrocatalyst Support

Establishing catalytically active species onto 2D layered materials is an efficient method of enhancing catalytic performance. Supporting 2D materials offers the advantages of increased active site exposure, less noble metal consumption, and special electrical interaction with the supported metal due to the larger lateral size [130]. Numerous studies have mentioned 2D layered materials as a cutting-edge electrocatalyst support and energy conversion application, which has caught researchers' attention.

1.4.9 Graphene Support

Since its discovery, the scientific world has shown great interest in graphene because of its attractive inherent qualities, including strong electrical conductivity, large surface area, and remarkable electrochemical stability, leading to the wide possibilities of graphene in electrocatalysis. Because of this, graphene has been thoroughly investigated as the perfect catalytic supporting material, with an emphasis on its usefulness as a conducting support. Based on the extent of loading material aggregation, catalysts deposited on graphene can be divided into three categories: single atoms, metal clusters, and nanoparticles. Nanoparticles greater than 1 nm and those less than or equal to 1 nm are commonly referred to as metal clusters or sub-nanoclusters. Catalysts with atoms dispersed singly are known as single-atom catalysts [131].

Graphene exhibits significant stability, a favorable electronic metal–support interaction, and good corrosion resistance. A wide range of metals, such as platinum, palladium, ruthenium, nickel, cobalt, zinc, and manganese, are used as sulfide or oxide as supporting metal/metal oxide/sulfide nanoparticles. Recently, Zhao et al. fabricated a nanocatalyst using ultra-fine platinum-cobalt with very low metal loading and great stability, that are shielded by graphene nanopockets. In addition to minimizing oxidative dissociation and diffusion by efficiently limiting catalyst aggregation and guaranteeing exceptional catalytic stability, the graphene shell also inhibited nanoparticle sintering and aggregation. Additionally, the platinum-cobalt nanoparticles' size was controlled by the space of the graphene pockets.

Catalytic efficiency of the metal clusters and individual atoms is quite different owing to their larger size. Unique geometric and electronic structural characteristics of metal clusters having an electronic structural resemblance to the molecules provide a large number of surface-active sites, which efficiently adsorb, activate, and convert the reactant into the desired products. At the moment, monometallic and multi-metallic catalysts are conjugated on the surface of graphene for the fabrication of the graphene-metal cluster composites. Mostly, the transition metals, for example, palladium, platinum, iron, cobalt, and nickel, were embedded earlier for this purpose. The multimetallic catalysts comprise platinum/nickel cluster-anchored graphene, and bimetallic cobalt/tungsten cluster-loaded graphene, etc.

Recent years have seen significant developments in the field of graphene-supported metal-cluster catalysts. The graphene-confined ultrafast radiant heating (GCURH) approach is a novel technique that Tian and his coworkers successfully developed. It has been demonstrated that this method works very well for creating metal cluster catalysts with remarkable loading capacities in a matter of microseconds. The GCURH technique makes use of graphene's special qualities, such as its high-temperature resistance, flexibility, and ultra-low permeability. They also used this technique to pyrolyze cobalt-based metal-organic frameworks (MOFs) to create sub-nano cobalt cluster catalysts with a metal loading of up to 27.1 weight percent. Because of its single-metal composition, this catalyst demonstrated exceptional activity in OERs and provided significant convenience in catalytic recovery and modifications. This innovative technique is well-positioned to address the intricate and ecologically sustainable challenges that metal cluster catalysts encounter.

Graphene is one of the potential substrates to be utilized in electrochemical energy conversion. Concerning pure graphene, heteroatom-doped graphene, or graphene composites of Au, polyethylene, or calcium oxide offer enhanced catalytic activity [132]. Pt deposited on nitrogen-doped graphene offered enhanced methanol oxidative capability as reported by Xiong and coworkers [133]. The effect of defects present on graphene-based catalysts due to the presence of active metal sites is one of the plausible subjects of discussion. Previous research has shown that although graphene vacancies can operate as active sites for metal catalysis, an excess of vacancies may dilute the catalytic sites. Hence, intensive research is highly desired to explore the optimal amount of vacancies to achieve the best catalytic performance. Additionally, there is disagreement over surface oxidation groups. Graphene surfaces with oxygen-containing functional groups have been shown in numerous studies to operate as active sites and increase the catalytic activity of metal catalytic sites. In terms of the adsorption of active species, alternative research findings have indicated that the oxygenated functional groups can compete with the metal catalytic sites, thus lowering catalytic activity. Therefore, catalytic activity can be tailored by controlling surface modifications or functional groups that include oxygen on the surface. The doping of heteroatoms can also elevate the activity of metal catalytic sites by introducing extra electronic states and defect sites. However, there is continuous discussion because of the complex effects of heteroatom type and concentration on catalytic characteristics. Graphene deals with various issues like (i) active site problem as commonly found functional moieties present on graphene surface such as carboxyl and hydroxyl are not ideal catalytic sites; (ii) reduction rate as the oxygen-containing functional groups often requires reduction treatment; and (iii) cyclic stability of the catalyst to increase recyclability of the is desired. Despite a great deal of research in this field, as of right now, very little is known about how heteroatom doping affects catalyst activity. Therefore, to improve the design of catalysts with graphene-based flaws, a thorough investigation is required due to the material's complexity.

1.4.10 Boron Nitride Support

Commonly known as white graphene, which is hexagonal boron nitride (h-BN), contains equal amounts of alternating B and N atoms that create a honeycomb pattern and extend up to 2D layers. The h-BN possesses excellent mechanical strength, chemical stability, and ultrahigh thermal conductivity, thus being researched extensively by scientists as lubricants, dielectric layers, and thermal conducting materials. However, the study of h-BN as a catalyst, owing to its electrochemical inertness and interlayer insulation due to the large band gap, is still in its infancy. Figure 1.11 shows the illustration of insertion of metal into B-N cavity.

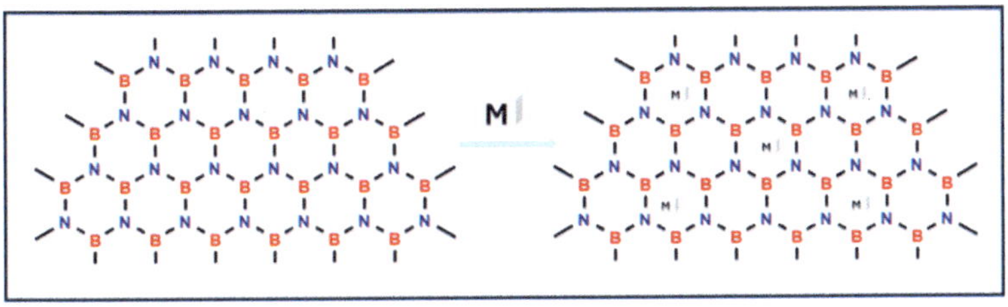

Figure 1.11 Schematic diagram of metal-BN support.

The catalysis mostly takes place on the surface of the material; hence, surface and interface engineering can alter inert support into an efficient electrocatalyst. With high site density on the 2D substrate surface, the support can control the electronic structure, affecting the active sites of the catalyst or catalytic efficacy through metal-support interactions [134]. Adsorption energy, reaction energy barriers, the number of exposed sites, the electronic configuration, and the microstructure of these active sites determine the performance of a material as a catalyst. To improve the active centers' inherent catalytic efficiency, optimization of the electrical structure is highly demanding.

Various methods have been implemented to activate h-BN in addition of conductive additives. With high N content and durability, h-BN can be considered as one of the preferred supports for attaching metal sites.

Through theoretical calculations, Taketsugu and coworkers (2013) anticipated that Ni-modified h-BN would function as an efficient catalyst for the ORR. Moreover, by theoretically creating a model with a single Mo atom, positioned on the top of a defective h-BN layer, Chen's group predicted an excellent catalyst for NRR. Based on the theoretical calculations, researchers used three experimental methods to activate BN by introducing defects, band structure, and interface engineering [135].

(i) **Defect Engineering:** It has been demonstrated that the incorporation of unsaturated nitrogen atoms resulting from boron vacancies in the h-BN layer leads to the attachment of metal atoms and thus develops M–Nx sites as active sites. For instance, Zhao and his colleagues added a single molybdenum (Mo) atom to the surface of an h-BN with a boron monovacancy. This defect-enriched hybrid composite exerted remarkable catalytic efficiency for nitrogen fixation at room temperature by stabilization of N_2H^* or destabilization of NH_2^* species. The creation of C–N and/or C=N bonds may also be accomplished by adding carbon to the BN lattice, enabling the attachment and dispersion of metal sites. Using pulsed laser ablation (PLA), Du et al. developed a special type of BN doped with carbon and oxygen (L-BN). On heteroatom fusion, the electrical conductivity of the L-BN increased, stemming from the dipolar interactions between interlayer B-B dipolar and offers a high rate of OERs. The strong interactions between the N-C=N bonds in L-BN and IrOx led to substantial catalytic activity when IrOx was loaded with L-BN. DFT simulations supported that after PLA, the carbon atom of N-C=N bonds in L-BN's lost 0.403 electrons, resulting in an unsaturated electronic structure.

(ii) **Band Structure Engineering:** The catalytic activity of the boron-BN is limited due to its wide conduction band structure associated with electronic regulation. Research has demonstrated that heteroatom doping, which produces B- and N-vacancies and defects, can successfully narrow the band structure of BN. Furthermore, the band structure of BN can be changed by metal components. Uosaki and his group reported that a slight increase of unoccupied BN states toward the Fermi level can diminish the band gap of BN by loading BN with a gold (Au) surface. The BN–Au composite material has been shown as an excellent electrocatalyst for ORR in both theoretical and experimental investigations. In addition to this, platinum (Pt) loading onto BN sheets can improve the catalytic activity of the HER by adjusting the band gap [136].

(iii) Interface Engineering: The microstructure and local electrical environment are essential for catalytic activity, as the mostly catalytic reactions take place on a material's surface. Sun and his coworkers developed a hetero-structured h-BN/Pd hybrid material with distinct surfaces to create a stable ORR catalyst. Strong interfacial interactions between the Pd nanoparticles and the BN sheets were evidenced in theoretical and experimental results, thus by decreasing the d-band center of Pd and improving the adsorption efficiency [137].

1.4.11 MXene Support

Layered ternary compounds, $M_{n+1}AX_n$, can be prepared from metal carbides and nitrides with space group $P6_3/mmc$. The ternary compounds are prepared from early transition metal "M", "A" stands for a main group element (often Si, Al, Sn, or Ga); and "X" stands for C and/or N; while "n" is one, two, or three. The stability of the MAX composites arises from the ionic, covalent, and metallic bond formations. HF readily breaks down the weaker-bonding layer, resulting in the production of MXene. MXene takes on distinct layers with an accordion-like shape once the Al layer is removed.

Since 2011, the MXene family has experienced tremendous growth [138] because of its superior electrical/thermal conductivity, ductility, hydrophilicity, and low-capacity layered structure. Till now, more than 70 ternaries, layered compounds have been reported with LTMC, nitrides, and carbonitrides. Various elements, for example, Ta, Nb, V, Cr, Zr, Ti, Al, C, and Hf, are among the constituent elements used for the preparation of multilayer, machinable LTMC, nitrides, and carbonitrides. The ternary compound, Ti_3AlC_2, has attracted a lot of interest and is becoming one of the most promising options for energy conversion [139]. Layered structures of MXene are depicted in Fig. 1.12.

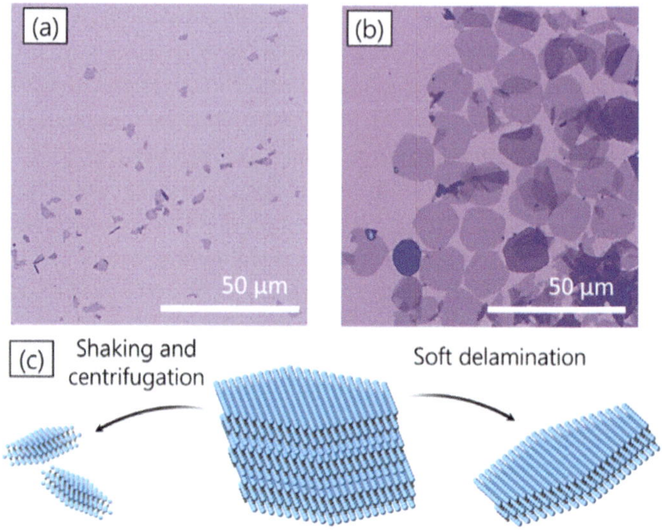

Figure 1.12 Comparison of single-layer MXene flakes prepared by regular delamination and soft delamination. Reproduced with permission from [140]. Copyright 2022, American Chemical Society.

Despite the removal of the A layer, MXene displays a distinctive 2D layered structure with a range of functional groups, such as -O, -F, and -Cl, on its surface. MXene is one of the great options to be used as catalytic electrodes because of its properties. The material properties of the MXene can be modulated by altering (i) surface functional group customization, (ii) engineering of the surface/interface, and lastly (iii) construction of heterostructure. MXene is created by etching a layer in a liquid phase stripping method that contains numerous functional groups (-Cl, -O, -F, -OH, -H) on the surface, resulting in enhanced hydrophilicity and adsorptive features. Activity modeling shows that the surface functional groups, specifically the terminal moieties anchoring to the surface of the MXene, can control catalytic efficiency. Strategic functionalizations of the MXene surface can easily improve the catalytic performance of targeted reactions. As it is thought, -O terminations increase HER, while -F terminations reduce HER but enhance NRR. Wang and coworkers reported that adding $TiOF_2$ nanospheres to the surface of $Ti_3C_2O_x^-$ MXene successfully stabilize the substrate's oxygen-containing terminations and generates remarkable oxidation resistance and HER catalytic efficacy. O-terminals, with almost zero Gibbs free energy ($\Delta G°_{H*}$), can serve as HER active sites, have been demonstrated by Gao and his group. According to certain published research, surface F terminations impede active sites, for which the efficiency of NRR reduces. By adding Fe and eliminating inactive F*/OH* terminations, a very effective NRR catalyst was prepared, which lowers the energy barrier, as demonstrated by Zhi and his colleagues. From the above discussion, it can be predicted that appropriate molecular function with interfacial interaction can increase or decrease the structure, activity, and catalytic performance of the catalyst for electrochemical applications. Moreover, the granularity, size of the crystallites, the solubility of the catalyst in the active solution medium, pH, and the support offered by the metals can regulate the productivity of the catalyst.

1.5 Challenges and Future Directions

Materials employed for electrocatalytic activities are discussed in this chapter. The choice of noble metals, their activity, and sustainability for commercial applications leads to the choice of cost-effective non-noble materials. The activity causes a material rise from d-band theory, the Fermi level, and Sabatier principle are explained to decide their fitness in the electrocatalytic reactions. Metal-based catalysts like metal oxides, chalcogenides, high entropy metals, POMs, perovskite, and spinel materials were discussed. The non-metal carbon-based materials are also considered in terms of their engineering in carbon skeleton for their importance in electrocatalytic applications. In such cases, the origin of carbon materials, their activity, microstructures, pores, interconnectivity between pores, and defectivity play crucial roles in electrocatalytic productivity.

Currently, the electrochemistry related research focuses on innovative electrocatalyst design for their sustainable applications in the future days not only for energy-related issues but also for better environment. In those directions, the development of perfect electrocatalyst for suitable applications is very essential. But certain factors like catalyst stability, durability, scalability, transitioning from laboratory-scale to industrial-scale applications, new synthesis methods, and eco-friendly processes are big challenges for their smooth way. In such cases, screening of materials and their selection are very vital for efficient activity. The following phenomena are very important for specific applications during the material's characterization to reach a successful goal.

i. Dissolution activity of some non-noble metals and their compounds are reported in the electrolyte during electrocatalytic operation or at open-circuit potential. Such type of activity leads to the formation of a new active catalytic phase, but the questions arise about their broad application.
ii. Post-mortem analysis is highly essential for highlighting the real stability of the materials. In such cases, the materials characterization and electrolyte's ICP-MS analysis are vital to check the material's stability.
iii. Intrinsic catalytic activity is a typical task that should be more focused on the measurement of catalytic efficiency as well as the related mechanisms.
iv. Surface reconstruction of precatalyst materials is observed during electrocatalytic investigation. The leaching of materials and generation of electrocatalytic phases leads to higher efficiency in the electrocatalytic applications. However, the fate of leaching elements and related waste treatment is a big concern to the scientific community.

In the material's selection for electrocatalytic applications, literature reports are very important to gain knowledge about the material's activity and nature. Trial and error methods are randomly used for selective purposes to gain maximum activity. The five main steps that can be considered for choosing materials are design, material characteristics, screening, assessment for maximum activity, and confirmation. Some methods, such as TOPSIS (technique of ranking preferences by similarity to the ideal solution), ELECTRE (elimination and choice translating reality), and AHP (analytical hierarchy process), have been chosen as a better final approach. In such cases, chart methods, computer-aided materials selection, knowledge-based systems, and fuzzy methods are also reported by researchers for material screening purposes.

References

1. C. G. Zoski, *Handbook of Electrochemistry* (Elsevier, 2007).
2. C. Breitkopf and K. Swider-Lyons, *Springer Handbook of Electrochemical Energy* (Springer, 2016).
3. A. P. O'Mullane, in *Ref. Modul. Chem. Mol. Sci. Chem. Eng.* (Elsevier, 2013).
4. X. Zhao, M. Liu, Y. Wang, Y. Xiong, et al., *ACS Nano* **16**, 19959 (2022).
5. P. Banoth, C. Kandula, and P. Kollu, in *Noble Met. Electrocatal. New Trends Electrocatal. Energy Appl. Vol.* 2 (American Chemical Society, 2022), p. 1.
6. S. Nitopi, E. Bertheussen, S. B. Scott, X. Liu, et al., *Chem. Rev.* **119**, 7610 (2019).
7. X. Wu, Y. Wang, and Z.-S. Wu, *Chem.* **8**, 2594 (2022).
8. C.-Z. Yuan, K. S. Hui, H. Yin, S. Zhu, et al., *ACS Mater. Lett.* **3**, 752 (2021).
9. S. Li, L. Shi, Y. Guo, J. Wang, et al., *Chem. Sci.* **15**, 11188 (2024).
10. J. Wu, N. Liu, F. Li, B. Jia, et al., *Coord. Chem. Rev.* **525**, 216343 (2025).
11. H. Xie, T. Wang, J. Liang, Q. Li, et al., *Nano Today* **21**, 41 (2018).
12. W. Hu, D. Yang, Y. Chang, K. Yu, et al., *Chem. Eng. J.* **496**, 154120 (2024).
13. C. Lucky and M. Schreier, *ACS Nano* **18**, 6008 (2024).
14. J. K. Nørskov, F. Abild-Pedersen, F. Studt, and T. Bligaard, *Proc. Natl. Acad. Sci.* **108**, 937 (2011).
15. Y. Yang, M. Luo, W. Zhang, Y. Sun, et al., *Chem.* **4**, 2054 (2018).
16. H. Ooka, J. Huang, and K. S. Exner, *Front. Energy Res.* **9**, (2021).
17. Y. Li, Y. Sun, Y. Qin, W. Zhang, L. et al., *Adv. Energy Mater.* **10**, 1903120 (2020).
18. Q. Zhang, M. Song, G. Luo, T. Shen, et al. *Mater.* **36**, 10967 (2024).
19. S. Kabir, K. Lemire, K. Artyushkova, A. Roy, et al., *J. Mater. Chem.* A **5**, 24433 (2017).
20. E. S. Davydova, F. D. Speck, M. T. Y. Paul, D. R. Dekel, et al., *ACS Catal.* **9**, 6837 (2019).
21. M. K. Adak, A. Rajput, D. Ghosh, and B. Chakraborty, *ACS Appl. Energy Mater.* **5**, 13645 (2022).
22. A. Rajput, M. K. Adak, and B. Chakraborty, *Inorg. Chem.* **61**, 11189 (2022).

23. G. Wang, W. Li, B. Huang, L. Xiao, et al., *ACS Appl. Energy Mater.* **2**, 3160 (2019).
24. S. Wang, L. Li, K. S. Hui, D. A. Dinh, et al., *Nano Energy* **113**, 108543 (2023).
25. E. S. Davydova, J. Zaffran, K. Dhaka, M. C. Toroker, et al. **8**, (2018).
26. C. Wang, H. Yang, Y. Zhang, and Q. Wang, *Angew. Chemie Int. Ed.* **58**, 6099 (2019).
27. A. Sajeev, A. Sathyaseelan, K. Serbara Bejigo, and S. Jae Kim, *J. Colloid Interface Sci.* **637**, 363 (2023).
28. H. Liu, C. Xi, J. Xin, G. Zhang, et al., *Chem. Eng. J.* **404**, 126530 (2021).
29. J. Zhang, K. Wang, L. Fan, H. Liu, et al. *Alloys Compd.* **933**, 167613 (2023).
30. V. Hasannaeimi, X. Wang, R. Salloom, Z. Xia, et al., *ACS Appl. Energy Mater.* **3**, 12099 (2020).
31. S. Liu, Y. Wang, T. Jiang, S. Jin, et al., *ACS Nano* **18**, 4229 (2024).
32. X. Zhou, L. Zou, H. Zhu, M. Yan, et al. *Feng, Small* **20**, 2310327 (2024).
33. P. Yang, M. Sun, J. Wang, J. Li, et al., *Particuology* **93**, 180 (2024).
34. Z. Cui, W. Jiao, Z. Huang, G. Chen, et al., *Small* **19**, 2301465 (2023).
35. H. Peng, in *Advances in Brazing*, edited by D. P. Sekulić (Woodhead Publishing, 2013), pp. 221–248.
36. S. Furukawa and T. Komatsu, *ACS Catal.* **7**, 735 (2017).
37. A. Samanta, S. Das, and S. Jana, *Nanoscale Adv.* **2**, 417 (2020).
38. C. Ma, H. Zhang, J. Xia, X. Zhu, et al., *J. Am. Chem. Soc.* **146**, 20069 (2024).
39. Y. Yang, R. Deng, C. Wang, J. Sun, et al. *Kanatzidis, Chem. Mater.* **36**, 2566 (2024).
40. R. Song, J. Han, M. Okugawa, R. Belosludov, et al., *Nat. Commun.* **13**, 5157 (2022).
41. V. K. Mariappan, K. Krishnamoorthy, P. Pazhamalai, R. Swaminathan, et al., *Small* **20**, 2402355 (2024).
42. H. Shi, T.-Y. Dai, X.-Y. Sun, Z.-L. Zhou, et al., *Adv. Mater.* **36**, 2406711 (2024).
43. Z. Jia, T. Yang, L. Sun, Y. Zhao, et al., *Adv. Mater.* **32**, 2000385 (2020).
44. A. Kormányos, Q. Dong, B. Xiao, T. Li, et al., *IScience* **26**, 107775 (2023).
45. A. Lavacchi, M. Bellini, E. Berretti, Y. Chen, et al., *Curr. Opin. Electrochem.* **28**, 100720 (2021).

46. Y. Fu, J. Wang, H.-Y. Yu, X. Li, et al., *Int. J. Hydrogen Energy* **42**, 20711 (2017).

47. K. B. Ghosh, W. Zhang, F. Tassinari, Y. Mastai, et al., *J. Phys. Chem. C* **123**, 3024 (2019).

48. T. Saha-Dasgupta, *Mater. Res. Express* **7**, 14003 (2020).

49. J. B. Falqueto, N. Hales, T. J. Schimidt, and E. Fabbri, *ACS Mater. Lett.* **6**, 5227 (2024).

50. C. Balamurugan, S. Song, H. Jo, and J. Seo, *ACS Appl. Mater. Interfaces* **13**, 2788 (2021).

51. C. L. McBean, H. Liu, M. E. Scofield, L. Li, et al., *ACS Appl. Mater. Interfaces* **9**, 24634 (2017).

52. S. Bhowmick, A. Dhankhar, T. K. Sahu, R. Jena, et al., *ACS Appl. Energy Mater.* **3**, 1279 (2020).

53. Y. Lu, A. Ma, Y. Yu, R. Tan, et al., *ACS Sustain. Chem. Eng.* **7**, 2906 (2019).

54. S. Malkhandi, P. Trinh, A. K. Manohar, A. Manivannan, et al., *J. Phys. Chem. C* **119**, 8004 (2015).

55. S. Ganguly, R. Datta, C. Loha, and S. Ghosh, *Energy & Fuels* **38**, 13196 (2024).

56. E. Tsuji, T. Motohashi, H. Noda, Y. Aoki, et al., *J. Phys. Chem. C* **122**, 7081 (2018).

57. Z. Cheng, J. Huang, G. Lu, Y. Wu, et al., *Chem. Mater.* **35**, 9669 (2023).

58. M. Bohra, V. Alman, and R. Arras, *Nanomaterials* **11**, (2021).

59. Z. Feng, P. Wang, Y. Cheng, Y. Mo, et al., *J. Electroanal. Chem.* **946**, 117703 (2023).

60. Y. Li, Z. Zhang, C. Li, Y. Zhou, et al., *Appl. Catal. B Environ. Energy* **355**, 124116 (2024).

61. J. Luo, X. Wang, Y. Gu, S. Wang, et al., *Chem. Eng. J.* **472**, 144839 (2023).

62. D. S. Nair, A. Anil, L. Elias, N. Satyanarayana, et al., *Int. J. Hydrogen Energy* **62**, 162 (2024).

63. M. Gopalakrishnan, W. Kao-ian, M. Rittiruam, S. Praserthdam, et al., *ACS Appl. Mater. Interfaces* **16**, 11537 (2024).

64. M. Ya, Y. Wang, J. Wang, G. Gao, et al., *ACS Sustain. Chem. & Eng.* **11**, 15451 (2023).

65. Y. Duan, B. Chakraborty, C. K. Tiwari, M. Baranov, et al., *ACS Catal.* **11**, 11385 (2021).

66. L. Mallick, A. Rajput, M. K. Adak, A. Kundu, et al., *Dalt. Trans.* **51**, 15094 (2022).

67. C. Dette, M. R. Hurst, J. Deng, M. R. Nellist, et al., *ACS Appl. Mater. Interfaces* **11**, 5590 (2019).
68. L. J. Enman, A. E. Vise, M. Burke Stevens, and S. W. Boettcher, *ChemPhysChem* **20**, 3089 (2019).
69. Z. Chen, X. Wang, S. Keßler, Q. Fan, et al., *J. Energy Chem.* **71**, 89 (2022).
70. L. Tang, M. Xia, S. Cao, X. Bo, et al., *Nano Energy* **101**, 107562 (2022).
71. X. Han, C. Yu, Y. Niu, Z. Wang, et al., *Small Methods* **4**, 2000546 (2020).
72. W. Bao, J. Yang, W. Yan, Y. Tang, et al., *J. Power Sources* **580**, 233307 (2023).
73. C. Wang, P. Zhai, M. Xia, W. Liu, et al., Adv. Mater. **35**, 2209307 (2023).
74. Q. Jiang, S. Wang, C. Zhang, Z. Sheng, et al., *Nat. Commun.* **14**, 6826 (2023).
75. C. Chauhan, T. Gupta, and B. Mondal, *Small* **21**, 2410228 (2025).
76. R. A. Marquez, E. Kalokowski, M. Espinosa, J. T. Bender, et al., *Energy Environ. Sci.* **17**, 2028 (2024).
77. P. H. S. Borges, J. M. Gonçalves, C. B. Breslin, and E. Nossol, *ACS Appl. Mater. & Interfaces* **16**, 53705 (2024).
78. X. Chen and K. M. Ok, *Chem. Sci.* **13**, 3942 (2022).
79. R. Li, D. Nordlund, and L. C. Seitz, *Chem. Mater.* **36**, 1299 (2024).
80. M. Banoo, R. S. Roy, M. Bhakar, J. Kaur, et al., *Nano Lett.* **22**, 8867 (2022).
81. M. A. Ghanem, M. S. Amer, A. M. Al-Mayouf, P. Arunachalam, et al., *Catalysts* **11**, (2021).
82. Y. Miyahara, T. Fukutsuka, T. Abe, and K. Miyazaki, *Chem. Mater.* **32**, 8195 (2020).
83. T. Kokulnathan, S. Ganesan, T.-J. Wang, and A. Palaniappan, *J. Environ. Chem. Eng.* **12**, 114049 (2024).
84. D. Singh, B. Joshi, and P. Poddar, *Langmuir* **39**, 11414 (2023).
85. Y. Zhou, P. Yan, J. Jia, S. Zhang, et al., *J. Mater. Chem. A* **8**, 13320 (2020).
86. M. Pal, R. Biswas, S. Barman, and A. Dutta, *Energy Adv.* **3**, 1562 (2024).
87. C. H. Wood and R. E. Schaak, *J. Am. Chem. Soc.* **146**, 18730 (2024).
88. M. K. Adak, A. Rajput, L. Mallick, and B. Chakraborty, *ACS Appl. Energy Mater.* **5**, 5652 (2022).
89. M. K. Adak, H. K. Basak, and B. Chakraborty, *ACS Org. Inorg. Au* (2023).
90. A. Radha and S.-F. Wang, *Environ. Sci. Nano* **10**, 3122 (2023).
91. S. Vinoth and S.-F. Wang, *ACS Appl. Nano Mater.* **7**, 25388 (2024).
92. W. Bao, J. Liu, T. Ai, J. Han, et al., *Adv. Funct. Mater.* **34**, 2408364 (2024).

93. S. M., Y. Vijay Rajyaguru, C. Kumar Rastogi, S. A. MP., et al., *Mater. Today Proc.* **73**, 316 (2023).
94. C. C. Mayorga-Martinez, Z. Sofer, D. Sedmidubský, Š. Huber, et al., *ACS Appl. Mater. Interfaces* **9**, 12563 (2017).
95. Y. Wang, Y. Zhao, X. Ding, and L. Qiao, *J. Energy Chem.* **60**, 451 (2021).
96. K. C. Majhi and M. Yadav, *J. Alloys Compd.* **852**, 156736 (2021).
97. M. Nath, H. Singh, and A. Saxena, *Curr. Opin. Electrochem.* **34**, 100993 (2022).
98. H. Yin, F. Rong, and Y. Xie, *Int. J. Hydrogen Energy* **52**, 350 (2024).
99. S. I. Mutinda, D. Li, J. Kay, and S. L. Brock, *J. Mater. Chem. A* **6**, 12142 (2018).
100. D. R. Liyanage, D. Li, Q. B. Cheek, H. Baydoun, et al., *J. Mater. Chem. A* **5**, 17609 (2017).
101. C. Tang, L. Gan, R. Zhang, W. Lu, et al., *Nano Lett.* **16**, 6617 (2016).
102. C. A. Downes, K. M. Van Allsburg, S. A. Tacey, K. A. Unocic, et al., *Chem. Mater.* **34**, 6255 (2022).
103. H.-M. Yang, C.-C. Weng, H.-Y. Wang, and Z.-Y. Yuan, *Coord. Chem. Rev.* **496**, 215410 (2023).
104. S. Wang, H. Pan, Y. Wang, H. Tang, et al., *Small Struct.* **5**, 2400031 (2024).
105. M. Guillen-Soler, N. V Vassilyeva, E. P. Quirós-Díez, J. M. Vila-Fungueiriño, et al., *Adv. Sustain. Syst.* **8**, 2300607 (2024).
106. C. Singh, A. Haldar, O. Basu, and S. K. Das, *Inorg. Chem.* **60**, 10302 (2021).
107. A. Ravi, S. Mulkapuri, and S. K. Das, *Inorg. Chem.* **62**, 12650 (2023).
108. S. Mulkapuri, A. Ravi, R. Nasani, S. K. Kurapati, et al., *Inorg. Chem.* **61**, 13868 (2022).
109. S. D. Bhoyate, J. Kim, F. M. de Souza, J. Lin, et al., *Coord. Chem. Rev.* **474**, 214854 (2023).
110. L. Li, S. Lu, L. Fang, Y. Wei, et al., *Energy & Fuels* **37**, 18485 (2023).
111. V. S. Sapner, P. P. Chavan, A. V Munde, U. S. Sayyad, et al., *Energy & Fuels* **35**, 6823 (2021).
112. V. S. Sapner, P. P. Chavan, R. V. Digraskar, S. S. Narwade, et al., *ChemElectroChem* **5**, 3191 (2018).
113. V. S. Sapner, P. P. Chavan, and B. R. Sathe, *ACS Sustain. Chem. Eng.* **8**, 5524 (2020).

114. R. Gutru, Z. Turtayeva, F. Xu, G. Maranzana, et al., *Int. J. Hydrogen Energy* **48**, 3593 (2023).
115. H. W. Kim, H. Park, J. S. Roh, J. E. Shin, et al., *Chem. Mater.* **31**, 3967 (2019).
116. M. P. Browne, Z. Sofer, and M. Pumera, *Energy Environ. Sci.*, **12**(1), 41–58 (2019).
117. P. Earis, N. Nugent, M. Martinez-fresno, S. Wilkes, et al., *Energy Environ. Sci.* (207890), 3–10 (2009).
118. Y. Kim, E. Coi, S. Kim, and H. R. Byon, *Chem. Sci.*, **14**(39), 10644–10663 (2023).
119. Z. Lu, H. Wang, D. Kong, K. Yan, et al., *Nat. Commun.*, **5** (1), 4345 (2014).
120. A. Kuc, N. Zibouche, and T. Heine, *Phys. Rev. B Condens. Matter*, **83**(24), 0–11 (2011).
121. H. Bao, Y. Huang, F. Ma, Z. Yang, et al., *J. Mater. Sci.*, **51**(14), 6850–6859 (2016).
122. X. Gao, Y. Zhou, Z. Cheng, Y. Tan, et al., *Appl. Surf. Sci.*, **547**, 149113 (2021).
123. D. Merki, H. Vrubel, L. Rovelli, S. Fierro, et al., *Chemi. Sci.*, **3**(8), 2515–2525 (2012).
124. J. Kibsgaard, Z. Chen, B. Reinecke, T. Jaramillo, *Nat. Mater.* **11**, 963–969 (2012).
125. P. Yang, X. Zou, Z. Zhang, M. Hong, et al., *Nat. Commun.* **9**, 979 (2018).
126. Y. Yu, S.-Y. Huang, Y. Li, S. N. Steinmann, et al., *Nano Lett.* **14**, 553–558 (2014).
127. C. C. Mayorga-Martinez, Z. Sofer, D. Sedmidubský, Š. Huber, et al., *ACS Appl. Mater. Inter.*, **9**(14), 12563–12573 (2017)
128. X. Zhang, Z. Luo, P. Yu, Y. Cai, et al., *Nat. Catal.*, **1**(6), 460–468 (2018).
129. J. Wang, X. Li, B. Wei, R. Sun, et al., *Adv. Funct. Mater.*, **30**(12), 1908708 (2020).
130. G. Shi, C. Yu, Z. Fan, J. Li, et al., *ACS Appl. Mater. Inter.*, **11**(3), 2662–2669 (2019).
131. C. Wang and D. Astruc, *Progress in Materials Science*, **94**, 306–383 (2018).
132. D. O. Idisi, J. A. Oke, I. T. Bello, *Int. J. Energy Res.* **45**, 19772–19788 (2021).
133. B. Xiong, Y. Zhou, Y. Zhao, J. Wang, X. et al., *Carbon N. Y.* **52**, 181–192 (2013).

134. Z. Li, Y. Chen, T. Ma, Y. Jiang, et al., *Adv. Energy Mater.*, **11**(25), 1–13 (2021).

135. J. Zhao, and Z. Chen, *J. Am. Chem. Soc.*, **139**(36), 12480–12487 (2017).

136. K. Uosaki, G. Elumalai, H. Noguchi, T. Masuda, et al., *J. Am. Chem. Soc.* **136**, 6542 (2014).

137. Y. Chen, J. Cai, P. Li, G. Zhao, et al., *Nano Lett.* **20**(9), 6807–6814 (2020).

138. M. Naguib, M. Kurtoglu, V. Presser, J. Lu, et al., *Adv. Mater.*, **23**(37), 4248–4253 (2011).

139. Z. Li, Y. Chen, T. Ma, Y. Jiang, et al., *Adv. Energy Mater.*, **11**(25), 1–13 (2021).

140. M. Shekhirev, J. Busa, C. E. Shuck, A. Torres, et al., *ACS Nano*, **16**, 13695–13703 (2022).

Chapter 2

Synthetic Techniques for Electrocatalytic Material Design

Dolly Singh, Uttam Sarkar, and Sagarika Bhattacharya
Department of Chemistry, National Institute of Technology Raipur, Chhattisgarh, India
sbhattacharya.chy@nitrr.ac.in

In this chapter, the development and the purposeful architecture of diverse classes of materials, specifically metallic and nonmetallic materials, employed for electrocatalysis processes, were discussed. Metastable two-dimensional (2D) metallic compounds have garnered considerable interest because of their unusual characteristics and multifunctional catalytic aptitude. There exists an array of materials, including metal organic frameworks (MOFs), transition metal dichalcogenides (TMDs), metal nanoparticles, and layered metal oxide materials, including perovskites, P2-layered oxides, and spinels, which have been earlier investigated as electrocatalysts. These materials are beneficial because of their tunable electronic structures and high catalytic activity. Additionally,

Materials Design for Electrocatalysis
Edited by Sagarika Bhattacharya and Biswarup Chakraborty

ISBN 978-981-5352-32-0 (Hardcover), 978-1-042-02558-9 (eBook)
www.jennystanford.com

DOI: 10.1201/9781042025589-2

these materials could carry out many key reactions in chemical transformations in energy storage and conversion. Materials such as layered double hydroxides (LDHs), metal nitrides, MXenes, and the new 2D materials, such as silicene and germanene, were actively being researched owing to their great electrocatalytic, structural, and chemical properties [1].

Nonmetallic carbonaceous materials were also employed for electrocatalytic applications. Mostly, literature reported on carbon-based materials, such as carbon black, carbon nanofibers, carbides, carbon nanotubes, carbon dots, and various catalysts with graphene and graphene oxide support [2]. Specifically, reduced graphene oxides (rGOs) are beneficial in enhancing the electrocatalytic activity due to their high electrical conductivity, high surface area, and ease of modification. The addition of rGOs lowers the material cost with acceleration of the reaction rate, thus the performance of numerous electrocatalytic processes was enhanced. Several techniques were implemented to grow the materials for research purposes, as discussed below. All the fabrication techniques are equally adopted for designing the electrocatalysts. Coupled with these fabrication technologies, some strategies help to offer increased control of the architecture, including shape, size, composition, and morphology of the material to optimize the electrocatalytic properties of the material for hydrogen evolution reaction (HER), oxygen evolution reaction (OER), carbon dioxide conversion, and so forth, which are vital to energy conversion and storage. Material design is one of the most important steps for electrocatalysis, as the stability of the catalysts and efficiency of the catalysts depend on the chemical composition. In the following sections, various techniques will be illustrated in detail [3, 4].

The catalysts can be prepared in liquid or solid state. It can either be atomic or in molecular in nature. Electrocatalysts can be used in the form of powder, pastes, emulsions, or many other forms.

2.1 Synthesis Techniques

2.1.1 Top-Down vs. Bottom-Up Approach

Synthesis of materials can commonly be classified under these two categories. In the top-down process, large materials are mostly sliced or cut successively to produce micro- or nano-sized materials. A top-down approach can be achieved by mechanical or lithographic techniques [5]. Ball-milling or grinding method produces micro- or nano-crystalline materials from the macrocrystalline structures, retaining the original integrity of the materials. Mostly, nanostructured metal oxides are prepared by grinding two different components in this method. The nonmetallic carbonaceous electrocatalysts were fabricated from the graphene oxide or graphitic rods. On the contrary, bottom-up techniques are being used to build micro-/nano-structured materials from atoms or molecules [6]. This method is extensively exploited for the strategic design of materials. Examples of bottom-up techniques are sol-gel technology, solvothermal treatment, microwave-assisted synthesis, physical and chemical vapor deposition (CVD), etc. Figure 2.1 shows a brief understanding of both pathways [7–9].

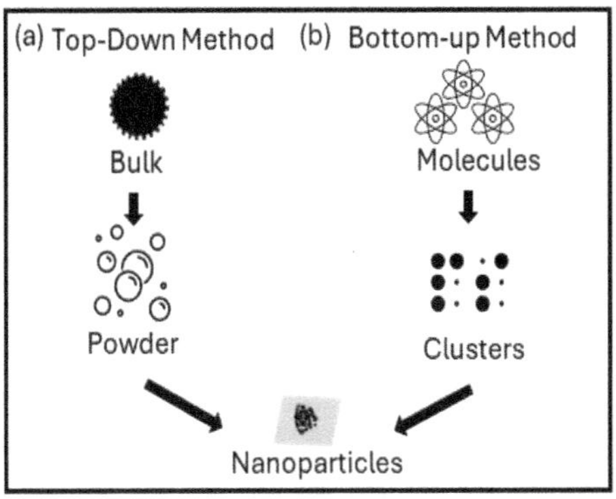

Figure 2.1 Schematic representation of two different approaches for nanoparticles synthesis.

2.1.2 Sol-Gel Techniques

This method comes under the bottom-up synthesis process. A sol is a colloidal phase of solution with very small particles where the gravitational force is inoperative, but van der Waals forces and surface charges are present, while the gel is a soft mass having a three-dimensional network encompassing the liquid. The colloidal particles begin to coagulate on evaporation of the solvent, and the gel is formed. Further, on grinding and sifting, the final product is obtained. The sol-gel technique is dependent on the sol's pH, the reaction's temperature, the reagent's concentration, the reaction time, and lastly, the nature of the catalysts employed. The sol-gel technique is inexpensive, needs low temperatures, and allows for precise control over product shape and morphology. This strategy anticipates that the rate of reactions and uniform dispersion of the dopants could be adjusted. The sol-gel procedure is regarded as a conventional chemical method capable of creating ceramic particles as small as 100 nm. It is a technique for achieving chemical homogeneity at the molecular level while fabricating ceramic structures. This technique enables the development of a high-quality product for use in sensitive industries by designing accurate trials. The selection of raw components is an important phase in the sol-gel process [10]. The structural and chemical characteristics of raw materials might have an impact on the product's quality. Figure 2.2 depicts the mechanism of the sol-gel method for materials design [11, 12].

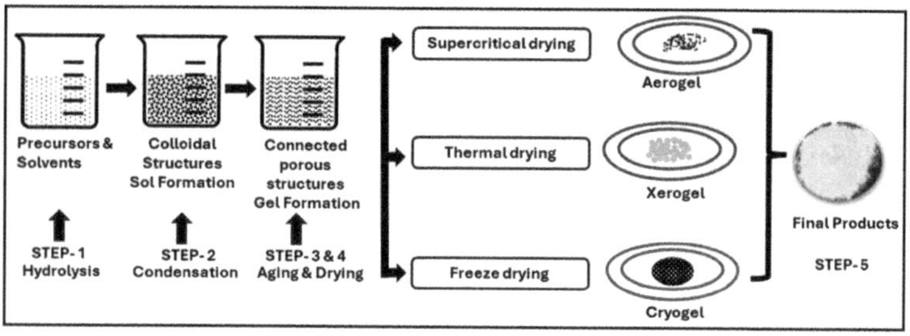

Figure 2.2 Schematic representation of the sol-gel method of nanoparticles synthetic approach.

2.1.3 Microwave-Assisted Synthesis

Microwave techniques are used to fabricate variety of inorganic materials. The frequency range of the microwave is 30 to 300 MHz [13]. Microwave techniques are clean, cheap, and faster with a high-efficiency rate. The fast heating in microwave is achieved by heating from within with continuous stirring. Metal salts, mostly silver (Ag) and gold (Au), or molecules, are placed inside the oven with a suitable concentration of reducing agents, which produce nanoparticles with a narrow distribution [14]. Besides, the microwave method has earlier produced Pt-, Cu-, and Ni-containing nanoparticles. Microwave irradiation is a popular approach for quickly producing nanomaterials, replacing traditional chemical processes, as it consumes more time. Microwave heating systems also provide advantages, such as great penetration and concentrated power, with the capacity to heat a system uniformly. Microwave synthesis produces more nucleation sites than standard heating techniques. Thus, for homogeneous nucleation, the products are formed with more purity. Additionally, by regulating the microwave's heating, the particle shapes can be modified. Particles with specific facets and superior form can significantly speed up the catalytic reactions [15–18].

2.1.4 Lithographic Process

A concentrated beam of light or electrons is employed in the lithographic technique to produce nanoarchitectures. It falls into two categories: masked lithography and maskless lithography. Nanopatterns are applied over substantial surface areas in masked lithography using a pre-made mask or template. This group of techniques includes nanoimprint lithography, soft lithography, and photolithography. However, maskless lithography uses techniques like focused ion beam lithography, scanning probe lithography, and electron beam lithography to print direct nanopatterns without using a mask [19]. The procedure for developing nanoarchitecture is shown in Fig. 2.3.

Maskless lithography prints random patterns straight onto the substrate. To build 3D freeform micro- and nanostructures, methods like focused ion beam lithography can also include wet chemical etching and ion implantation [20–21]. Vacuum-based and solution-based nanolithography techniques are commonly used to build 1D and 0D nanostructures with extremely fine line widths, which can be produced using these technologies [19]. These developments in nanolithography are essential to produce nanomaterials for a variety of uses [22–23].

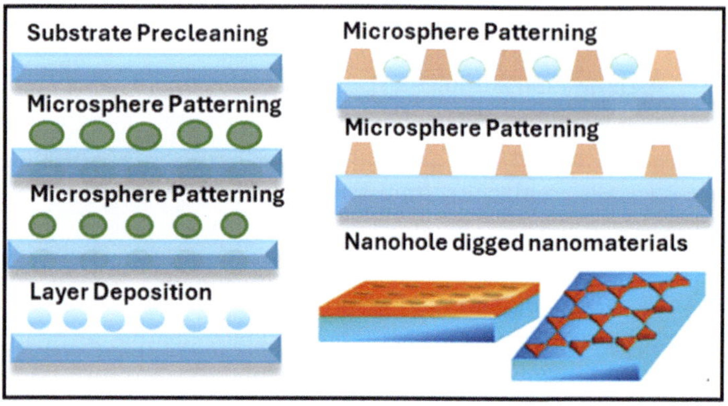

Figure 2.3 Typical lithography procedure for developing nanoarchitectures.

2.1.5 Chemical Vapor Deposition

The chemical components in the vapor phase were deposited on the solid substrate to make a thin film of micro or nanostructures as shown in Fig. 2.4. It shows how the shape of the thin film can be varied by adjusting the concentration and conditions of the component in the gas phase. Chemical vapor deposition (CVD) produces solid material with high purity and high performance [24]. The CVD process is associated with two steps: (i) the transportation of gas phase materials into the reaction zone and (ii) the deposition of the reactant onto the solid substrate. Mostly, the gas phase reactions are homogeneous, whereas the surface reactions are heterogeneous for fabricating thin films [25, 26]. The CVD process is categorized into three parts depending on the characteristics of the vapor:

Aerosol-Assisted CVD: It incorporates aspects from both aerosol and CVD methods. In this process, precursor materials are injected as aerosols into a heated reaction chamber, where chemical reactions take place to deposit the required substance on the substrate.

Direct Liquid Injection CVD: The liquid starting materials are injected into a vaporization chamber, and the vapors are carried out to the substrate. This technique is associated with a high growth rate.

Metalorganic CVD: It combines CVD with metal-organic precursors to produce high-quality films or structures with precise control over composition, thickness, and crystallinity.

CVD is a manageable approach for producing nanotubes with pre-ET characteristics (electron transfer processes) [27]. The CVD approach is the result of substantial work to develop a controllable route for synthesizing nanotubes by breaking down gaseous molecules as shown in Fig. 2.4. However, the arc discharge process produces more nanotubes, but they are not filtered [28].

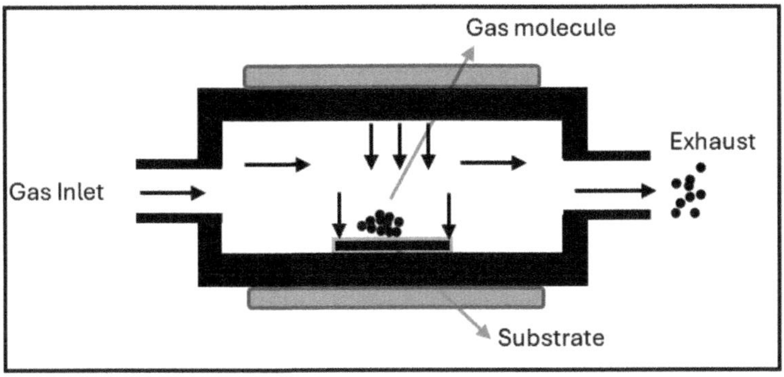

Figure 2.4 Schematic diagram of chemical vapor deposition technique.

2.1.6 Physical Vapor Deposition

It is a vacuum-based coating process in which material is vaporized from a liquid or solid source and then condensed on a substrate to form a thin layer. The process involves the vaporization of the target material, often through heating or bombardment with ions, and its subsequent travel through a vacuum environment to be deposited on a substrate surface. Physical vapor deposition (PVD) is commonly used in the semiconductor, aerospace, and tool-coating industries for producing high-quality films with excellent adhesion, uniformity, and purity. It offers versatility in the deposition of metals, ceramics, and composite materials, making it ideal for applications in electronics, optical coatings, decorative finishes, and hard coatings for wear resistance. The primary benefits of PVD are precise control over film characteristics, environmental friendliness, and the ability to create high-performance coatings without the need for chemical precursors.

2.1.6.1 Two-dimensional layered material

Researchers have focused significant emphasis on the various uses of two-dimensional (2D) structured materials in disciplines like materials science, nanotechnology, and electrocatalysis. In the past, metal chalcogenides, graphite, and boron nitride were well-known examples of 2D inorganic layered materials. The efficiency of molybdenum disulfide (MoS_2) and layered double hydroxides (LDHs), sometimes known as hydrotalcite, in electrocatalytic applications was acknowledged in the 1970s and 1980s. These layered materials are easily exfoliated into nanosheets due to their stronger coplanar covalent interactions and active van der Waals interactions across the 2D planes inside each layer. Several 2D nanomaterials are being used in a variety of applications today, including MXenes (including silicene and germanene), layered metal oxides (including layered perovskites and P2-layered oxides), nonmetallic layered materials (like graphene and boron nitride), LDHs, and 2D transition metal composites [29].

2.1.6.2 Synthesis of layered structural materials

In general, top-down and bottom-up methods are implemented for layered materials synthesis. The exfoliation technique can easily remove the uppermost layer of the 2D nanomaterials and produce nanosheets with mono- or multi-layers. Exfoliation methods can modulate the physicochemical properties of the catalyst. Exfoliation techniques are classified as mechanical (i.e., Au-initiated and traditional mechanical exfoliation) and liquid phase exfoliation (oxidation and ion intercalation/exchange). Moreover, chemical vapor and physical vapor deposition can also produce layered materials with good thicknesses [30].

2.1.7 Mechanical Exfoliation

Mechanical exfoliation, which was utilized in the initial attempt to isolate graphene, is one of the most successful procedures for this purpose as depicted in Fig. 2.5. Single- and few-layered graphene from graphite were fabricated by new ribbon-based exfoliation processes, as reported by Geim and co-workers in 2004 [31]. In addition to this, transition metal dichalcogenides (TMDs), hexagonal boron nitride, and MoS_2 were prepared by this method. The conventional method is to peel off the larger particles with glued tape. After repeating this step multiple times, a few-layered or monolayer nanosheets are obtained. The exfoliation technique is advantageous as it does not require any toxic chemicals. The inherent nature of the crystal remains conserved, and a material with a clean surface is produced. On the contrary, manual handling sometimes lacks accuracy and reproducibility.

Oxygen plasma-triggered exfoliation was reported by Sutter et al. from the University of Nebraska–Lincoln, USA, for the developing $Bi_2Sr_2CaCu_2O_x$ nanosheets [32]. The undesired molecules absorbed on the surface of the layered structures can easily be wiped off by oxygen plasma treatment. Gold-assisted exfoliation techniques were put into use for generating single-layer TMD nanosheets on different substrates, for example, SiO_2/Si or quartz. Once again, such a method was adopted by Magda and colleagues on the mica surface [33].

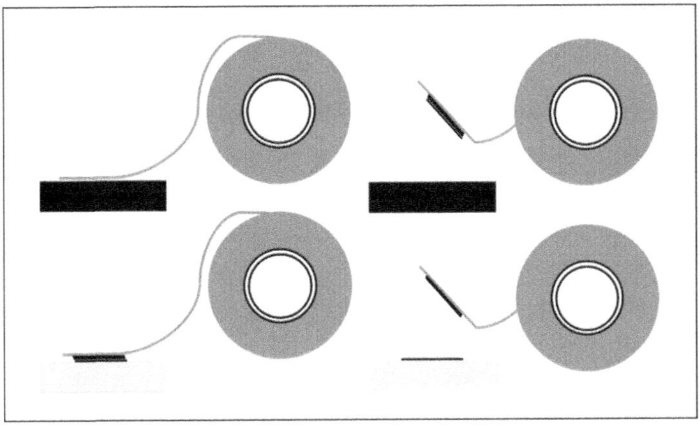

Figure 2.5 Schematic diagram of mechanical exfoliation technique.

2.1.8 Liquid-Phase Stripping

A widely accepted method for preparing 2D nanosheets is mechanical exfoliation, although it is associated with low yield as shown in Fig. 2.6. Nanosheets are dispersed in a surfactant-containing aqueous or organic solution on an industrial scale in the liquid-phase exfoliation method. Such methods produce a large surface area of 1000 m^2. Three-step techniques are adopted for fabricating layered materials. First, oxidative sonication is performed, followed by ion intercalation, and lastly, ion exchange. In 2006, Ruoff and co-workers prepared thick, monolayer graphene oxide by the oxidation of graphite [34]. The most common method for graphene oxide preparation is Hummer's method, where graphite powder is used as the raw material and sulfuric acid/potassium permanganate are used as strong oxidizing agents to incorporate –OH and epoxide functional moieties on the surface of the substrate. The increment in the percentage oxygen content on the surface enhances the extent of hydrophilicity, with a more probability of water intercalation. Direct ultrasonication was employed on layered crystals in N, N'-dimethylformamide (DMF), dimethyl sulfoxide (DMSO), and acetone to synthesize various 2D nanosheets, for example, graphene, BN, $TaSe_2$, MoS_2, and $MoSe_2$. The oxygenated functional group on the surface can control electrical conductivity and resistivity [35].

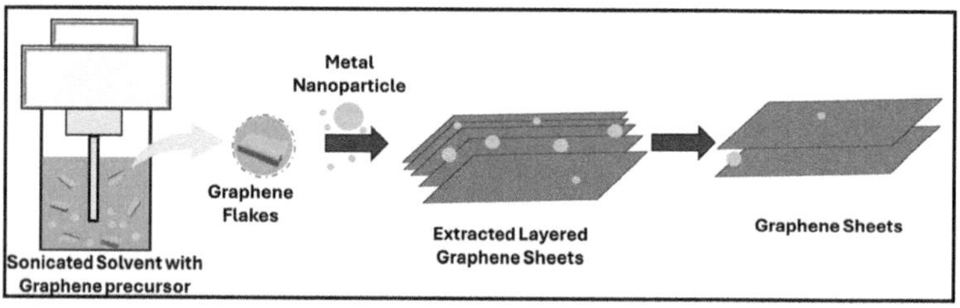

Figure 2.6 Block diagram of liquid-phase stripping.

2D-layered materials are used to develop lithium-ion batteries using the intercalation exfoliation method. Highly pure and multi-layered WS_2 nanosheets were prepared by the lithium-ion intercalation exfoliation method on a large scale. In addition to the intercalation exfoliation method, the ion exchange-enabled liquid-phase exfoliation process is also well known. Adachi-Pagano and co-workers utilized an anionic surfactant dodecyl sulfate with a dispersing agent butanol for the successful exfoliation of LDHs in ion exchange method [36].

2.1.9 Single-Source Precursor Approach

The single-source precursor (SSP) approach is often used to design the catalysts for electrocatalytic reactions. In this technique, the electrocatalysts are synthesized using the molecules, and that specific molecules contain all the required atoms required for nanoparticle nucleation from corresponding metal salt. Besides, on proper modification of SSP techniques, different chalcogens can be embodied into the targeted phase of a metal to yield desired metal chalcogenide phases. The SSP approach is preferred over conventional solid-state methods as it required lower temperatures with less reaction time as higher temperature often fostered larger particles with mixed phases or with undesired impurity. In addition, the final products obtained in SSP techniques are of high purity with homogenous elemental composition. Under identical reaction conditions, the precursor will produce desired metal chalcogenides with high yield. Using SSP for thin film deposition eliminates the need for pre-reaction in the vapor phase, which is a common issue when using dual precursors. Hence, maintaining phase purity and reproducibility is one of the significant difficulties in dual-/multi-source routes. The SSP method effectively addresses these issues and offers numerous advantages [35].

Compared to multi-source synthetic methods, adopting an SSP provides significant advantages in the synthesis of semiconductor nanomaterials. Chemical linkages between constituent elements may minimize the nucleation activation energy, allowing for low-temperature access to the final material. The breakdown temperature of SSP depends on the molecular complexes' kinetic and thermal stability, which is determined by the metal, auxiliary ligand, and chalcogen atom. Driess and Menzes group reported the SSP approach to design nickel chalcogenide precatalysts for electrocatalytic oxidations of water, alcohol, and aldehydes [37]. Chakraborty and co-workers prepared mineral digenite Cu_9S_5 for OERs [38].

2.1.10 Hydrothermal and Solvothermal Approach

Hydrothermal or solvothermal technique involves the preparation of materials under ambient temperature and pressure in a sealed stainless-steel container. On utilization of aqueous media, the reaction is termed as hydrothermal, while for nonaqueous solvent it is solvothermal. It is one of widely accepted route for inorganic material synthesis for multiple applications, such as CO_2 reduction (CO_2RR), oxygen evolution, hydrogen evolution reaction (HER), and many more [39–41]. This process allows for better control of catalyst's size, shape, and crystallinity, thus enhancing their stability and efficiency. Bhattacharya and group developed Tb-doped graphene quantum dots, achieving enhanced multicolor emission for advanced optoelectronic applications [42]. Weinstock and co-workers fabricated an anatase TiO_2 core adhered with polyoxometalate clusters solvothermal for photocatalytic HER. The metal polyoxometalate composite was prepared with titanium (IV) tetra-isopropoxide in isopropanol and $Na_7[\alpha\text{-}PW_{11}O_{39}]\cdot 12H_2O$ in water of pH 6.5 at 170°C for 20 h [43]. Chakraborty and his group synthesized a γ-FeO(OH) hydrothermally for electrocatalytic OERs from Ferric nitrate nonahydrate in KOH (1 M, pH ~12) solution at 85°C for 24 h in an autoclave. Moreover, on alteration of concentration of iron salt and reaction temperature, a series of pure phase α-FeO(OH) (goethtite) and γ-FeO(OH) (lepidocrocite) materials can be developed [45]. This group also devised a hexagonal covellite phase CuS in a hydrothermal approach for OER in alkaline medium for overall water splitting and electrochemical alcohol oxidation. The CuS material was derived from copper (II) chloride with thiourea in deionized water and kept in an oven for 5 h at 150°C. These reported materials served as efficient electrocatalysts for multiple electrochemical transformations picture of hydrothermal autoclave is shown in Fig. 2.7 [46]. Bhattacharya and his group developed sulfanilamide-functionalized carbon dots synthesized from citric acid exhibit zwitterionic behavior and selective antibacterial activity and selective detection of Cr(VI) static quenching and its electrocatalytic application in HER [44, 47–49].

Figure 2.7 Picture of an autoclave reactor.

2.2 Conclusion

To design an electrocatalyst, some issues need to be taken into account such as, its thermal and chemical stability, dispersibility in the solvent the media, inertness, simple fabrication method, and lastly the ease of surface modification. In this chapter, the structure, design, and the techniques associated with the fabrication of the metallic and nonmetallic materials commonly engaged for electrocatalytic reaction were reviewed. The methodology adopted for preparation of metallic materials were discussed. The most commonly used techniques—including top-down, bottom-up, sol-gel process, lithography, CVD, physical vapor deposition, mechanical exfoliation, single precursor approach, hydrothermal/solvothermal methods, and liquid-phase stripping—were thoroughly considered.

The chapter concludes by discussing the wider range of metallic and nonmetallic materials in electrocatalysis, and more specifically, the areas of metal nanoparticles, MOFs, TMDs, RGO, and 2D metallic complexes, including carbon-based materials. These materials are critically important in energy conversion and storage because of their high catalytic activity and tailoring of electronic structure. Modern synthesis techniques such as mechanical exfoliation, solvothermal synthesis, and CVD facilitate electrocatalytic effectiveness by enabling material properties to be controlled. The development of nanomaterials such as metal nanoparticles, MOFs, TMDs, RGO, and 2D metallic complexes represents a transformative shift in the design of advanced electrocatalytic systems. These materials offer distinct advantages in electrocatalysis due to their high surface-to-volume ratios, tunable electronic structures, and abundant active sites. MOFs, with their inherent porosity and structural flexibility, can serve as precursors for the synthesis of porous carbon-based catalysts with embedded metal centers. Electrocatalytic applications such as OER, HER, carbon dioxide reduction reaction (CO_2RR), and nitrogen reduction reaction (NRR) greatly benefit from the preparation of these nanomaterials. The structural tunability, doping strategies, and surface modification techniques available for these classes of materials allow to achieve the targeted enhancement of catalytic activity, selectivity, and long-term stability. Meanwhile, 2D metallic complexes and heterostructures open new avenues for synergistic effects by combining multiple functionalities at the atomic scale.

These developments in research not only enable the development of better and more efficient and sustainable energy systems but also emphasize the importance of further electrocatalytic research advancement.

Looking ahead, the future of electrocatalysis depends on the rational material synthesis, theoretical modeling, and in situ characterization techniques. Sustainable and scalable production methods, coupled with a deeper understanding of reaction mechanisms at the nanoscale, will be crucial in translating these materials from the laboratory scale to technology development with fabrication of commercial energy devices. Thus, the interplay between nanostructure design and electrocatalytic function will continue to be a cornerstone of innovation in energy conversion and environmental remediation technologies.

References

1. Suzuki, S., Iwasaki, T., De Silva, K. K. H., Suehara, S., et al. (2021). Direct growth of germanene at interfaces between van der Waals materials and Ag(111). *Adv. Funct. Mater.* 31(5), 1–34.
2. Burkholder, M. B., Rahman, F. B. A., Chandler, E. H., Regalbuto, J. R., et al. (2022). Metal supported graphene catalysis: A review on the benefits of nanoparticular supported specialty sp^2 carbon catalysts on enhancing the activities of multiple chemical transformations. *Carbon Trends* 9.
3. Chen, P., Tong, Y., Wu, C., and Xie, Y. (2018). Surface/interfacial engineering of inorganic low-dimensional electrode materials for electrocatalysis. *Acc. of Chem. Res.* 51(11), 2857–2866.
4. Chia, X. and Pumera, M. (2018). Characteristics and performance of two-dimensional materials for electrocatalysis. *Nat. Catal.* 1(12), 909–921.
5. Yu, H. D., Regulacio, M. D., Ye, E., and Han, M. Y. (2013). Chemical routes to top-down nanofabrication. *Chem. Soc. Rev.* 42(14), 6006–6018.
6. Biswas, A., Bayer, I. S., Biris, A. S., Wang, T., et al. (2012). Advances in top-down and bottom-up surface nanofabrication: Techniques, applications & future prospects. *Adv. Colloid Interface Sci.* 170(1–2), 2–27.
7. Chattopadhyay, K. K. (2009). *Introduction to Nanoscience and Nanotechnology*. PHI Learning Pvt. Ltd.

8. Choi, W. K., Liew, T. H., Chew, H. G., Zheng, F., et al. (2008). A combined top-down and bottom-up approach for precise placement of metal nanoparticles on silicon. *Small* 4(3), 330–333.
9. de Oliveira, P. F., Torresi, R. M., Emmerling, F. and Camargo, P. H. (2020). Challenges and opportunities in the bottom-up mechanochemical synthesis of noble metal nanoparticles. *J. Mater. Chem. A* 8(32), 16114–16141.
10. Pradeep, T. (2007). NANO: *The Essentials: Understanding Nanoscience and Nanotechnology*. 1st ed. McGraw-Hill: New Delhi.
11. Mahmoudi, Z., Soleymani, F., Mesgari Abbasi, S., Khalaj, G., et al. (2024). Sol-gel synthesis and characterization of mesoporous TiB_2 nano powder via Titanium isopropoxide/trimethyl borate precursors. *Ceram. Int.* 50(11), 18081–18089.
12. Dubey, R. S., Rajesh, Y. B. R. D., and More, M. A. (2015). Synthesis and characterization of SiO_2 nanoparticles via sol-gel method for industrial applications. *Mater. Today: Proc.* 2(4–5), 3575–3579.
13. Zhu, Y. J. and Chen, F. (2014). Microwave-assisted preparation of inorganic nanostructures in liquid phase. *Chem. Rev.* 114(12), 6462–6555.
14. Law, V. J. and Dowling, D. P. (2023). Microwave-assisted Au and Ag nanoparticle synthesis: An energy phase-space projection analysis. *Am. J. Anal. Chem.* 14(04), 149–174.
15. Kundu, S., Ma, L., Chen, Y., and Liang, H. (2017). Microwave assisted swift synthesis of $ZnWO_4$ nanomaterials: material for enhanced photo-catalytic activity. *J. Photochem. Photobiol. A: Chem.* 346, 249–264.
16. Dahal, N., Garcia, S., Zhou, J., and Humphrey, S. M. (2012). Beneficial effects of microwave-assisted heating versus conventional heating in noble metal nanoparticle synthesis. *ACS Nano* 6(11), 9433–9446.
17. Motshekga, S. C., Pillai, S. K., Sinha Ray, S., Jalama, K., et al. (2012). Recent trends in the microwave-assisted synthesis of metal oxide nanoparticles supported on carbon nanotubes and their applications. *J. Nanomater.* (1), 691503.
18. Li, D. and Komarneni, S. (2006). Microwave-assisted polyol process for synthesis of Ni nanoparticles. *J. Am. Ceram. Soc.* 89(5), 1510–1517.
19. Hector, S. and Mangat, P. (2001). Review of progress in extreme ultraviolet lithography masks. *J. Vac. Sci. Technol., B: Microelectron. Nanometer Struct. Process. Meas., Phenom.* 19(6), 2612–2616.
20. Henry, D., Gemmink, J. W., Pain, L., and Postnikov, S. V. (2006). Status and future of maskless lithography. *Microelectron. Eng.* 83(4–9 SPEC. ISS.), 951–955.

21. Raza, A., Saeed, Z., Aslam, A., Nizami, S. M., et al. (2024). "Advances, Application and Challenges of Lithography Techniques" in 2024 5th International Conference on Advancements in Computational Sciences (ICACS)., 1–6.

22. Baig, N., Kammakakam, I., Falath, W., and Kammakakam, I. (2021). Nanomaterials: A review of synthesis methods, properties, recent progress, and challenges. *Mater. Adv.* 2(6), 1821–1871.

23. Modaresialam, M., Chehadi, Z., Bottein, T., Abbarchi, M., et al. (2021). Nanoimprint lithography processing of inorganic-based materials. *Chem. Mat.* 33(14), 5464–5482.

24. Cai, Z., Liu, B., Zou, X., and Cheng, H.-M. (2018). Chemical vapor deposition growth and applications of two-dimensional materials and their heterostructures. *Chem. Rev.* 118(13), 6091–6133.

25. Sun, L., Yuan, G., Gao, L., Yang, J., et al. (2021). Chemical vapour deposition. *Nat. Rev. Methods Prime.* 1(1), 5.

26. Choy, K. L. (2003). Chemical vapour deposition of coatings. *Prog. Mater. Sci.* 48(2), 57–170.

27. Dai, H., Rinzler, A. G., Nikolaev, P., Thess, A., et al. (1996). HiPCO-Rice University. *Chem. Phys. Lett.* 260, 471–475.

28. Nizamuddin, S., Abbasi, S. A., Jatoi, A. S., Siddiqui, M. T. H., et al. (2019). Microwave-assisted synthesis for carbon nanomaterials. *Nanomater. Synth. Des. Fabr. Appl.* 121–147.

29. Shi, Y., Li, H., and Li, L. J. (2015). Recent advances in controlled synthesis of two-dimensional transition metal dichalcogenides via vapour deposition techniques. *Chem. Soc. Rev.* 44(9), 2744–2756.

30. Cai, Z., Liu, B., Zou, X., and Cheng, H. M. (2018). Chemical vapor deposition growth and applications of two-dimensional materials and their heterostructures. *Chem. Rev.* 118(13), 6091–6133.

31. Novoselov, K. S., Geim, A. K., Morozov, S. V., Jiang, D., et al. (2004). Electric field in atomically thin carbon films. *Science* 306(5696), 666–669.

32. Huang, Y., Sutter, E., Shi, N. N., Zheng, J., et al. (2015). Reliable exfoliation of large-area high-quality flakes of graphene and other two-dimensional materials. *ACS Nano* 9(11), 10612–10620.

33. Magda, G. Z., Pető, J., Dobrik, G., Hwang, C., Biró, L. P., and Tapasztó, L. (2015). Exfoliation of large-area transition metal chalcogenide single layers. *Sci. Rep.*, 5(1), 14714.

34. Dreyer, D. R., Park, S., Bielawski, C. W., and Ruoff, R. S. (2010). The chemistry of graphene oxide. *Chem. Soc. Rev.* 39(1), 228–240.

35. Hadi, A., Zahirifar, J., Karimi-Sabet, J., and Dastbaz, A. (2018). Graphene nanosheets preparation using magnetic nanoparticle assisted liquid phase exfoliation of graphite: The coupled effect of ultrasound and wedging nanoparticles, *Ultrason. Sonochem.* 44, 204–214.

36. Adachi-Pagano, M., Forano, C., and Besse, J. P. (2003). Synthesis of Al-rich hydrotalcite-like compounds by using the urea hydrolysis reaction—control of size and morphology. *J. Mater. Chem.*, 13(8), 1988–1993.

37. Karmakar, G., Atomic, B., Tyagi, A., Atomic, B., et al. (2021). Exploring the gravity of tin chalcogenides as potential energy materials. *SMC Bulletin* 12(1).

38. Dasgupta, B., Yao, S., Mondal, I., and Mebs, S., et al. (2024). A knowledge-based molecular single-source precursor approach to nickel chalcogenide precatalysts for electrocatalytic water, alcohol, and aldehyde oxidations, *ACS Nano* 18, 33964–33976.

39. Chakraborty B., Kalra S., Beltrán-Suito R., Das C., et al. (2020). A low-temperature molecular precursor approach to copper-based nano-sized digenite mineral for efficient electrocatalytic oxygen evolution reaction. *Chem Asian J.* 15(6), 852–859.

40. Muralikrishna, S., Manjunath, K., Samrat, D., Reddy, V., et al. (2015). Hydrothermal synthesis of 2D MoS_2 nanosheets for electrocatalytic hydrogen evolution reaction. *RSC Adv.* 5(109), 89389–89396.

41. Wang, X., Kolen'ko, Y. V., and Liu, L. (2015). Direct solvothermal phosphorization of nickel foam to fabricate integrated Ni_2 P-nanorods/Ni electrodes for efficient electrocatalytic hydrogen evolution. *Chem. Commun.* 51(31), 6738–6741.

42. Sarkar, U. and Bhattacharya, S. (2025). White light emission from luminescent lanthanide molecular complex/terbium doped graphene quantum dots/PMMA Composites, *Small* 2503419.

43. Raula, M., Gan Or, G., Saganovich, M., Zeiri, O., et al. (2015). Polyoxometalate complexes of anatase-titanium dioxide cores in water, *Angew. Chemie Int. Ed.* 54, 12416–12421.

44. Mishra, A. and Bhattacharya, S. (2024). Switching photoluminescence "On-Off-On" for detection of chromium (VI) ion and ascorbic acid with N, S-Codoped carbon dots, *ChemistrySelect* 9, e202403960.

45. Mallick, L., Rajput, A., Adak, M. K., Kundu, A., et al. (2022). γ-FeO(OH) with multiple surface terminations: Intrinsically active for the electrocatalytic oxygen evolution reaction, *Dalt. Trans.* 51, 15094–15110.

46. Kundu, A., Adak, M. K., Kumar, Y., and Chakraborty, B. (2022). Electrochemically derived crystalline CuO from covellite CuS nanoplates: A multifunctional anode material, *Inorg. Chem.* 61, 4995–5009.

47. Mishra, A., Lzaod, S., Dutta, T., and Bhattacharya, S. (2024). Selective bacterial growth inactivation by pH-sensitive sulfanilamide functionalized carbon dots, *ACS Appl. Bio Mater.* 7, 2752–2761.

48. Mehta, I, Rajput, A., Yewale, P., Chakraborty, B., and Bhattacharya, S. (2025). The electrocatalytic hydrogen evolution reaction using a heteroatom-doped hollow onion-like fullerene across a wide pH range, *Nanoscale*, 17, 27947–27958.

49. Mishra, A., Basak, H. K., Chakraborty, B., and Bhattacharya, S. (2026). Hydrogen evolution reaction using a sulfanilamide/citric acid derived N, S-doped carbon dot, *ACS Appl. Energy Mater.*, 9, 1771–1781.

Chapter 3

Spectroscopic Techniques for Material Characterization

Astha Gupta, Anakshi Boruah, Prekshi Gupta, Samya Suvra Datta, and Surojit Pande
Department of Chemistry, Birla Institute of Technology and Science, Pilani, Rajasthan-333031, India
spande@pilani.bits-pilani.ac.in

3.1 Introduction

Over the past three decades, the research field of nanomaterials has experienced exponential growth. The distinct physicochemical characteristics such as size, shape, surface area to volume ratio, and surface properties of nanomaterials are often markedly different from their bulk counterparts. These characteristics make them highly versatile and suitable for a range of applications in various fields such as energy, medicine, electronics, optical engineering, etc. Since nanomaterials are gradually getting integrated into advanced technologies, a comprehensive understanding of their fundamental properties becomes crucial to exploit their full

Materials Design for Electrocatalysis
Edited by Sagarika Bhattacharya and Biswarup Chakraborty

ISBN 978-981-5352-32-0 (Hardcover), 978-1-042-02558-9 (eBook)
www.jennystanford.com

DOI: 10.1201/9781042025589-3

potential. A thorough physicochemical characterization of the nanomaterials is essential for an in-depth understanding that can affect their properties. Characterization of nanomaterials can be done by two different methods: microscopic and spectroscopic, both providing different information about the nanomaterials. While microscopic techniques give information about the size, shape, and morphology of nanomaterials, spectroscopic techniques are used to obtain the structural and electronic information. Both methods are indispensable for a holistic understanding of nanomaterials.

The objective of this chapter is to give a comprehensive overview of the key spectroscopic techniques used in the characterization of nanomaterials. It explores the fundamental principles, sample preparation, and applications of different spectroscopic techniques that are essential in analyzing nanomaterials. Herein, we have widely discussed the techniques for crystal structure information such as X-ray diffraction analysis (XRD). It generates detailed information about the atomic arrangement, phase composition, and crystallinity of nanomaterials. The data obtained gives information about lattice parameters, crystal size, and strain, which are vital for understanding how the material's structure influences its properties. Fourier Transform Infrared (FTIR) spectroscopy is utilized to acquire molecular-level structural information about nanomaterials. FTIR spectra provide insights into the chemical structure of surface-bound species, specifically for nanomaterials, it helps to understand how the surface chemistry influences their reactivity and interactions with other materials. In the nanoscale, the nature of the material influences the properties, which makes surface study utmost important. Raman spectroscopy offers complementary information to FTIR and is sensitive to vibrational modes of molecular bonds, making it valuable for studying the atomic vibrations and defects in nanomaterials. The information about the chemical and electronic state, composition of elements, etc., of nanomaterials can be obtained from X-ray photoelectron spectroscopy (XPS). This technique is predominantly useful for studying oxidation states, bonding configurations, and surface modifications of nanomaterials. The state-of-the-art technique, X-ray absorption spectroscopy (XAS), gives information at the atomic level and provides insight into the distance and types of the different species from the absorbing atom. UV-Vis spectroscopy is particularly useful for characterizing metal nanoparticles, which exhibit unique plasmonic properties that are size- and shape-dependent, and also provides information about their optical properties. In this chapter, we will explore the theoretical basics of these spectroscopic techniques and their applications for characterization of nanomaterials.

3.2 Spectroscopic Characterization of Materials

For the characterization of materials, there are many useful spectroscopic techniques. For example, XRD, FTIR, Raman, XPS, XAS, and UV-Vis. Herein, we will focus on the basics of each spectroscopic technique.

In 1895, Wilhelm Conrad Roentgen made a major breakthrough by discovering X-rays when working with a discharge tube. In 1912, Max von Laue discovered the wave-like properties of X-rays by demonstrating that they could be diffracted by crystals. Laue, Friedrich, and Knipping further explored X-rays, which led to developing XRD techniques for studying crystals. X-rays, with wavelengths in the range of 0.01 nm to 10 nm, interact strongly with atomic structures due to their short wavelengths, which are comparable to atomic radii and bond lengths. With the increase in atomic number of an atom, the number of electrons increases, resulting in enhanced absorption and scattering of the X-rays, making it possible to differentiate the elements across the periodic table. X-rays are generated in X-ray tubes, which consist of a cathode and an anode in a vacuum or low-pressure environment. When a high-voltage field accelerates high-energy electrons and these electrons strike the metal target, usually copper, molybdenum, or silver, their kinetic energy is converted into X-ray radiation. The interaction of accelerated electrons with the anode material results in two types of X-rays: characteristic X-rays and bremsstrahlung radiation. The energy released during the transition (from a higher-energy shell to the lower-energy vacancy) is emitted as an X-ray photon, known as a characteristic X-ray, with an energy that is unique to the element of the anode material.[1–3]

FTIR spectroscopy is supplementary to XRD, which uses infrared (IR) radiation range of the electromagnetic spectrum for analysis. FTIR spectroscopy provides valuable information about the functional groups in a material, which helps to predict the molecular structure. FTIR is widely used due to its sensitivity and non-destructive nature. The technique is based on the interaction of IR waves with a sample, causing vibrational transitions in the molecular bonds. These transitions produce absorption bands unique to the sample, creating a vibrational spectrum. FTIR spectroscopy, developed after the Fourier Transform algorithm, was introduced in the 1960s, allows for simultaneous measurement of all IR wavelengths, making it an essential tool for both academic research and industrial applications.[4,5]

Complementary to the above-discussed techniques, Raman spectroscopy examines the inelastic scattering of light scattered by molecules, which was found by C. V. Raman in 1928. In 1930, he won the Nobel Prize in Physics for this groundbreaking discovery. Raman scattering occurs when an inelastic collision between light and matter takes place, followed by a reduction or increment of the frequency of the scattered radiation with reference to incident light, resulting in Stokes and anti-Stokes shift, respectively. It provides vibrational and structural insights into materials, including symmetry, interlayer interactions, and defects. Improvements in lasers and detectors have enhanced the technique's sensitivity for analyzing molecular vibrations and structural fingerprints.[4]

For surface analysis, XPS is essential. The surface of a solid material interacts with its surroundings, making it the most critical region of the material. Many properties such as catalysis, adhesion, friction, optical, electrical properties, and corrosion are dependent on the interaction of the surface with the environment. Thus, the study of the surface to understand these specific phenomena becomes extremely important, especially in the field of materials in the nano domain. XPS is a surface-sensitive analytical technique for the analysis of chemical composition, oxidation state, etc. In 1960, Kai Siegbahn and co-workers developed XPS at the University of Uppsala, Sweden, for which he was awarded the Nobel Prize in Physics, 1980. In XPS, the sample is bombarded with X-rays ($E > 1000$ eV), and the kinetic energy of the released electrons, which are known as photoelectrons, is analyzed. XPS analysis is performed under high vacuum conditions, typically around 10^{-6} Pascal. A range of solid samples, such as thin films, powder samples, glasses, ceramics, fiber, etc., can be analyzed with the help of XPS.[6]

In order to further study, the local atomic structure, valence state, and symmetry of the absorbing atom, XAS is discussed. It involves the analysis of attenuation in the intensity of a monochromatic X-ray beam when it passes through a sample and reaches the detector. This process of absorption of X-rays by the target atoms depends on the quantized energy levels of the atoms. The energy of the X-ray utilized depends on the specific element that is analyzed.[7]

Another crucial analytical tool is the UV-Vis spectroscopy, which operates on the UV-Vis region of the spectrum that ranges from 190 to 400 nm and 400 to 800 nm, respectively. Molecules absorb radiation in the UV and visible regions, causing transitions between electronic energy levels, especially in the valence shells. As a result, electronic-absorption spectroscopy is another name for UV-visible spectroscopy. It is the most extensively utilized analytical technique to measure the extent of light absorbed by an analyte as a function of frequency or wavelength.[8] Diffuse reflectance spectroscopy (DRS) is used to study the spectral characteristics of solid opaque samples. Surface-reflected electromagnetic radiation is collected and examined as a function of wavelength or frequency to produce a reflectance spectrum.[9]

3.3 Powder X-Ray Diffraction Analysis

3.3.1 Fundamental Concepts

Bragg's Law: Bragg's law explains how X-rays diffract off a crystal's atomic planes, leading to constructive or destructive interference. When the path difference is equivalent to an integral multiple of the X-ray wavelength, it results in constructive interference, which produces strong reflections. Using the basic concept of Bragg's law, we can get the information of Miller indices (hkl), interplanar spacing (d), etc.[1]

Reciprocal Lattice: It is a mathematical concept used to analyze the diffraction of X-rays or electrons by a crystal's periodic atomic structure. It is a 3D grid perpendicular to the real-space lattice and helps visualize the scattering patterns of X-rays. While the real-space lattice represents the physical positions of atoms, the reciprocal lattice represents diffraction reflections. Directions in real and reciprocal space are indexed as lattice as r_{uvw} with unit vectors *a*, *b*, and *c* in real space, and *u*, *v*, and *w* are integers for indexing crystallographic directions, and the direction in the reciprocal lattice is defined by d^*_{hkl} in reciprocal space, represented by Eqs. 3.1 and 3.2, respectively.[3]

$$r_{uvw} = ua + vb + wc \tag{3.1}$$

$$d^*_{hkl} = \mathrm{h}a^* + \mathrm{k}b^* + \mathrm{l}c^* \tag{3.2}$$

Ewald Sphere: This method is a graphical way that uses the reciprocal lattice to illustrate the conditions of Bragg's law. It is an imaginary sphere having a diameter of $2/\lambda$, where λ represents the wavelength of the incident X-ray or electron beam in reciprocal space. The incident beam (OA) passes through the crystal, centered at the inception of the reciprocal lattice. The diffracted beam (AB) connects the sphere's center to a lattice point in the reciprocal lattice. When the lattice point coincides with the Ewald sphere, and the diffraction angle (θ) must be twice the angle between the incident and diffracted beams (OA and AB), Bragg's law is fulfilled. This condition leads to constructive interference, predicting a diffraction peak in that direction. From Fig. 3.1,

$$\mathrm{OA} = \frac{1}{\lambda}\ ,\ \mathrm{OC} = \frac{d^*_{\mathrm{hkl}}}{2} \tag{3.3}$$

Hence,

$$\sin\theta = \frac{\mathrm{OC}}{\mathrm{OA}} = \frac{\frac{d^*_{\mathrm{hkl}}}{2}}{\frac{1}{\lambda}} \tag{3.4}$$

Recall,

$$d^*_{\mathrm{hkl}} = \frac{1}{d_{\mathrm{hkl}}} \tag{3.5}$$

Thus,

$$\lambda = 2d_{\mathrm{hkl}}\sin\theta \tag{3.6}$$

This proves the correlation between the Ewald Sphere and Bragg's law. Thus, the Ewald sphere is just a mathematical tool to visualize the conditions for diffraction, which helps to determine the points in the reciprocal lattice that yield a diffraction peak after interacting with the incident beam based on the orientation of the crystal and the λ of the incident beam.[3,10]

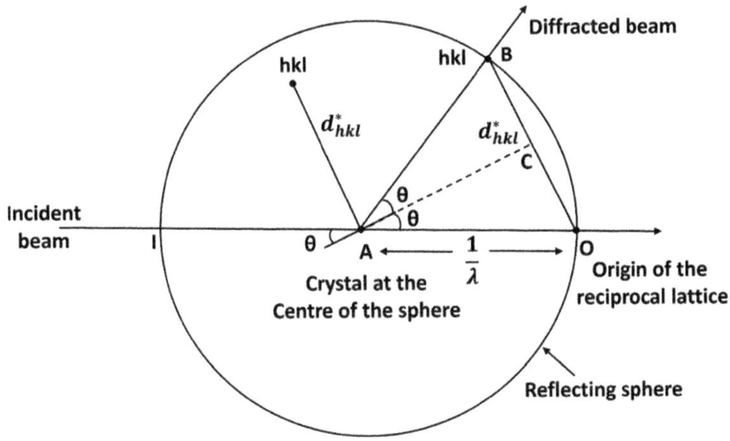

Figure 3.1 Construction of Ewald sphere.

Diffraction Intensity: The XRD peak intensity depends not only on the conditions given by Bragg's law but also on the scattering factors of the atoms in the crystal. The diffraction intensity is dependent on the combined scattering by all the electrons of the constituent atoms in the crystal. When X-rays interact with a crystal, they are scattered by the electrons surrounding the atoms, not by the atomic nuclei. The scattering intensity scattered by an electron is dependent on the angle between the incident and the scattered direction 2θ and is given by Eq. 3.7:

$$I(2\theta) = \frac{I_o}{r^2} K \frac{1+\cos^2(2\theta)}{2} \tag{3.7}$$

where I_o is the incident beam intensity, r is the distance from the electron to the detector, and K is a constant for atomic properties. The atomic scattering factor quantifies the scattering intensity and depends on the X-ray wavelength and the direction of the scattered beam. If the X-ray wavelength is constant, the scattering intensity is inversely proportional to 2θ. The atomic scattering factor (f) is given in Eq. 3.8:[2,3]

$$f = \frac{\text{Amplitude of the wave scattered by one atom}}{\text{Amplitude of the wave scattered by one electron}} \quad (3.8)$$

Structure Extinction and Structure Factor: Structure extinction refers to the absence or reduced intensity of some diffraction peaks, even when they are theoretically expected based on the crystal structure. The summarized rules for the structure extinction for a crystallographic plane (hkl) are given in Table 3.1. The structure factor F_{hkl} for a plane (hkl) is a mathematical expression used to predict the intensity of a diffraction peak. Assuming per unit cell, there are n atoms and the locations are given by x_n, y_n, z_n. Consider, f_n as the atomic structure factor. So, the structure factor F_{hkl} for a plane (hkl) is given in Eq. 3.9. A higher value of the structure factor represents higher intensity:[1]

$$F_{\text{hkl}} = \sum_{n}^{N} f_{\text{n}} e^{\left[2\pi i\left(hx_{\text{n}} + ky_{\text{n}} + lz_{\text{n}}\right)\right]} \quad (3.9)$$

Table 3.1 Structure extinction rules for crystallographic plane (hkl)

Lattice form	Diffraction probably will occur	Diffraction certainly will NOT occur
Simple	All	None
Base-centered	Unmixed h and k (applicable to a cell with (001) face centered)	Mixed h and k (applicable to a cell with (001) face centered)
Body-centered	Sum of (h+k+l) is even	Sum of (h+k+l) is odd
Face-centered	h, k, and l is unmixed	h, k, and l is mixed

3.3.2 Sample Preparation

Preparing samples for XRD is a crucial step as accuracy of results and quality of the diffractogram depends on it. Proper preparation ensures that the sample provides clear, interpretable diffraction peaks, prevent irregularities which decrease the probability of artefacts, improves peak resolution, enhances reproducibility, and also reduces instrumental errors. The procedure for preparing powder samples and thin-film samples differ significantly, and different considerations need to be kept in mind while preparing the samples to get reliable diffraction data.

Preparing Powder Samples: While preparing powder samples, it should be safeguarded that it has a uniform, well-distributed layer on the sample holder to obtain clear diffraction patterns. Initially, the sample must be carefully selected to represent the material being analyzed and should be free from contamination, thoroughly cleaned, and dried. If the sample is not already in powder form, it must be ground into fine particles, typically less than 10 µm, using tools like a mortar and pestle or a ball mill. Homogeneity is essential, so the powder should be carefully mixed to avoid segregation by particle size or composition. Subsequently, the sample must be loaded into an appropriate holder, typically flat or shallow, ensuring it forms a thin, even layer across the surface. The powder should be gently leveled using a spatula or other tool, and, if necessary, slightly compacted to improve the diffraction pattern without over-compressing, which could alter the sample structure. It is important to clean all equipment to avoid contamination and to prevent oils or residues from hands affecting the results. The powder sample should then be inspected to ensure it is free from clumps or irregularities. Finally, the sample should be carefully mounted in the XRD instrument and properly aligned. Calibration of the equipment may also be necessary to obtain precise results. Following steps ensure that the powder sample is well-prepared for accurate and reliable XRD analysis.[11]

Preparing Thin-Film Samples: While preparing a thin film sample for XRD, handling should be done diligently to ensure that the film is well-prepared so that it provides accurate, high-quality diffraction data. The first step is selecting a compatible substrate that is clean, flat, and free from defects. The substrate should be cleaned thoroughly using solvents like acetone, ethanol, or isopropanol, followed by rinsing with deionized water, and in some cases, ultrasonic baths or plasma cleaning for more rigorous cleaning. The film's thickness and evenness are very essential during the thin film preparation, and annealing may be performed to improve film quality and promote crystallization. The annealing process must be carefully controlled to avoid damaging the film or altering its composition. Before XRD analysis, the thin film should be inspected for smoothness and absence of defects. If contamination is present, gentle cleaning with solvents or deionized water is necessary, avoiding abrasive methods that could damage the film. The film should be mounted properly on the sample holder, ensuring it is securely aligned for accurate measurements. For thin films, grazing incidence X-ray diffraction (GIXRD) is often used, where the X-ray beam is directed at a shallow angle to enhance sensitivity to the film. The θ-2θ scan method is also common for collecting diffraction data. Proper preparation ensures that thin films are well-characterized, providing valuable insights into their crystal structure, orientation, and other properties.

3.3.3 Characterization

Crystal-Phase Identification: To identify the crystalline substance and its phase, the diffraction spectrum of the sample is compared with the XRD data stored in a Powder Diffraction File (PDF) database, which is maintained and revised by the International Centre for Diffraction Data (ICDD). Most PDFs use CuKα radiation, and the most current files should be referenced for phase identification. To compare the spectra with PDF data, primarily the corresponding 2θ values for each peak need to be located in the PDF database. Attention must be paid to the intensity and relative positions of the peaks, to confirm the phase. Ideally, at least three most intense peaks should match to ensure the presence of a specific phase. Additionally, the d-spacing of the peaks of the sample should align with those in the PDF, and potential peak broadening due to crystallite size or strain must also be considered, as this may cause shifts in peak positions. Thamaphat and co-workers differentiated anatase and rutile phases of TiO_2 nanopowders by comparing diffraction patterns with the JCPDS database.[12] Similarly, Verma and co-workers analyzed the XRD patterns of synthesized $In_2S_3/In_2O_3/Au$ materials, confirming the presence of pure tetragonal In_2S_3 and In_2O_3 phases by matching with JCPDS references. Their analysis also showed how the deposition time of gold nanoparticles (Au NPs) affected the intensity of Au peaks, verifying the successful decoration of the In_2S_3/In_2O_3 substrate with Au NPs.[13]

Crystallite Size: XRD patterns typically display peaks corresponding to specific crystal planes. Ideally, these peaks should be sharp, narrow, and symmetrical, but in reality, they exhibit broadening due to factors such as crystallite size, internal strain, and instrumental effects. The broadening of peaks occurs because the Bragg's law applies only to perfect crystals, which have infinite lattice repetition. In actual crystals, the presence of "coherent scattering domains" or "crystallites," regions where the crystal structure is periodic, affects the diffraction pattern. As crystallite size decreases, the ability of these domains to diffract X-rays coherently diminishes, leading to broader peaks in the XRD diffractogram. The broadening can also be attributed to incomplete destructive interference caused by X-rays with incident angles deviating slightly from the ideal Bragg's angle. This leads to diffraction intensities around the Bragg's angle, widening the diffraction peak. Consequently, the width of the diffraction peaks is inversely related to the crystallite size. To quantify the crystallite size, the Scherrer equation is used, which relates the crystallite size to the peak broadening, specifically the full width at half maximum (FWHM). The Scherrer equation is given by:

$$D = \frac{K\lambda}{\beta \cos\theta}$$

where K is the Scherrer constant which is dimensionless having value about 0.9, D is the crystallite size, i.e., average diameter of the coherent diffraction domain, β represents the FWHM of the diffraction peak that is corrected for instrumental broadening, and θ is the Bragg's angle, that is half of the diffraction angle 2θ, whereas λ accounts for the wavelength of the X-ray radiation utilized.[14]

Rathore and co-workers used the Debye Scherer equation to calculate the crystallite size of bare $Ni_{0.85}Se$ and Ce-doped $Ni_{0.85}Se$ and reported it as 32.1 and 25.7 nm, respectively. The doping with Ce decreases the crystallite size.[15] Theivasanthi and co-workers reported the grain size of Cu nanoparticles for each 2θ values using the FWHM values and obtained the particle size to be ~24 nm.[16] Nevertheless, the Scherrer formula has some limitations. The Scherrer equation assumes that the crystallites are perfect, homogeneous, and spherical in shape. In reality, crystallites may have irregular, elongated, or anisotropic shapes that may cause the equation to overestimate or underestimate the actual size of the crystallites. Muniz and co-workers reported that the Scherrer equation remains justifiable with an acceptable margin of error, up to 600 nm in size for crystallites. The same group also reported that the equation performs better at higher diffraction angles, where the crystal size impact is more apparent. Instrumental effects on peak broadening can also interfere with accurate size determination, especially if they are larger than the crystal effects. Additionally, the Scherrer equation may not be reliable for thin films where crystal grain size exceeds the film's thickness. In such cases, alternative methods are needed.[17]

Residual Stress: Residual stress in crystals refers to internal stress that remains in a crystal lattice after external forces are removed, often resulting from the synthesis process, deformation, or defects. This stress causes strain in the crystal, altering lattice parameters and potentially shifting diffraction peak positions in XRD analysis. An increase in lattice spacing due to tensile stress shifts the peak to lower 2θ values, while compression stress decreases the spacing and shifts the peak to higher 2θ values. The Williamson-Hall (W-H) equation is an extension of the Scherrer equation for analyzing XRD peak broadening. It considers both crystallite size and strain effects (Fig. 3.2), unlike the Scherrer equation, which only accounts for crystallite size. The W-H equation (Eq. 3.10) expresses the total broadening as the sum of size and strain contributions:

$$\beta_{\text{total}} = \beta_{\text{size}} + \beta_{\text{strain}}$$

$$\beta_{\text{total}} . \cos\theta = \frac{K\lambda}{D} + 4\epsilon \sin\theta \qquad (3.10)$$

where ϵ is the microstrain, that is, the strain in the crystal lattice. The $\frac{K\lambda}{D}$ term signifies the broadening effect caused by the finite size of the crystallites and $4\epsilon \sin\theta$ represents the broadening due to lattice strain.

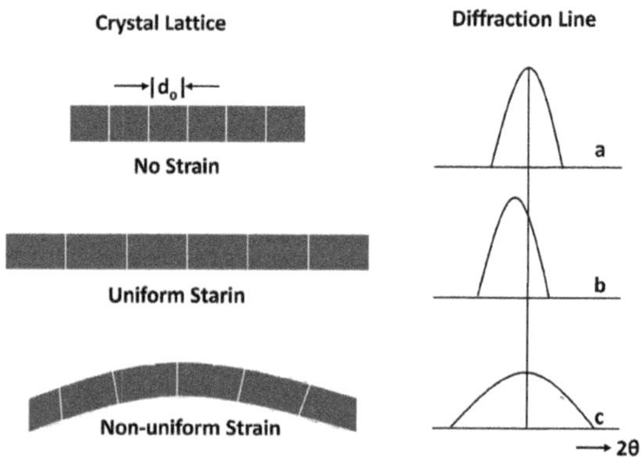

Figure 3.2 Effects of residual stress and strain on diffraction peak position and shape: (a) no strain, (b) uniform strain, and (c) nonuniform strain.

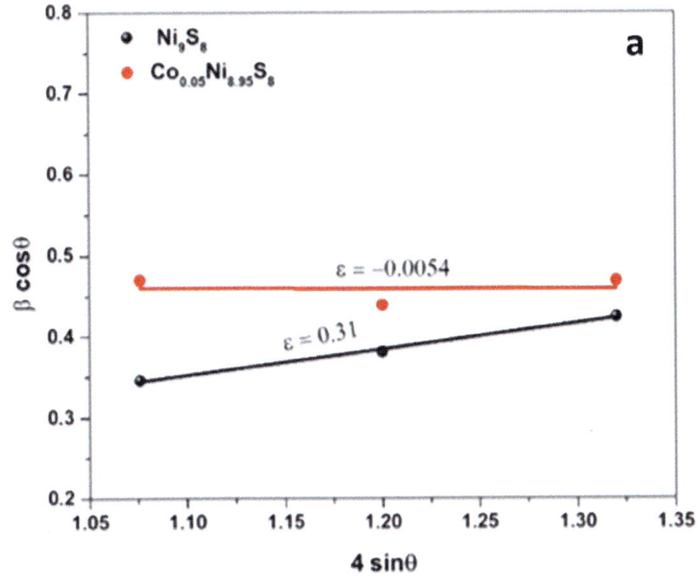

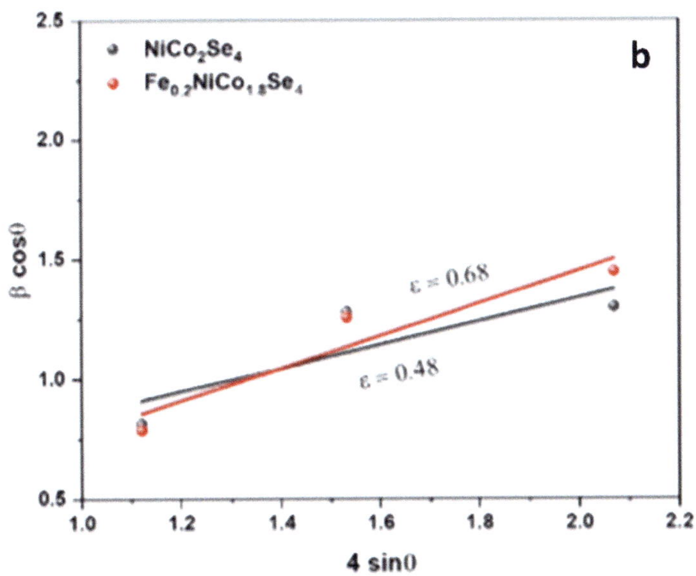

Figure 3.3 W-H plot of (a) Co-doped Ni_9S_8 and bare Ni_9S_8 materials. Reprinted with permission from [19]. Copyright 2022, American Chemical Society. (b) W-H plot of bare $NiCo_2Se_4$ and $Fe_{0.2}NiCo_{1.8}Se_4$ materials. Reprinted with permission from [20]. Copyright 2023, American Chemical Society.

By plotting $\beta_{total}.\cos\theta$ vs. $4\sin\theta$ for multiple diffraction peaks, the line's slope gives microstrain (4ε), and the intercept provides crystallite size. This method helps to determine both the crystallite size and internal strain of a sample. Modified versions of the W-H equation, such as the Uniform Deformation Model (UDM) which postulates that the strain is distributed evenly throughout the material, the Uniform Stress Deformation Model (USDM) which presumes the stress is evenly spread across the material, and Uniform Deformation Energy Density Model (UDEDM) which considers the energy density due to deformation is uniform, are also used for more refined analysis.[18] Rathore and co-workers calculated the strain in the Co-doped Ni_9S_8 using the Williamson–Hall equation. 31.3°, 35.2°, and 38.8° were the diffraction angle (2θ) values chosen to calculate the strain in both the pristine and doped samples. From the plot, 4 $\sin\theta$ was considered to get the value of the slope, which accounts for the increased strain (ε) in the system. In the Co-doped Ni_9S_8 sample, the negative slope value indicates that the strain is of a compressive nature (Fig. 3.3a).[19] Similarly, the same group has reported, from the W-H plot that $NiCo_2Se_4$ and $Fe_{0.2}NiCo_{1.8}Se_4$ samples have the resulting values of the strain to be 0.48 and 0.68, respectively (Fig. 3.3b) showing that the positive value of the strain corresponds to tensile strain in both the systems.[20] Zak and co-workers prepared ZnO nanoparticles (ZnO-NPs) via a sol gel combustion method and applied W-H plot to all the XRD peaks of ZnO having a wurtzite hexagonal phase. The data was collected for a range of angles from 20° to 100°. To make the analysis more accurate, the researchers compared the different models: UDM, USDM and UDEDM. By using these models and methods, they clarify thoroughly how strain, stress, and energy affect the properties of ZnO nanoparticles.[18]

Rietveld Refinement: Rietveld refinement is a powerful crystallographic technique used to derive accurate structural information from PXRD and neutron diffraction data. Developed by Hugo Rietveld in the late 1960s, this method allows for the determination of crystal structures, particularly when single-crystal data is unavailable. It is mainly useful for analyzing complex materials like multiphase systems or less crystalline substances. Unlike conventional methods that focus on individual diffraction peaks, Rietveld refinement uses the entire diffraction profile, improving the analysis of these materials. The technique depends on the scattering of X-rays or neutrons by the atoms in the crystal, producing a diffraction pattern that provides valuable insights into the crystal structure, with the structure factor given in Eq. 3.9, describing the intensity of each diffraction peak. This method is used to refine crystal structures from X-ray or neutron diffraction data through least squares fitting. The process involves adjusting structural model parameters to diminish the difference between observed and calculated diffraction intensities. Initially, a model is created based on known structures, including atomic coordinates, thermal parameters, and scale factors. Background noise is removed using polynomial fitting, and a peak shape function (e.g., Gaussian or Lorentzian) is chosen to model diffraction peaks accurately. Software tools like GSAS, Full Prof, or TOPAS are used to refine parameters such as atomic coordinates, thermal displacement, scale factors, and background. To perform the refinement by iteratively adjusting parameters to minimize the weighted residual given by Eq. 3.11, where I_{obs} and I_{calc} are the observed intensity and the calculated intensity, respectively.

$$R = \frac{\sum \left| I_{obs} - I_{calc} \right|}{\sum I_{obs}} \tag{3.11}$$

The refinement process continues until convergence is achieved, indicated by minimal improvement in the fit. The imaginary and real components of the structure factor for various Bragg planes, attained through Rietveld refinement, are utilized in the maximum entropy method (MEM) calculations to generate the electron density maps. The quality of refinement is assessed using R-factors, and residual electron density maps to help visualize inconsistencies. The final output includes refined structural parameters, atomic positions, bond lengths, angles, and electron density maps, which aid in understanding the material's crystal structure and properties. Abinaya and co-workers used Rietveld refinement of XRD data of manganese-doped cobalt ferrite ($CoFe_2O_4$) to understand the material's atomic structure and the subtle structural changes that occur due to doping. By performing Rietveld refinement, the authors mapped the electron density in the crystal structure, which helped to understand how the Mn ions influence the bonding and coordination of the metal ions. The electron density map provides detailed information about the distribution of electrons around each atom in the crystal lattice, revealing how the bonding environment around the Co^{2+}, Fe^{3+}, and Mn^{2+} ions changes with doping. The electron density maps were also used to infer the bonding characteristics in the Mn-doped $CoFe_2O_4$ structure. For example, if Mn ions substitute for Co or Fe, they could create distortions in the crystal lattice that affect the local electron density, which can influence both the bonding strength and the bonding angle between neighboring ions.[21, 22]

3.4 FTIR Spectroscopy

3.4.1 Fundamental Concepts

For a molecule to be IR active, it must experience a change in dipole moment when exposed to IR radiation. Homonuclear diatomic molecules are IR inactive as there are no electronegativity differences between their atoms, so their vibrations do not alter the dipole moments. In contrast, heteronuclear diatomic molecules (e.g., HCl, CO, etc.) possess dipole moments due to electronegativity differences, making their stretching vibrations IR active. Polyatomic molecules exhibit changing dipoles during vibrations, allowing them to absorb IR radiation and undergo vibrational–rotational transitions. The number of vibrational modes depends on their degrees of freedom—for linear ones $3N$-5 and for nonlinear polyatomic molecules, it is $3N$-6—where N is the number of atoms. IR activity ultimately depends on the ability of a molecule to experience dipole changes during these vibrations. The vibrational frequency at which a molecule absorbs IR radiation can be calculated using Eq. 3.12:

$$\omega = \frac{1}{2\pi}\sqrt{\frac{K}{\mu}} \text{ Hz or } \varpi = \frac{1}{2\pi c}\sqrt{\frac{K}{\mu}} \text{ cm}^{-1} \qquad (3.12)$$

where the force constant of the bond between the atoms is represented by K, μ is the reduced mass of the atoms, and the velocity of light is represented by c. This formula assumes a simple harmonic oscillator model, where the vibrational frequencies are directly proportional to the bond strength and inversely proportional to the mass of the vibrating atoms. The quantized vibrational energy of an anharmonic oscillator is given by Eq. 3.13:

$$E = \left(v + \frac{1}{2}\right)h\omega \text{ joules or } E = \left(v + \frac{1}{2}\right)\varpi \text{ cm}^{-1} \tag{3.13}$$

where v is the vibrational quantum number, Planck's constant is represented by h, and ϖ is the wave number.[4,5]

3.4.2 Sample Preparation

Different states of samples like solids, liquids, and gases can be studied by using IR spectroscopy. However, sample handling in IR is quite sophisticated as a particular inert cell may not be transparent over the entire IR range. NaCl is commonly used as a window material, because it transmits approximately to 650 cm^{-1}, while KBr transmits to about 400 cm^{-1}. Both materials are water-soluble and can be easily affected by exposure to atmospheric humidity or moist samples; when used under such conditions, these require frequent maintenance.

Gas Cells: Gas cells are typically 100 mm in length, making them suitable for measuring gases and vapors at standard partial pressures. They usually have 40 mm as diameter and a volume of approximately 125 cm^3. Two tubes with stopcocks connect the gas cell to a gas-handling system. While glass cells often feature a uniform shape, a tapered design is more effective, because it aligns better with the radiation beam and reduces the required sample amount. Longer path lengths are necessary for studying weak absorption bands, but fitting these into the instrument can be challenging. In such cases, the use of mirrors and a beam condenser can aid in capturing the spectra of trace components.

Liquid Cells: Most liquid hydrocarbons strongly absorb IR light, as shown by their extinction coefficient. When using a pure sample with a thickness between 0.01 and 0.05 mm, the absorption spectrum is clear and useful for analysis, with transmittance values between 15% and 70%. There are two primary types of cells used for testing liquid samples, namely, sealed cells which are designed for liquids with high vapor pressure, and demountable cells for other types of liquids. Both cell types are available in various path lengths and have IR-transparent windows. The cells have an outside diameter of 30 mm, and approximately 100 μL of the sample is required for every 0.1 mm of path length. The IR-transparent windows are separated by thin copper and lead gaskets treated with mercury. These are securely clamped together to ensure a proper and effective seal.

Variable Path Length Cells: These cells are utilized to determine the absorption coefficient and concentration of a test substance and to generate calibration curves. These cells typically contain CaF_2, sodium chloride, and KBr windows. The standard dimensions of the cell are 120 mm in length and 75 mm in diameter. To maintain uniform cell thickness, it is important to measure the path length accurately. This can be achieved by placing the empty cell in the spectrometer and analyzing the interference pattern formed by radiation reflecting off the internal surfaces of the cell. The path length can be calculated using Eq. 3.14:

$$\text{Path length (in cm)} = \frac{n}{2(v_1 - v_2)}' \qquad (3.14)$$

where n is the number of fringes between the wavenumbers v_1 and v_2. The method is not dependable if the cells are in a bad condition.

Pressed Pellet Technique: The pressed pellet technique involves finely grinding a solid sample and mixing it with KBr. This mixture is then compressed into a disc shape for instrumental analysis. The pressing is performed in a specialized die that can be evacuated under high pressure, and the resulting pellet is used for analysis. For measurements at longer wavelengths, alternative alkali halides like CsI and CsBr can be used.

Mull Technique: The IR analysis of powder solid materials can be performed by mixing the powder directly with Nujol to develop a thick slurry, where Nujol is a mixture of liquid paraffin (mineral oil). However, for C-H bending and stretching analysis, Nujol cannot be used. In that case, other liquids like perfluorokerosene or hexachlorobutadiene can be used. The liquid is chosen according to the refractive index of the sample to reduce energy loss from scattered light. Finally, the slurry is placed in the IR equipment.

Micro-sampling: When only a small amount of sample is available for IR analysis, standard methods may not work due to the need for a sufficient amount of absorbing material. To address this, micro-sampling techniques are used. One solution is to use cells with smaller openings, such as the M-OX series microcells, ensuring the spectrophotometer beam covers the entire sample. If the sample is extremely small, a beam condenser can be employed to shrink and focus the beam size onto the sample. Beam condensers can condense the beam by four to six times and are effective for liquids, solids, or gases. For liquid samples, ultra-micro cells with volumes around 2 μL are used, while solid samples can be analyzed by mixing the powder with mineral oil or KBr to create micro-disks for placement in a micro solid sample holder. The micro KBr disk method is particularly popular for handling very small amounts, ranging from 1 to 100 μg.[23]

3.4.3 Characterization

Functional Group Identification: FTIR can identify functional groups in nanoparticles by detecting characteristic absorption bands. This helps in determining the chemical composition of the materials used in the synthesis of nanoparticles. Adhikari and co-worker reported by comparing the FTIR spectra of Ag nanocluster-DHLA nanoconjugate and dihydrolipoic acid (DHLA) (Fig. 3.4) that the characteristic band at 2548 cm^{-1}, which corresponds to the ν of S-H vibration of DHLA, disappeared in Ag nanocluster-DHLA nanoconjugate. This change suggests that the thiol groups in DHLA exist as thiolates within the nanoconjugate. Furthermore, in both the Ag nanocluster-DHLA and DHLA nanoconjugate, a band at 1564.3 cm^{-1} is observed, indicating that the carboxylate ion is in free state. From this observation, it has been reported that freshly reduced DHLA comprises per molecule two free thiol groups. These thiol groups are highly sensitive to the surfaces of Ag nanoclusters and play a crucial role in effectively stabilizing the fluorescent Ag nanoclusters.[24] Similarly, Yang and co-workers characterized the copper nanoclusters (Cu NCs) using FTIR and confirmed that L-Cys has been successfully attached to the copper nanoparticles. The spectral analysis illustrates that the 2511.32 cm^{-1} peak vanishes specific to –SH stretching vibrations, indicating that Cu atoms of CuNCs have formed Cu-S bond with L-cysteine.[25]

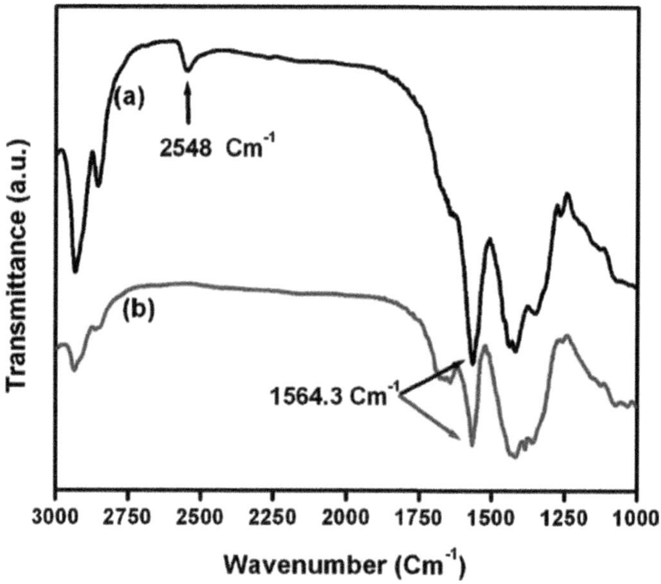

Figure 3.4 Comparative FTIR spectra obtained from DHLA and Ag nanocluster-DHLA nanoconjugate. Reprinted with permission from [24]. Copyright 2010, American Chemical Society.

Surface Chemistry Characterization: FTIR can examine the surface chemistry of nanoparticles by studying the absorption bands related to functional groups attached to their surfaces. For example, the adsorption of biomolecules or surfactants onto nanoparticle surfaces can be monitored using FTIR, allowing researchers to determine how these modifications affect the nanoparticles' properties and interactions. Wang and co-workers used FTIR spectroscopy to study the functional groups on Fe_3O_4@TiO_2@PEG and the precursor Fe_3O_4@TiO_2 nanoparticles. The authors observed both the spectra of Fe_3O_4@TiO_2@PEG and Fe_3O_4@TiO_2 between 500 cm^{-1} and 700 cm^{-1} exhibited a broad band, which was assigned to the stretching modes of Ti-O-Ti. Several new vibrational absorption peaks were observed after the functionalization with thiol-polyethylene glycol (SH-PEG). The C-H stretching vibration peak at 2882 cm^{-1}, $-CH_2$ bending vibration the peak at 1468 cm^{-1}, and a strong peak of the C-O-C asymmetric vibration at 1106 cm^{-1}. Through these findings, the group confirmed the successful attachment of SH-PEG to the Fe_3O_4@TiO_2 nanomaterials.[26] Zuoli He and co-workers, employed

FTIR spectroscopy to characterize the surface chemistry of the carbon-doped TiO_2 nanomaterial. The spectral analysis allowed the researchers to observe characteristic absorption bands associated with various chemical groups. The FTIR spectra helped in examining the nature of carbon species incorporated into the TiO_2 matrix. This included identifying peaks related to C-O and C=C bonds, which indicated successful carbon doping and the existence of functional groups that may improve photocatalytic activity. By comparing FTIR spectra before and after thermal treatment, the study assessed how the thermal process affected the chemical structure of the photocatalysts. Position of absorption peaks or changes in the intensity provided insights into the stability and transformation of functional groups under different conditions.[27]

Quantitative Analysis: FTIR can also be used for quantitative analysis of the components in a nanoparticle material. By applying Beer's law, the principle of constant ratio method and for the computation part software like Essential FTIR software v3.10.016 can be used to calculate the concentration of specific functional groups based on the intensity of their absorption bands. Initially, the key peaks associated are identified. To quantify the functional groups, the intensity of the absorption peaks in the FTIR spectra is analyzed. This concentration is proportional to the intensity of the corresponding functional group present on the nanoparticle surface. By determining the area under the peaks of interest and comparing them to standard curves created from known concentrations of the functional groups, the researchers can determine the amounts of each functional group present on the nanoparticles. Doolaanea and co-workers used FTIR to compute the loading efficiency of Nigella sativa oil (NSO) directly from the intact nanoparticles, eliminating the requirement to extract the NSO and dissolve the nanoparticles. The quantification was based on the Beer-Lambert law, which was applied after selecting specific combinations of wavenumbers for the analysis.[28]

Identification of Contaminants and Impurities: This technique can help to identify unwanted contaminants or impurities in nanoparticle samples by comparing their spectra to those of pure materials. This is particularly important in applications where the purity of nanoparticles is critical, such as in biomedical applications. Xiong and co-workers examined the chemical composition and structural characteristics of zinc oxide nanoparticles, focusing on identifying the impurities and defects and their correlation with photoluminescent properties. The analysis focused on detecting changes in the position or intensity of absorption peaks, which could be linked to the presence of impurities such as organic contaminants or defects like oxygen vacancies. By comparing the FTIR spectra of pure ZnO nanoparticles with those containing impurities, the researchers could infer the effects of these factors on the material's properties.[29]

3.5 Raman Spectroscopy

3.5.1 Fundamental Concepts

When light encounters with matter, it can be absorbed, scattered, or diffracted, or it can eject electrons from the surface of the matter. The elastic or inelastic light-matter collision results in the scattering of a relatively small portion of the incident light by the molecules present in the matter. In the case of inelastic collision, the molecule gets excited to an excited energy state by absorbing energy from the incident beam of radiation and then drops down to a higher or lower energy state with respect to its initial energy state. When the scattered radiation is of lower frequency compared to the incident radiation, the excited molecule drops down to a higher energy state than its initial state; this scattering phenomenon is referred to as Stokes scattering. On the other hand, anti-Stokes scattering takes place if the scattered radiation is of higher frequency than the incident radiation, and the excited molecule drops down to a lower energy state than its initial state. Raman scattering phenomena are the collective result of both the Stokes and anti-Stokes scattering. The scattered light by the analyte molecule is recorded as a function of frequency or wave number. According to the classical theory of Raman scattering, for a molecule to be Raman active, there should be a deviation in polarizability in the molecule due to its rotational or vibrational motion. Raman scattering intensity depends on both the change of polarizability of the vibrational motion of the bonds and the symmetry of the vibrational motion with respect to the incident radiation.[30,31]

3.5.2 Sample Preparation

Raman spectroscopic analysis can be done for samples with various sizes, shapes, and physical states. Raman spectroscopy can analyze a very low amount of sample as the laser source can focus accurately on a very small sample area, and the radiation can easily be transferred through the entrance slit of the spectrometer. Different kinds of sampling techniques are discussed as follows:

Gas Samples: A glass tube having a diameter of 1 to 2 cm and a thickness of 1 mm is used as a gaseous sample chamber. A sealed capillary tube filled with the gaseous sample can be analyzed. A pressure of 10 torr is enough when the multi-pass system is used specifically for weak scattering elements.

Liquid Samples: For liquid samples, capillary cells with a bore of 0.5 to 0.1 mm and length 1 mm are used. Glass tubes, ampoules, capillaries, and even NMR tubes can be used as sample holders for liquids. Water is considered as a compatible solvent for Raman analysis as it has a weak Raman scattering effect. Therefore, water-soluble biomolecules and inorganic molecules, as well as polluted water samples, can be analyzed.

Solid Samples: Solid samples in powder form can be irradiated directly with Raman laser. Even a small amount of powder sample (5–10 mg) can be analyzed and give adequate results. In the case of colored samples, some more amount will be better to reduce localized heating. Polymers can also be used without any typical sample preparation before irradiation. Sometimes, KBr pellets can be used, which further minimizes local heating and prevents the decomposition of the sample.

Fiber Optic Sampling: Raman spectroscopy mostly uses visible or near IR radiation, which can be transmitted through optical fiber. These fibers can be dipped into liquid samples or employed to irradiate the solid sample. Optical fiber probes are very effective to analyze locations far from sample sites; therefore, they can be used in maleficent environments like molten salts or reactors containing hazards, or in biological samples like tissues or arterial walls or samples belonging to the environment like seawater or groundwater.

Microprobe Sampling: Raman microprobe is an important equipment of Raman spectrometer, where the laser is focused in a selected ~1 μM area of the sample that is placed on the stage of the microscope and then the lamp is switched off and only the irradiating laser is focused. With this technique, just a picogram range of sample is required to provide a highly resolved Raman spectra having ~1 μM spatial resolution.[31]

3.5.3 Characterization

Superconductors: Superconductors are those materials that typically show no resistance below a particular temperature (T_c) and pass maximum current. High T_c superconductors can be investigated by using Raman spectroscopy. Zhang and co-workers developed $YBa_2Cu_3O_{7-x}$ superconductor with different thickness and analyzed the samples. The film with a thickness less than 60 nm showed an unusual enhanced mode at 145 cm^{-1}, indicating that phases have oxygen deficiency resulting in more superconductivity with reduced T_c as oxygen content is an important parameter in superconductivity.[32] Furthermore, Li and co-workers reported how Raman spectral analysis can be used to determine the oxygen content for the superconducting $YBa_2Cu_3O_{6.87}$ film while the film was having a laser written pattern in O_2 or N_2 atmosphere by varying pattern power density and scanning speed of writing.[33] Zhou and co-workers used a variation of temperature while using polarized Raman spectra for the characterisation of single crystal $HgBa_2Ca_2Cu_3O_{8+x}$. They observed that the $A^1{}_g$ mode shifted with respect to frequency and line width, whereas the $B^1{}_g$ mode was unshifted across T_c.[34]

Semiconductors: Semiconductors made of Ga, Si, Zn, and Cd can be characterized by Raman spectroscopy to have an idea of the physical and chemical mechanisms occurring at their surface. In this context, Velterhoeffer and co-workers isolated and detected molecular hydrogen and deuterium in GaAs,[35] On the other hand, the quantum site effect of porous nano-structure of Si is reported by Champagnen and co-workers.[36] The compensation due to N doping in p-type ZnSe semiconductor was observed by Towsley and co-workers using spin flip Raman scattering.[37]

Carbonaceous Materials: The optical, vibrational, and electronic characteristics of carbonaceous materials can be effectively described by Raman spectra. To differentiate between different carbonaceous materials by Raman spectroscopy, the relative intensity and position of D and G modes are analyzed. The D mode is associated with a breathing vibration of sixfold aromatic rings, whereas the G band results from the stretching of a bond between two sp^2-hybridized carbons arranged in either olefinic chains or as aromatic rings. Graphite and diamond are two allotropes of natural carbon, where graphite is a conducting substance but brittle in nature, whereas diamond is a non-conducting substance but crystalline in nature. Merging the advantages of both allotropes, carbon nanotube (CNT) has been developed, which is a conducting substance with high mechanical strength. Bokobza and co-workers reported Raman bands of CNT at 1342 cm^{-1} for D band, which generates due to carbon system's disorder and at 1580 cm^{-1} for G band that signifies in-plane C-C bond vibration.[38] Similarly, Raman shifts at 1350 cm^{-1} for D band and 1593 cm^{-1} for G band for CNT were observed by Lopez-Lorente and co-workers.[39]

Nanomaterials: Various nanomaterials such as graphene, metal chalcogenides, etc., can be characterized with the help of Raman analysis. Some are discussed as follows:

I. Graphene:

Graphene is one of the emerging 2D materials made of carbon in which carbon atoms are present in sp^2 hybridization. It is a good conductor of heat and electricity, and it also provides high optical transparency. The prominent Raman peaks in graphene arise from the interaction between electrons and phonons due to the incidence of the incident radiation. The G band occurs due to the interaction of iTO and iLO phonons' interaction. The ratio of the D and G bands' intensity I_D/I_G gives information about the defect density in graphene. If G and D bands are absent in any sample's Raman spectrum, then it is not considered graphene. Raman analysis provides information about different properties of graphene, such as the thickness of different layers, strain, and doping, explored below in detail.

Thickness: The physical and chemical properties of the 2D materials depend on the number of layers as well as the interlayer interaction. Raman spectroscopy gives crucial information about the number of layers and the stacking phenomenon of the material. Bayle and co-workers measured the number of layers in a few layers' graphene

(FLG) by combining optical contrast with Raman spectroscopy to remove the limitation caused by stacking order.[40] Silva and co-workers calculated the I_{2D}/I_G value for the tested graphene sample to be 0.43, indicating the number of layers more than 5.[41]

Strain and Doping: Detection of the strain and doping in graphene can also be done by Raman spectroscopy. To understand the extent of doping and strain, Muller and co-workers separated the G and 2D band of Graphene along with their shifting.[42] The lattice disorder due to external strain was quantitatively determined by Sampathkumar and co-workers. They used the D band of graphene to calibrate lattice disorder.[43] By using confocal Raman spectroscopy, Belyaeva and co-workers discovered that liquid has more advantages than solid as substrate for graphene for lowering strain. They used water instead of any solid substrate and observed lower amounts of strain as well as more uniformity in doping.[44] Judek and co-workers observed increment of strain and defect when they inserted hydrogen during the cooling process for graphene-germanium interaction using Raman spectroscopy.[45]

II. Transition Metal Chalcogenides:

Most of the transition metal chalcogenides (TMCs) are depicted symbolically as MX_2, where M stands for metal atom such as Sn, Re, W, Mo, etc., and X stands for chalcogenide such as O, S, Se, Te, and collectively they form X-M-X sandwich-like structure. TMCs have applications in multiple research areas such as semiconductors, transistor, catalysis, and medicine. Raman spectroscopy has a huge role in the characterization of these TMC materials. TMCs are classified as 1T, 2H and 3R based on their crystal structure, which are characterized using Raman spectra along with their other physical and chemical properties. Raman spectroscopic characterization of TMCs based on various parameters is discussed in greater detail in the following sections:

Thickness: Firstly, it is very important to detect the number of layers of TMC materials. Raman spectroscopy is mostly utilized due to its non-destructive and fast characterizing facility. In MX_2-type materials, the common Raman active modes are E^1_{2g} (in-plane vibration of M and X) and A_{1g} (out-of-plane vibration of X atom). Usually, as the layer number increases, the red shift of E^1_{2g} mode and blue shift of A_{1g} mode is observed. Zhao and co-workers reported a blue shift of A_{1g} mode and red shift of E^1_{2g} mode with increasing layer number for WS_2. With an increment of thickness, atomic vibration is suppressed by interlayer van der Waals force due to which the blue shift of A_{1g} was observed. On the other hand, structural variation along with long range coulombic interaction between the layers resulted to red shift of E^1_{2g}. Same group has also reported similar phenomena for WSe_2.[46] Also, when number of layers is less than 6, we can identify few layer MX_2 by difference in frequency of A_{1g} and E^1_{2g} modes using Eq. 3.15 where $\Delta\omega$ is frequency difference and N is number of layers.

$$\Delta\omega(\text{A-E}) = P_{OS}(A_{1g}) - P_{OS}(E^1_{2g}) = (25.8\text{-}8.4)/N \ (1<N<6) \quad (3.15)$$

Yu and co-workers had shown that the difference of frequency of A_{1g} and E^1_{2g} mode increases with increasing thickness of MoS_2 films.[47] High-frequency Raman spectra was reported by Li and co-workers for MoS_2 up to 120 layers, and they observed the appearance of a new peak for ~85-layered MoS_2.[48] However, variation of low-frequency Raman modes with thickness is more prominent than high frequency. Low-frequency Raman modes are generated due to the perpendicular or parallel movement of a complete layer. The parallel oscillation of layer plane is shear (S) mode which is also symbolized as C mode in some literature, and the perpendicular oscillation of layer plane is layer-breathing (LB) mode or compression vibrational mode. The frequency of these LB and S modes are lower than 100 cm^{-1} due to weak van der Waals interaction of interlayer coupling. In bulk MX_2 materials, the optically inactive LB (B^2_{2g}) mode and the Raman-active S (E^2_{2g}) mode are present. For 2D TMC, more than one layer are observed and the type of mode differs depending whether the number of layers is odd or even.

The equations to calculate frequency are:

$$\text{Pos}(S_{NN-i}) = \sqrt{2}\,\text{Pos}(S_{21})\cos(i\pi/2N),\ i = 1,2,...,N-1 \quad (3.16)$$

$$\text{Pos}(LB_{NN-i}) = \sqrt{2}\,\text{Pos}(LB_{21})\cos(i\pi/2N),\ i = 1,2,...,N-1. \quad (3.17)$$

where Pos (S_{21}) and Pos (LB_{21}) are frequencies of S and LB modes, respectively. Using Eqs. 3.16 and 3.17, Zhang and co-workers reported for 19 layers MoS_2; C mode or S mode stiffens (blue shift) from 22.6 cm^{-1} for 2 layers to 32.5 cm^{-1} for 19 layers, but the LB mode softens (red shift) from 40 cm^{-1} for 2 layers to 4.7 cm^{-1} for 19 layers.[49] Similar trends in the softening of LB modes and the stiffening of S modes with an increasing number of layers are observed in other 2D materials, such as WSe_2, $MoTe_2$, $NbSe_2$, $MoSe_2$, and $Mo_{0.5}W_{0.5}S_2$ alloy. It is seen, various low-frequency Raman modes are influenced by different stacking orders. For "N" layers ReS_2, as the number of layers increases, the LB modes become stiffer due to both isotropic and anisotropic stacking effects. In contrast, the S mode exhibits different behavior which depends on the type of stacking, it softens under isotropic stacking but stiffens under anisotropic stacking.

Strain and Doping: In ultrathin TMC materials, strain can arise from factors such as doping, orbital interactions, electronic band dispersion, and mechanical properties. When strain reaches to a critical value, the Raman active $E^2{}_g$ mode splits into two modes, indicating the breaking of symmetry. Currently, uniaxial and biaxial strain can be created externally by forcing substrate having 2D material, with the resulting strain analyzed using Raman spectroscopy. Desai and co-workers reported splitting of $E^1{}_{2g}$ mode for WSe_2 due to tensile strain to be ~2%, and this splitting is directly proportional to increasing strain.[50] Wang and co-workers also reported the same splitting along with red shifting of $E^1{}_{2g}$ mode caused by ~1% uniaxial strain.[51] Zhu and co-workers also reported the breaking of isotropic symmetry for MoS_2 in the XY plane, and it resulted the splitting of E' and $E^1{}_{2g}$ mode into A'(1)/A'(2) and $A_g(1)/A_g(2)$, respectively. The vibration along the C axis is not very effective for tensile strain, and it can be easily explained from the minimum shifting of A' and A_{1g} modes.[52] Hui and co-workers reported blue shift of A_{1g} and E'_{2g} for biaxial compressive strain induced by piezoelectricity. The observed blue or red shifts in the Raman bands is resulted due to the increase or decrease in the force constants, respectively, resulting from changes in bond lengths under compressive or tensile strain.[53] Sharma and co-workers reported red shift in Raman spectra for MoS_2 after doping with Cu and V.[54] Similar phenomenon was also observed by Rathore and co-workers with Ce-doped $Ni_{0.85}Se$ and Co-doped Ni_9S_8. From Fig. 3.5, it can be observed that the 508 cm^{-1} Raman mode, which is a characteristic Ni-S vibrational mode, was shifted to 527 cm^{-1} after doping with Co in Ni_9S_8 ($Co_{0.05}Ni_{8.95}S_8$). Definitely, Co doping generates the tensile strain in the Ni_9S_8 lattice that leads to red shifting of its phonon band in Raman spectra.[15,19]

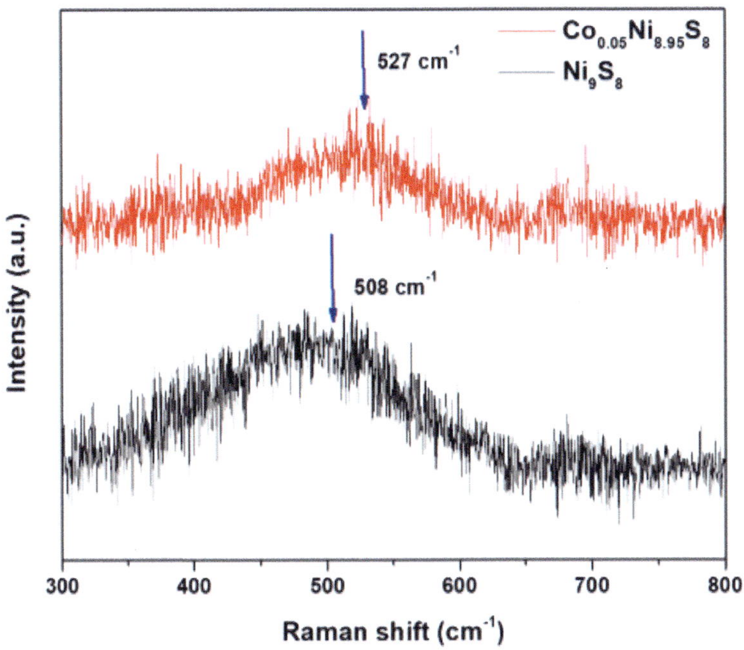

Figure 3.5 Raman analysis of Ni_9S_8 and $Co_{0.05}Ni_{8.95}S_8$. Reprinted with permission from [19]. Copyright 2022, American Chemical Society.

Anisotropy: Splitting of intralayer S mode as well as its angular dependency is proportional to in-plane anisotropy in 2D materials. Vandalon and co-workers reported that the polarized Raman response causes variation in the ratio of $A^1{}_g/E^1{}_{2g}$ modes in 2D materials.[55] Froehlicher and co-workers showed that excellent imaging of N layer TMC can be done by significant anisotropy.[56] From the Raman spectra of MoS_2, Kim and co-workers reported the relativity of $E^1{}_{2g}$ phonon band and anisotropic behavior for polarized.[57a]

III. MXenes:

MXenes are a class of nanomaterials with the general composition $M_{n+1}X_nT_x$, x = 1, 2, 3. They are classified into three types: M_2X, M_3X_2, and M_4X_3. These materials consist of alternating layers of transition metals (M such as Ti, V, Nb, Mo, etc.) and carbon or nitrogen (X) layers, with terminal groups (T such as O, F, Cl, etc.) situated on the outermost surfaces of the metal layers. This arrangement forms a distinctive sandwich-like layered structure. With the increment of the distance between layers and the surface disorder, certain characteristic Raman bands exhibit broadening while their intensity decreases. Hence, Raman spectroscopy can be taken as a standard characterization technique for the characterization of MXenes.

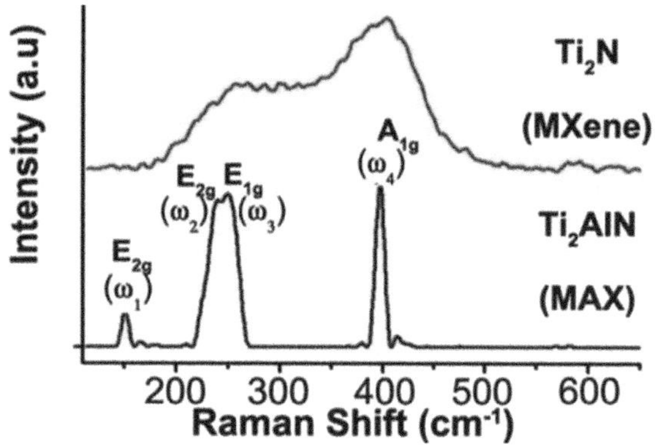

Figure 3.6 Raman spectra of Ti_2ALN and Ti_2N. Reprinted with permission from [57b]. Copyright 2017, American Chemical Society.

With the help of Raman Spectroscopy, the structural deviation was observed when Ti_2N was synthesized after etching of Ti_2AlN with KF–HCl mixture. In the Raman spectrum of Ti_2AlN, peaks were observed at 145 cm^{-1} (E_{2g}), 225 cm^{-1} (E_{2g}), 230 cm^{-1} (E_{1g}), and 360 cm^{-1} (A_{1g}), as shown in Fig. 3.6. In Fig. 3.6, the E_{2g} modes were resulted by the shear vibrations of Ti and Al atomic planes, whereas the parallel and perpendicular vibrations of the Ti atomic planes are the reason for E_{1g} and A_{1g} modes, respectively. The Raman spectra of Ti_2N after treatment with the KF-HCl showed only two wide peaks at 205 cm^{-1} and 335 cm^{-1}, correspond to the E_{1g} and A_{1g} modes, respectively. Exfoliation in the Ti_2AlN phase is shown by the red shift of E_{1g} and A_{1g} modes alone with the removal of the E_{2g} modes in the Raman Spectrum of Ti_2N.[57b] Aakyiir and co-workers determined the composition of MXenes V_2CT_x and elastomers, and after analyzing the composite with Raman spectroscopy, an increased presence of reaction sites and a higher number of =O groups were observed. V_2CT_x was characterized by Raman spectroscopy at different temperatures and pressures to understand its thermal stability. It was observed that below 600°C at non-oxidizing atmospheric conditions, its catalytic activity was not varied. Above 300°C in an oxidizing atmosphere, V_2CT_X undergoes dehydration followed by different degrees of oxidation. Thus, optimization of reaction conditions using V_2CT_X as a thermal catalyst has an important role in the Raman spectroscopic technique.[58] Hu and co-workers examined the charge–discharge system of $Ti_3C_2T_X$ electrode using Raman spectra for the higher-level understanding of the capacitance property of MXenes. The negative electrode at sulfuric acid medium has shown variation in Raman spectra. The Raman peak at 726 cm^{-1} was shifted remarkably when discharged $Ti_3C_2T_X$ was inserted by hydronium ions, confirming the transfer of $Ti_3C_2O_2$ into $Ti_3C_2O(OH)$.[59]

3.6 X-ray Photoelectron Spectroscopy

3.6.1 Fundamental Concepts

The photoelectric effect is the foundation of the XPS technique. When photons bombarded on a solid surface, the energy of the photon is transferred to the electrons, If the photon has sufficient energy, electrons can eject from the atom as photoelectrons; this phenomenon is called photoemission. Photoelectrons can be emitted from the valence and core orbitals of an atom. Core orbitals have discrete energy values and do not participate in chemical bonding, so they are less affected by their surrounding chemical environment. In contrast, valence electrons have broad energy values and are involved in chemical bonding, called valence band. Electrons present in the core shell have a characteristic binding energy (BE). The BE is the amount of energy required to dislodge the core electrons; these core electrons are of interest in XPS. Each element has a characteristic BE; thus, by determining the value of BE, we can identify the element. For example, the BE of the core level of C 1s is 284.5 eV, while for O 1s, the value is shifted to 530 eV. The BE of core level photoelectrons can be expressed as in Eq. 3.18:

$$BE = h\nu - KE - \phi_{spec} \tag{3.18}$$

where hν is the photon energy that induces photoemission, KE is the photoelectron's kinetic energy, and ϕ is the spectrophotometer's work function.[60] Kinetic energy gives the information about the surface element, chemical state, and BE of the electron. Various factors affect the BE of an element, such as orbital from which electrons are emitted and chemical environment of the element. For example, the BE of C 1s varies from 284.5 to 287.6 eV, depending on its surrounding or chemical environment. In the C-N bond, the BE of C 1s is 286.7 eV, while for carboxyl or ester linkage, it is 287.6 eV, furthermore, for C-F, BE shifted to 288.5 eV.[61]

3.6.2 Sample Preparation

XPS examines the sample's top 10 nm of surface. Any contamination during the handling of the sample and exposure to the surroundings can affect the result. Two important aspects should be considered during the analysis: (1) handling of the sample before analysis and (2) sample mounting for the analysis. During the time of analysis, to avoid surface contamination and protect the electron and ion interaction with the surroundings, an ultra-high vacuum ($\leq 10^{-9}$ mbar) is employed on the sample surface during XPS. Such high pressure restricts the analysis of samples with a hydrated surface, as it can lose the fraction of their $-OH/H_2O$ attached. Most systems accommodate samples ranging from $\sim 1 \times 1$ cm^2 to less than 1 cm in height. Powder samples can be fixed to the double-sided adhesive tape. Both powder and thin film samples can be analyzed in XPS. There are certain limitations during XPS analysis: (1) organic substances cannot be studied using XPS because they can be easily damaged by either the energy of X-rays or the heat from non-monochromatic X-ray sources. (2) XPS recognizes all components with an atomic number greater than 3 ($Z > 3$). It cannot detect H ($Z = 1$) or He ($Z = 2$) due to the short diameter of these orbitals, which reduces the catch probability of the photoelectrons to nearly zero.[6,62]

Position and Shape of XPS Spectra: XPS spectra are plotted as photoelectron intensity vs. BE. The background of an XPS spectra consists of (1) photo emission by core electron levels, (2) photo emission from valence levels, and (3) Auger emission. For example, the XPS spectra of clean Ag surface (Fig. 3.7) shows a stepped background, resulting from photoelectrons' inelastic scattering.[63]

The core and valence-level photoelectron peaks are represented as 3s, 3p, 3d, and 4s and 4p, respectively, whereas the marked MNN is the Auger peak. The primary peak for elemental analysis is core-level photoelectron peaks, whereas the valence-level peaks are useful for the study of the electronic structure of material. The Auger electrons are useful for chemical analysis that are emitted during a core-hole decay process.[64]

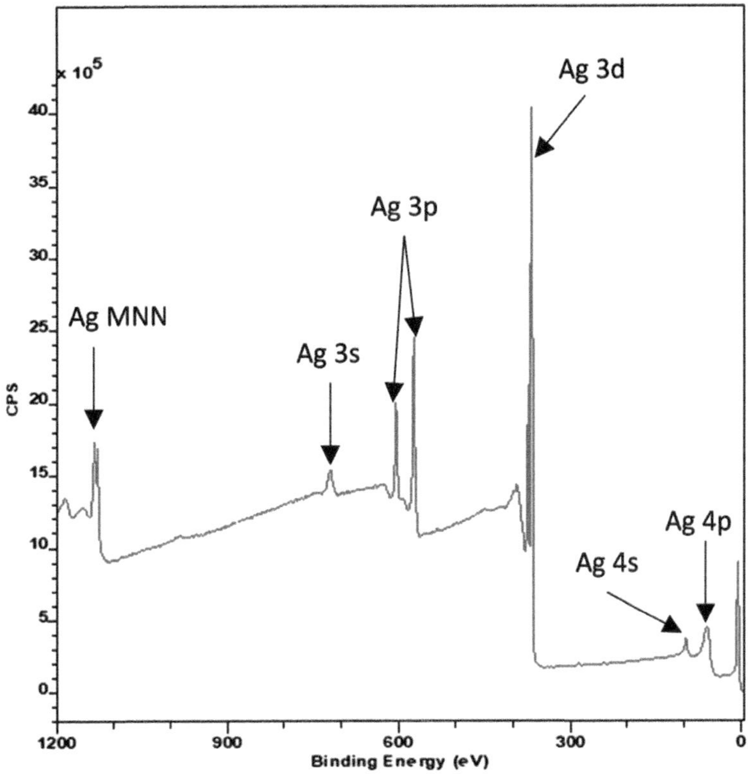

Figure 3.7 XPS spectrum of a clean silver Ag surface. Reprinted with permission from [63]. Copyright 2021, Royal Society of Chemistry.

Satellite Peak: During the photo ejection of core-electrons, the electrostatic force experienced by the outer shell electrons may induce a transition that involves the excitation of a valence electron. This process takes place simultaneously with the photo-ejection of core electrons. That results in the appearance of satellite peak.[65]

These satellite peaks can be seen in most of the transition metal oxides. For example, the XPS spectra of MnO shows two satellite peaks at a position 647 and 658.7 eV along with each main peak.[66] Furthermore, the satellite peaks are visible in samples containing d^9 Cu(II), while absent in samples having d^{10} Cu(I). For example, two satellite peaks appear at ~944 and ~963 eV along with the main peak of Cu $2p_{3/2}$ (933.4 eV) and $Cu_{1/2}$ (955 eV), respectively for CuO, whereas these satellite peaks are absent for Cu_2S.[67]

Multiplet Splitting: The core-level peak may undergo multiplet splitting if the valence level of the atom has unoccupied electrons. When photoionization causes a core electron vacancy, the remaining unpaired electron in the core may interact with the unpaired electrons in the outer shell. This interaction can lead to numerous final states, which can be observed in the photoelectron spectrum as a multi-peak envelope. Multiplet splitting is observed in rare earth metals with unfilled f orbitals and transition metals with unfilled p and d orbitals.[68,69] For example, in the XPS spectrum of NiO, the peak corresponding to the $2p_{3/2}$ state splits into two, corresponding to the Ni^{+2} and Ni^{+3} species at 853 and 856 eV, respectively.[70] Another example of splitting of core-level peaks can be observed for $NiCo_2O_4$ where core-level peak for both Ni and Co splits into two. The core-level peak for Ni $2p_{3/2}$ (855.81 eV) and $2p_{1/2}$ (873.71 eV) splits into two peaks corresponding to the Ni^{+3} and Ni^{+2} at 854.1, 872.1 eV and 855.81, 873.71 eV, respectively. On the other hand, the Co core-level peaks also split and confirm the presence of Co in a different oxidation state.[71]

3.6.3 Characterization

Elemental Identification: A typical XPS spectrum features a broad structure along with the sharp peaks; the broad structure is due to the inelastically scattered electrons from deeper in the solid that forms the background. The sharp peaks appear due to the elastically scattered electrons that are generated within the few atomic layers. For the elemental analysis, the sharp core-level photoelectron peaks are primary peaks.

Chemical Shift: Change in the Electronic Environment: The shift in the value of BE of an element depends on the bonding with other elements and the oxidation state of the element; for instance, if an element is present in higher oxidation state, its BE value will be higher compared to its lower oxidation state due to increased nuclear shielding. A shift in the BE of elements under the influence of surrounding is reported by Fageria and co-workers for AuPd nanoparticles decorated on C_3N_4.[72] Deconvoluted C1s BE spectrum fitted into four peaks corresponding to the different surrounding of C in AuPd-C_3N_4, appeared at 284.6, 286.2, 288, and 289.2 eV, which can be attributed to the C–C, C–N (or C–O), C=N, and C=O bonding, respectively. The value of C 1s BE increases as it is surrounded by the more electronegative elements. Similarly, the high-resolution Pd 3d spectra splits into two sets of doublets. One pair, which is cantered at the low BE side corresponds to the Pd(0) state of $3d_{5/2}$ and $3d_{3/2}$ cantered at 335.3 and 340.6 eV, respectively. Contrary, the other doublet's higher BE side corresponds to the higher oxidation state of Pd(+2), as it appeared at 337.4 and 342.6 eV. Similarly, Zhai and co-workers reported a shift in the BE of Ni and Fe in NiFe-LDH after the doping of Ru single atoms.[73] For Ni^{+2}, the peaks appeared at 856.2 and 874.0 eV corresponding to the Ni $2p_{3/2}$ and Ni $2p_{1/2}$. The core-level Fe 2p XPS spectrum is located at 712.1 and 725.7 eV, corresponding to the Fe $2p_{3/2}$ and Fe $2p_{1/2}$ of Fe^{+3}. Compared to the pristine NiFe-LDH, a positive shift in the B.E. of 0.5 and 0.3 eV is observed for Ni and Fe, respectively. This shift in the BE values indicate the change in the chemical state of Ni and Fe caused by the electronic coupling of Ru single atoms with NiFe-LDH.

Formation of the Active Sites: In many electrocatalytic reactions, the electrode material undergoes phase transition. This phase transition occurs on the surface layer of heterogeneous electrocatalysts, especially under anodic oxidation potential. This phase transformation before and after the electrocatalytic reactions can be investigated by XPS. For example, Lin and co-workers reported active phase transformation of NiO during oxygen evolution reaction (OER).[74] It is found that, NiO undergoes phase transformation and changes to NiOOH. XPS of Pt single atom decorated NiO (0.5 wt% Pt/NiO) were recorded before and after electrocatalysis. Figure 3.8a and b shows the high-resolution spectra of Ni $2p_{3/2}$ for before and after the 500 electrocatalytic cycles of OER, respectively. In Fig. 3.8a, peak A is assigned to the Ni^{+2} species, while the peaks B and C are assigned for the Ni^{+3}. The intensity of the peaks B and

C also reflects the amount of Ni present in the higher valence state. It is apparent from Fig. 3.8b that after the electrocatalytic cycle, peak A vanishes, suggesting that Ni(II)O transforms to the Ni(III) OOH during the OER and serve as an active site. Li and co-workers applied XPS analysis to know the role of the ketonic C=O group on CNTs during electrocatalytic OER.[75] With the help of O1s spectra, it was revealed that the ketonic C=O group acts as an active site for the electrocatalytic OER, as more ketonic C=O formed on the CNT surface, activity of the material enhanced for the electrocatalytic activity.

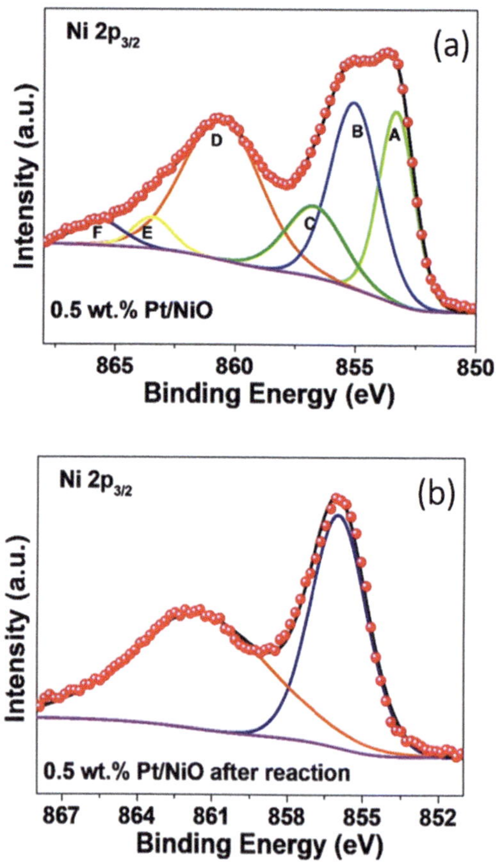

Figure 3.8 XPS spectra of Ni $2p_{3/2}$ of (a) as prepared Pt single atom decorated NiO and (b) after 500 electrocatalytic OER cycles. Reprinted with permission from [74]. Copyright 2018, Royal Society of Chemistry.

Depth Profiling: XPS depth profiling combined with the ion beam etching method can provide information regarding a material's composition uniformity as a function of depth. Though the X-ray used, penetrate the sample up-to several micrometers, however, the photoelectrons generated in the deep layer lose their energy due to inelastic collisions while escaping toward the surface.[76] In order to analyze the buried interface located at some micrometers from the surface, the surface layer of the material etched out with the help of ion sputtering. In a typical XPS depth profiling combined with the ion etching process, the soft X-ray is initially radiated on the sample to gather XPS spectra, followed by ion beam sputtering to etch the surface layer, subsequently by alternate spectra collection. Monoatomic ion beams such as Ar^+, O^{2+}, and cluster ion beams (C_{60}^+, C_{24}^+, Ar_n^+) are the ion sources that are generally used for the etching technique. Rathore and co-workers reported the synthesis of Ce-doped $Ni_{0.85}Se$ nanosheets. With the help of XPS depth profile, presence of Ni and Se bond throughout both pristine and doped sample was confirmed. The depth profile of both Ni 2p and Se 3d confirmed up to 135 nm etching.[15]

3.7 X-ray Absorption Spectroscopy

3.7.1 Fundamental Concepts

Extended X-ray absorption fine structure spectroscopy (EXAFS) and X-ray absorption near edge spectroscopy (XANES) are the two regions into which the X-ray absorption spectrum can be separated. On the absorption of X-ray, if the transitions are taking place from the core electronic state of the atom to the excited state (LUMO), then the spectra corresponding to this transition is known as XANES, while EXAFS is corresponding to the transition from the core electronic state to the continuum of the atom. Despite having the same physical origin, the differentiation is useful for the interpretation. XANES gives information about the oxidation state and coordination or symmetry of the metal sites, while EXAFS is useful to determine the distance and types of the different species from the absorbing atom. Thus, these two methods give complementary electronic state and structural information. XAS is element specific and it can be performed essentially to all the elements on the periodic table. The disordered samples such as powder samples, solutions, and amorphous samples can be characterized by XAS. Unlike the other structural probes such as X-ray crystallography, crystallinity of the sample is not required for EXAFS, which makes it more advantageous.[77,78] The wavelength of absorption peaks is characteristic of an element and depends on the chemical state of the element. There are three regions in the XAS spectrum.

First, the absorption edge region, where absorption of X-ray suddenly increases at a distinct point. This occurs due to the excitation of inner electrons to the upper electronic shell; these transitions are relatively weak. The core level from which the transition takes place is reflected in the nomenclature of the XAS edges. For example, if the transition originates from 1s core orbital, then the edge corresponds to K-edge, similarly transition from 2s and 2p represents L-edge, and M-edge represents the transition from 3s, 3p, and 3d levels (Fig. 3.9). Second, XANES, the low energy portion just above of the edge. Third, EXAFS, the higher energy portion (hundreds of eV) from the edge.

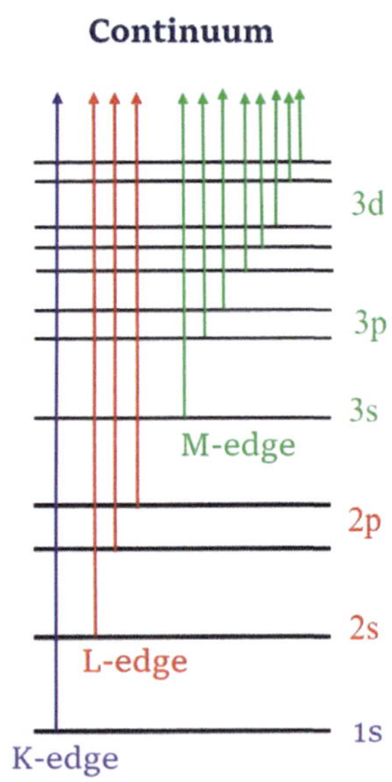

Figure 3.9 Transitions from the core level and corresponding edges.

When a quantum of the incident beam of X-rays is absorbed by an atom, its energy is transferred to the atom, resulting in the ejection of an electron, called a photoelectron, from the innermost energy level. The energy of the incident beam is divided between the kinetic energy of the ejected photoelectron and the potential energy of the excited ion. The probability of absorption of the X-ray photon is highest when the energy of the incident photon is comparable to the energy required to remove the electron just to the periphery of the atom. For example, the energy required to eject the K-electron from an atom of lead (Pb) corresponds to a wavelength of 0.14 Å of the incident X-ray. As the wavelength increases towards 0.14 Å, the probability of absorption rises to a maximum. Once the wavelength exceeds 0.14 Å (less energy), the photon no longer has enough energy to eject the k-electron, causing an abrupt drop in the absorption curve. A similar absorption edge occurs at the longer wavelengths corresponding to the lower energy L-shell electrons. The L-shell has three distinct subshells (LI, LII, LIII) with different binding energies corresponding to which separate absorption edges appear in that region of the spectra.[79]

The absorption of X-rays by the target atom provides useful information for the qualitative and quantitative analysis. The absorption edges are characteristic of an element, because each element has its own characteristic set of K, L, and M absorption edges. The wavelength corresponding to the adsorption edge, thus, can be used to identify the element present in a sample. The absorption of a part of the incident beam of X-rays, as it passes through the sample, leads to attenuation of the intensity of the transmitted beam that reaches the detector. The attenuation of the intensity can be related to the amount of the target element in sample using Beer-Lambert as in Eq. 3.19 as:

$$I_t = I_0 \, e^{\varepsilon c l} \tag{3.19}$$

where I_0 is the intensity of the incident beam of monochromatic X-rays, I_t is the intensity of the transmitted beam after passing through a distance of l cm to through the sample, and ε is the molar extinction coefficient.[7]

3.7.2 Sample Preparation

Any kind of sample, including powders and solutions, can be analyzed. Solid samples can be compressed into pellets for analysis after being diluted with an appropriate binder, such as boron nitride (BN), to provide the required thickness and absorption. To ensure homogeneity, the sample and binder are grinding together in the first phase of sample preparation. The homogenous sample is then pressed to create a self-supported pellet using a pellet press or sandwiched between an adhesive tape.

3.7.3 Characterization

Oxidation State and Symmetry of the System, XANES: XANES region of XAS is strongly sensitive for oxidation state or valence state and symmetry of the system. In XAS spectra, the oxidation state of the absorbing atom is often indicated by the edge's position. The edge shifts toward the higher energy with increased oxidation state of the absorbing atom. With the increase in effective nuclear charge on the absorbing atom, higher energy is required to convert a core electron into ions, which results in the shift of edge position. Kuo and co-worker[80] reported that on increasing the oxidation state of Mn, edge position shifted toward higher energy. Furthermore, the pre-edge features are sensitive toward the symmetry of the system. The transition corresponding to the pre-edge involves the 1s to nd transitions for the K-edge of a first-row transition metal. These transitions are forbidden as they do not follow the dipole selection rule. Despite not following the selection rule, the 1s → 3d transitions are observed due to both 3d-4p mixing and direct quadrupolar coupling. The extent of 3d-4p mixing reflects to the intensity of 1s → 3d transition, and that can be used as a probe of geometry. When the site is distorted from a centrosymmetric environment, the extent of 3d-4p mixing increases and thus the intensity of transition also enhances. The enhanced intensity can be used to distinguish between the centrosymmetric system, that is, tetrahedral < square pyramidal < octahedral and to distinguish between square planar (centrosymmetric) and tetrahedral sites. Thus, the details of pre-edge feature can be used to explore the electronic structure of the absorbing atoms. For example, in CuO, Cu is coordinated with the six O atoms and forms octahedral geometry (symmetrical), while in Cu_2O, Cu favors tetrahedral hybridization. Thus, the pre-edge feature is more intense for Cu_2O than CuO.[81]

Local Coordination Structure, Extended Fine Structure: In XAS spectrum, signal beyond 50 eV considered in the EXAFS region. The initial and final state of the electron influence the probability of the absorption of an X-ray photon. The initial level is represented by the core-level electrons, whereas the final state varies, and depends on the surrounding or neighboring atom. When there are no neighboring atoms, the final state of the electron is simply the emitted photoelectron wave. However, when there is a neighboring atom, an incoming wave is generated by the backscattering of the photoelectron (Fig. 3.10). This incoming wave is influenced by the outgoing wave, distance from the absorbing and scattering atom, as well as the type of scattering atom. Thus, the final wave is the combination of both the outcoming, and incoming waves and the interaction between the outcoming and incoming wave can be explained as an interference behavior. In case of constructive interference, the final electron wave will intensify, which results in an increased absorption probability and a maximum in the oscillating fine structure. On the other hand, for destructive interference, the electron wave will be reduced, and compared to that without a neighboring atom, the final state will be less probable. Thus, the probability of absorption will decrease, resulting in a minimum in the oscillating fine structure. As the surrounding of the absorbing atom may have different atoms and variable distances, several final waves resulting from these distinct surroundings can emerge, and EXAFS represents the total of these final waves.[82–85]

Consequently, from this information, the distance to adjacent atoms, coordination numbers, and lattice dynamics can be obtained. EXAFS plays an important role in the synthesis and understanding of single atom catalysts. Kuang and co-workers reported an efficient strategy to synthesize and stabilize Pt single atoms supported on N-doped mesoporous hollow carbon spheres, with the help of findings of EXAFS results. They reported the formation of Pt single atoms via the bonding of Pt with N of the carbon support, while aggregation occurred of Pt single atoms in the absence of N atoms on the carbon support matrix.[86] These EXAFS studies inspired numerous efforts to synthesize and stabilize the single atom catalysts.

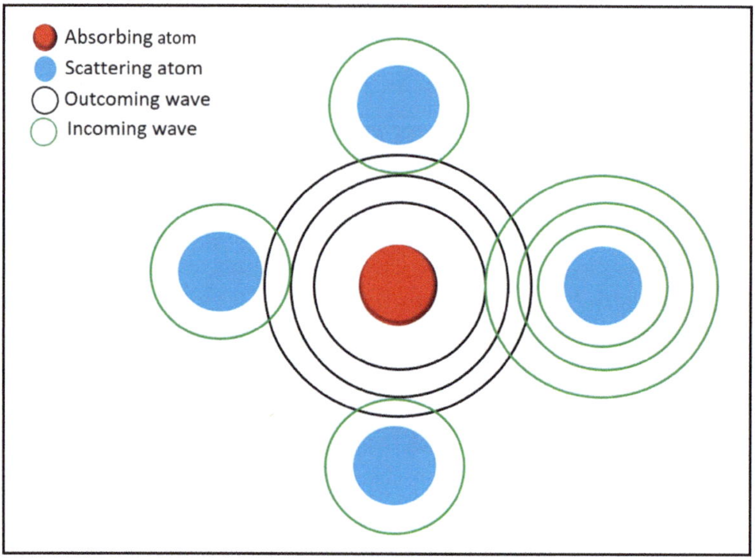

Figure 3.10 Schematic representation of the effect of neighboring atoms.

3.8 Ultraviolet-Visible Spectroscopy

3.8.1 Fundamental Concepts

Ultraviolet-visible (UV-Vis) spectroscopy is based on the Beer-Lambert law. This law states that light absorption is directly proportional to the sample's concentration and the path length of the sample cell:[87]

$$A = \log_{10}(I_o/I) = \varepsilon\, c\, l \qquad (3.20)$$

where A = absorbance, I_o = intensity of incident light, I = intensity of transmitted light, ε = molar absorption coefficient, c = concentration of analyte, and l = path length.

In the 190–800 nm range, the amount of radiation absorbed depends on the electronic structure of the absorbing species, such as atoms, molecules, ions, or complexes. An electronic energy diagram consists of a few vibrational energy levels, and each vibrational level comprises several rotational energy levels. If the photon's energy is equal to the energy difference between these levels, it can induce a transition between these energy levels when interacting with a moiety. To obtain the spectrum, the amount of radiation absorbed by the analyte is measured and plotted against the wavelength of the light. Consequently, a typical UV-Vis spectrum is a plot of wavelength or frequency against absorption intensity.[88]

3.8.2 Sample Preparation

The solution is placed in a standard 10-mm path length cuvette for liquid samples with a volume of between 100 μL and 3.5 mL. The cuvette and solvent selection are crucial, because they have absorption cut-off wavelengths below which the solvent or cuvette will absorb all light and become unusable for measuring sample absorption. Quartz cuvettes are more suited for examining materials that absorb at UV wavelengths, because their absorption cut-off wavelength is 160 nm, compared to around 300 nm for glass.[88]

3.8.3 Characterization

UV-Vis spectroscopy is a straightforward technique for investigating metal NPs, semiconducting materials such as quantum dots, CNTs, and 2D materials like graphene. It is used to study the optical characteristics of nanomaterials. In addition, it also helps to characterize attributes like shape, size, and stability that affect their properties.[8]

Size and Shape of the Nanoparticles: Metal nanoparticles exhibit optical properties due to surface plasmon resonance (SPR). In SPR, the electron clouds on the surface of nanoparticles oscillate at the same frequency as the incident radiation when exposed to light of that frequency. The SPR frequency of a nanoparticle depends on size and shape. By modifying these factors, nanoparticles with desired optical properties can be obtained.[8] For example, Njoki and co-workers observed that Au NPs exhibit a color transition from dark red to pink as the size varied from 30 to 90 nm. The particle size affects the wavelength of the surface plasmon band. As the size of nanoparticles increases, the λ_{max} of the surface plasmon band shifts from 519 nm to 569 nm, as shown in Fig. 3.11. This red shift is in conjunction with a little broadening of the SPR band in the long-wavelength region.[89]

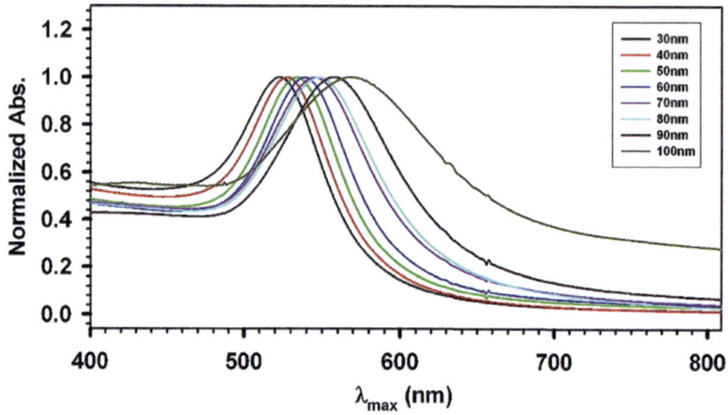

Figure 3.11 UV-Vis spectra for gold nanoparticles with different particle sizes. Reprinted with permission from [89]. Copyright 2007, American Chemical Society.

Localized surface plasmon resonance (LSPR) is similar to SPR. This technique generates amplified plasmon resonances on the surface of nanostructures, in contrast to the unstructured metal surfaces. A localized surface plasmon is produced when a surface plasmon is contained within a nanoparticle that is the same size or smaller than the wavelength of the light. For instance, Haiss and co-workers studied how the LSPR peak position of NPs depends on particle size. An empirical relation represented by Eq. 3.21 can be used to calculate the LSPR peak's position for the Au NPs with a diameter larger than 25 nm:

$$\lambda_{LSPR} = 512 + 6.53 \exp(0.0216d) \tag{3.21}$$

For particles with diameters ranging from 35 to 100 nm, the peak position can be calculated by using Eq. 3.22:

$$d = \frac{\ln(\lambda_{spr} - \lambda_o / L_1)}{L_2} \tag{3.22}$$

The particle diameter (in nm) can be specifically calculated using the A_{LSPR}/A_{450} ratio. The absorbance at the LSPR peak (A_{LSPR}) and that at $\lambda = 450$ nm (A_{450}) for particles with diameters, between 5 and 80 nm were found to follow the relationship provided by Eq. 3.23:

$$d = \exp((3.35(A_{LSPR}/A_{450})) - 3.11) \tag{3.23}$$

Equation 3.23 does not hold true for those particles which show agglomeration.[90] Huang and co-workers reported that the

SPR band of Au nanosphere splits into longitudinal and transverse bands when the shape changes to nanorod. The longitudinal band shows strong intensity in the NIR region represents the oscillation of electron through the long axis. On the other hand, the weak intense transverse band in visible region signifies similar wavelength to the Au nanosphere. Change in size has no effect on the transverse band, while the longitudinal band gets red-shifted as the ratio of length/ breadth increases.[91]

Using the UV-Vis spectroscopy, Qian and co-workers studied the splitting of the plasmon band with respect to the morphology of Ag nanoparticles in different solvent media. In ethanol, most nanoparticles were spherical with a size of 32 nm, while few had triangular geometry with a single SPR at 420 nm. On the other hand, in pyridine, spherical Ag nanoparticles with an average size of 8 nm were formed which showed only one sharp asymmetric peak at 403 nm. The absorption of silver nanoparticles synthesized in pyridine exhibits a narrower absorption band and a blue shift than those formed in ethanol. This phenomenon may be related to the size effect of the silver nanoparticles. They have also found out that with increased irradiation time, the plasmon resonance maximum and the band get sharpened in ethanol and pyridine, respectively. On the other hand, in DMF, the formation of nanoprisms takes place which shows two distinct quadrupole plasmon resonances, while in NMP (N-methyl-2-pyrrolidone) solvent, the nanoprisms exhibit three peaks.[92] Wei and co-workers studied the correlation between the LSPR peak and the morphology of copper nanoparticles. The structure of Cu nanoparticles changed from aggregated cubes to hemispheres by adding 5 mM sodium citrate, which led to the formation of more uniform-size nanoparticles on the substrate by inhibiting the re-nucleation. Aggregated Cu_2O/Cu cubes show an absorption band at 461 nm, while hemispheres show an absorption band at 601 and 608 nm, respectively, in the LSPR spectrum. Agglomerated NPs decrease the intensity of the LSPR peak, while the uniform-sized nanoparticles increase the intensity.[93]

Determination of Band Gap in Semiconductors: Semiconducting nanomaterials like quantum dots and 2D materials comprise a single layer of molecules and absorb light by excitation of their electrons to higher energy levels. Between these levels that make up the lower valence band and the upper conduction band, there is a significant gap in the energy levels that electrons in semiconductor materials can adopt. This is referred to as the band gap. Shorter light wavelengths have more energy, therefore light below a particular threshold can excite electrons in semiconducting materials to move from the valence band to the conduction band by providing them with enough energy to cross this gap.[8] In this context, Soltani and co-workers calculated the optical band gap of CdS QDs from the absorption band. The interband transition between extended states in the valence and conduction bands is the cause of the absorption in the Tauc region (high absorption zone), and the absorption coefficient is dependent on the optical band gap, which is determined by Eq. 3.24:

$$\alpha(\upsilon) = K\,(h\upsilon - E_g)^n/h\upsilon \qquad (3.24)$$

where n depends on the number of transitions, that is, 1/2, 2, 3/2 for direct allowed, indirect allowed, and direct forbidden; K is a constant; and E_g is called the optical band gap. By plotting $d(Ln(\alpha h\upsilon))/d(h\upsilon)$ from UV-Vis spectra, one may calculate the energy of the band gap. This plot indicates the discontinuity point given to the band gap value (at $h\upsilon - E_g = 0$) according to Eq. 3.25:

$$d(Ln(\alpha h\upsilon))/d(h\upsilon) = n/(h\upsilon - E_g) \qquad (3.25)$$

As the microwave exposure period increases, the band gap in the 2.54–2.65 eV region decreases, and particles with hexagonal structures have greater band gaps than those with cubic structures.[94] Sakthivel and co-workers determined the optical band gap of ZnS and Mn-doped ZnS by using the Tauc equation. The calculated band gap values for Mn = 0, 2, 4, 6% doped ZnS QDs were found to be 3.93, 3.91, 3.84, and 3.79 eV, respectively. A red shift has been observed in the band gap values compared to the bulk ZnS (3.6 eV) due to quantum confinement and the direct energy transfer between semiconductor-excited states and the 3d Mn^{2+} ion levels.[95]

3.9 Diffuse Reflectance Spectroscopy

3.9.1 Fundamental Concepts

When the particle dimensions are small compared to the beam cross-section yet substantial compared to the wavelength of light, diffraction phenomena arise due to rays striking upon the crystal and overcoming it, leading to interference among elementary waves. A part of the incident light reflects at all angles into the source's hemisphere in powders with randomly oriented particles. Diffuse reflection mainly results from refraction, absorption, and diffraction of particles oriented in all directions. The angular distribution of reflected light does not depend on the angle of incidence and obeys Lambert Cosine's law for ideal diffuse reflection. In the case of powder samples, near Lambertian behavior is observed, unlike that of an ideal diffuse reflector.[9] Kubelka-Munk (K-M) theory is commonly used to study diffuse reflectance spectra. It is analogous to Beer's law for transmission spectroscopy.[96] This theory is based on the following assumptions:

(i) the specular reflection component should not be taken into consideration, (ii) the sample should be of finite thickness, (iii) the sample should be considered homogeneous and isotropic, (iv) no external and internal reflection should occur from the sample's surface, and (v) scattering and absorption are the only interactions taken into consideration with the sample.

In transmission spectroscopy, the transmittance is the ratio of the intensity of transmitted to incident light provided by Eq. 3.26:

$$T = I/I_o \tag{3.26}$$

where I_o = intensity of incident light and I = intensity of transmitted light, and T = transmittance. Similarly, Eq. 3.27 determines the remittance of a diffusely reflected sample which is defined as the ratio of the reflected to incident light:

$$R_\infty = J/I_o \tag{3.27}$$

where I_o is the intensity of incident light, J is the intensity of reflected light, and R_∞ is absolute remittance. Practically, the value of R_∞ cannot be equal to 1 for an ideal diffuse reflector. Therefore, absolute remittance is measured in terms of relative remittance (R'_∞) which is represented by Eq. 3.28:

$$R'_\infty = R_{\infty(\text{sample})}/R_{\infty(\text{standard})} \tag{3.28}$$

In transmission spectroscopy, the data is expressed in terms of absorbance (A), where A = -log T. This approach is useful because it obeys Beer's law, which states that the absorbance is directly proportional to the concentration of the absorbing medium. Similarly, data is expressed in apparent absorbance units in DRS, which is obtained by plotting log $(1/R'_{\infty})$ against frequency or wavelength. However, this does not mean that Beer's law applies to DRS.[96]

The function derived by Kubelka and Munk to linearly relate the concentration of analyte species with the reflectance of the sample as Beer's law in the case of transmission spectroscopy is given as:

$$F(R_{\infty}) = \frac{(1 - R_{\infty})^2}{2R_{\infty}} = \frac{K}{S} = \frac{2.303\varepsilon C}{S} \tag{3.29}$$

where $F(R_{\infty})$ = K-M function, ε = molar extinction coefficient, C = analyte concentration, S = twice the scattering coefficient of the sample, and K = twice the true physical absorption coefficient of the diffusely reflecting medium. From Eq. 3.29, we can say that K-M theory is a two-constant theory since the reflectance properties depend upon the ratio of the absorbing to scattering coefficients, which are constants.[96]

3.9.2 Sample Preparation

During sample preparation, the scattering coefficient must be kept constant as it varies from sample to sample. To accomplish this, two important parameters, that is, particle size and particle packing, must remain constant for different samples. Particle size can be controlled by either sieving or ball and milling the sample into powder. On the other hand, sufficient particle packing can be achieved by gently tapping the sample in the sample cup or applying a known amount of pressure for a particular period. High pressure should be avoided otherwise, it will result in glossing of the surface, which is a sign of Fresnel reflection.[96]

The following key points must be kept in consideration while preparing the sample:

- Sample selection: The sample should be dry and free from contaminants.
- Grinding: The sample should be crushed into a fine powder.
- Sifting: For particle size to be uniform, the sample must be sieved.
- Solid samples can be handled as KBr pellet, Nujol mull, solution, or film.

(a) Mix the powdered sample with a suitable non-absorbing matrix in the ratio of 1:10. The most often utilized matrix is potassium bromide (KBr). The matrix helps to dilute the sample and provides a uniform background. The mixture is pressed into a pellet in an evacuated die set by applying a pressure of about 3000–5000 kg cm^{-3}.

(b) By adding a drop of Nujol film (liquid paraffin: transparent to mid-IR-region), the powdered solid sample is converted into a paste or mull and then spread on the KBr sample window cell.

(c) Suitable solvents like CS_2, $CHCl_3$, or CCl_4 are used to dissolve solid samples to form the solution. About 0.1 mL of solution is placed in the liquid sample cell.

(d) By allowing the solvent to evaporate from a sample solution and leaving a sample film on the window, films of polymers and amorphous solids can be deposited on the sample cell window.

- Drying: Keep the sample in the desiccator if it is moist; otherwise, it will affect the reflectance properties.
- Measurement setup: Place the prepared sample in the instrument and calibrate it properly before taking the readings.

3.9.3 Characterizations

DRS is an essential tool for characterizing nanomaterials, allowing researchers to explore their optical properties and surface features. By examining the diffuse light reflected from a sample, DRS yields valuable details about the electronic structure and band gap of nanomaterials, which are vital for applications in catalysis, sensors, and photovoltaics. This technique is especially beneficial for analyzing materials that are challenging to study with traditional spectroscopy methods, as it requires minimal sample preparation and can be used with various forms, like powders and thin films.

Determination of Optical Band Gap: Aydin and co-workers studied the optical properties of undoped and Fe-doped ZnO using DRS. In this case, the DRS spectrum of weakly absorbing materials is analyzed by using the K-M theory.[97] Aydin and co-workers observed the substantial change in the absorption edge as the concentration of Fe dopant increases, suggesting a strong bond between zinc and iron molecules. The diffuse reflectance of undoped and doped ZnO increases with increasing wavelength. However, there is a decrease in the diffuse reflectance upon the addition of Fe dopant into ZnO. The fundamental absorption, which indicates the excitation of an electron from the valence band to the conduction band, can be used to calculate the optical band gap (E_g). For DRS calculations, the equation for the determination of E_g is:

$$\alpha h\vartheta = \frac{F(R)h\vartheta}{t} = A(h\vartheta - Eg)^{n} \tag{3.30}$$

where F(R) is the Kubelka-Munk function, A is an energy-independent constant, and n is a constant such that its value is equal to 2, 3, 1/2, and 3/2 for indirect allowed transitions, indirect forbidden transitions, direct allowed transitions, and direct forbidden transitions. The undoped ZnO sample's optical band gap is 3.19 eV, lower than the bulk ZnO sample's 3.37 eV. However, the E_g value agrees with diffuse reflectance's 3.22 eV. As Fe doping increases, the optical band gap drops from 3.19 eV for the undoped sample to 2.75 eV for the 20% doped sample. This suggests that Fe-dopants significantly alter the ZnO optical band gap. The Burstein-Moss effect, resulting from a rise in carrier concentrations, is anticipated to cause changes in the band gap of ZnO doped with Fe.[97] Singh and co-workers characterized Nd^{3+} ion-doped cadmium sulphide nanoparticles by DRS using K-M theory. They observed that with an increase in the molarity of dopant (Nd^{3+} ions) from 0.1 mol % to 0.3 mol% in CdS, there is a decrease in the band gap energy from 4.57 to 4.49 eV. This is due to the substitution of CdS by Nd^{3+} ions, which adds electron states into the band gap of CdS, leading to the formation of new lowest unoccupied molecular orbitals.[98]

To Study the Influence of Shape and Size of Nanoparticles on Diffuse Reflectance Properties: Khadga and co-workers reported how the shape and size of nanomaterials influence the diffuse reflectance properties. According to K-M theory, initially, the value of diffuse reflectance for thick powder is unity when absorption is zero, and then it declines monotonically as the absorption to scattering ratio (K/S) increases. The complex refractive index of the material increases with increasing real part for lowest order scattering (S) and with rising imaginary part for absorption (K). Equation 3.31 gives the complex refractive index as:

$$m = n + ik \tag{3.31}$$

where i is the imaginary number and n and k are real and imaginary parts of the complex refractive index (m), respectively. Figure 3.12 shows NIR diffuse reflectance curves for different samples. CuO nanorods (a, b) demonstrated strong near-infrared (NIR) diffuse reflectance, as demonstrated by the comparison investigation. This is explained by their complicated refractive index's substantial real (mathematical) component. Conversely, Cu nanoparticles (f, g, i) and conventional micrometre-scale CuO particles (e, h) with various morphologies exhibited low NIR diffuse reflectance. Larger CuO particles have lower NIR reflectance because of their rough surfaces, which enhance NIR absorption. Conversely, the substantial imaginary (mathematical) part of the complicated refractive index of copper metal accounts for the low diffuse reflectance in Cu nanoparticles.[99]

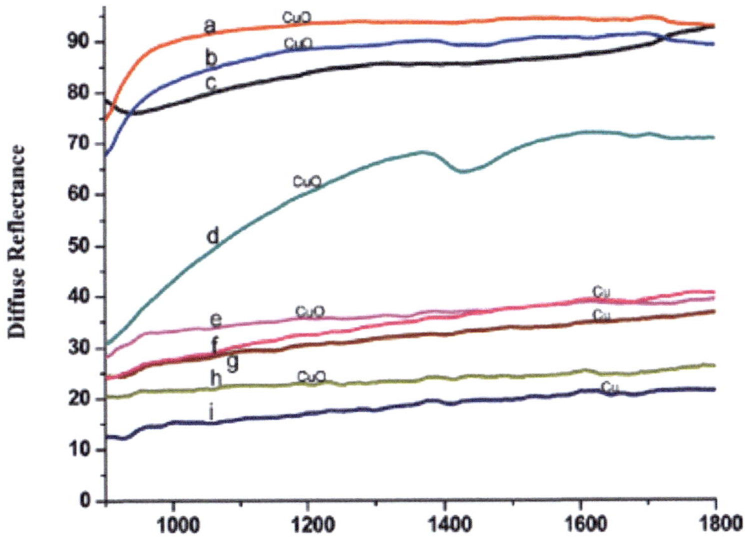

Figure 3.12 Diffuse reflectance curves at NIR of different Cu and CuO samples. Reprinted with permission from [99]. Copyright 2010, American Chemical Society.

Biju and co-workers characterized different shapes and sizes of α-Fe_2O_3 nanoparticles with the help of DRS. The absorption band edge shifts to lower wavelength values with the decrease in the size of α-Fe_2O_3 NPs. As a result, the value of the optical band gap also increases for G1-G5 (by varying the molar ratio of $FeCl_2$ to glycine-1:1, 1:3, 1:9, 1:12, 1:15) as 1.775(3), 1.889(2), 1.959(1), 1.949(2) and 1.936(2) eV, respectively. The Burstein-Moss effect causes the band gap to increase as the size decreases in the case of n-type semiconductors.[100]

3.10 Conclusion

In this chapter, an overview of the most relevant spectroscopic techniques, namely XRD, FTIR, RAMAN, XPS, XAS, and UV-Vis spectroscopy is discussed. It includes the fundamental concepts, sample preparation methodology, and the application of these spectroscopic techniques for the characterization of nanomaterials, correlating with the scientific literature. From this chapter, we learned how these spectroscopic techniques could be utilized for the profound understanding of the different characteristics of the nanomaterials. For instance, the structure, size, and residual stress of the crystal can be determined by XRD. Similarly, different functional groups and bonding types in a material can be identified by FTIR. Furthermore, Raman analysis is discussed for the exhaustive analysis of new age materials such as graphene, transition metal chalcogens. Additionally, the surface chemistry of materials can be elaborately comprehended by XPS analysis. This chapter also focuses on advanced characterisation techniques like XAS and XPS. Finally, emphasis was given on UV-Vis spectroscopy and DRS to know the optical and diffused reflectance properties of liquid and solid samples, respectively. Together, these techniques give comprehensive information required to understand and analyze the physical, chemical, and optical characteristics of nanomaterials that can guide the researchers to optimize their performance in a wide range of applications.

References

1. Epp, J., X-ray diffraction (XRD) techniques for materials characterization. In *Materials Characterization Using Nondestructive Evaluation (NDE) Methods*, Elsevier: **2016**, pp. 81–124.

2. Giannini, C., Ladisa, M., Altamura, D., Siliqi, D., et al., X-ray diffraction: A powerful technique for the multiple-length-scale structural analysis of nanomaterials. *Crystals* **2016,** *6* (8), 87.

3. Leng, Y., *Materials Characterization: Introduction to Microscopic and Spectroscopic Methods.* John Wiley & Sons: **2013**.

4. Banwell, C. N. and McCash, E. M., *Fundamentals of Molecular Spectroscopy.* Indian Edition: **2017**.

5. Blinder, S. M., *Introduction to Quantum Mechanics.* Academic Press: 2020.

6. Watts, J. F. and Wolstenholme, J., *An Introduction to Surface Analysis by XPS and AES.* John Wiley & Sons: **2019**.

7. Koningsberger, D., Mojet, B., Van Dorssen, G., Ramaker, D., XAFS spectroscopy; fundamental principles and data analysis. *Top. Catal.* **2000,** *10*, 143–155.

8. Heath, J. and Poggel, C., *UV-Vis-NIR Spectroscopy for Nanomaterials Research.* John Wiley & Sons Ltd., Chichester, West Sussex: **2021**.

9. Torrent, J., Barrón, V., Diffuse reflectance spectroscopy. *Methods Soil Anal., Pt 5: Mineral. Methods* **2008,** *5*, 367–385.

10. Ali, A., Chiang, Y. W., and Santos, R. M., X-ray diffraction techniques for mineral characterization: A review for engineers of the fundamentals, applications, and research directions. *Minerals* **2022,** *12* (2), 205.

11. Bish, D. L., Reynolds, R., and Post, J., Sample preparation for X-ray diffraction. *Mod. Powder Diffr.* **1989,** *20*, 73–99.

12. Thamaphat, K., Limsuwan, P., and Ngotawornchai, B., Phase characterization of TiO_2 powder by XRD and TEM. *Agric. Nat. Resour.* **2008,** *42* (5), 357–361.

13. Verma, R., Rani, R., Choubey, P., Sharma, A., et al., Au nanoparticles on In_2S_3/In_2O_3 nanopyramids increase photoanodic activity in photoelectrochemical water splitting. *ACS Appl. Nano Mater.* **2024**.

14. Holzwarth, U. and Gibson, N., The Scherrer equation versus the 'Debye-Scherrer equation'. *Nat. Nanotechnol.* **2011,** *6* (9), 534–534.

15. Rathore, D., Ghosh, S., Gupta, A., Chowdhury, J., et al., Ce-doped NiSe nanosheets on carbon cloth for electrochemical water-splitting. *ACS Appl. Nano Mater.* **2024,** *7* (8), 9730–9744.

16. Theivasanthi, T. and Alagar, M., Nano sized copper particles by electrolytic synthesis and characterizations. *Int. J. Phys. Sci* **2011,** *6* (15), 3662–3671.

17. Muniz, F. T. L., Miranda, M. R., Morilla dos Santos, C., and Sasaki, J. M., The Scherrer equation and the dynamical theory of X-ray diffraction. *Acta Crystallogr. A: Foundations and Adv.* **2016,** *72* (3), 385–390.

18. Zak, A. K., Majid, W. A., Abrishami, M. E., and Yousefi, R., X-ray analysis of ZnO nanoparticles by Williamson–Hall and size–strain plot methods. *Solid State Sci.* **2011,** *13* (1), 251–256.

19. Rathore, D., Ghosh, S., Chowdhury, J., and Pande, S., Co-doped Ni_9S_8 nanostructures for electrocatalytic water splitting over a wide pH range. *ACS Appl. Nano Mater.* **2022,** *5* (8), 11823–11838.

20. Rathore, D., Ghosh, S., Chowdhury, J., and Pande, S., Fe-doped $NiCo_2Se_4$ nanorod arrays as electrocatalysts for overall electrochemical water splitting. *ACS Appl. Nano Mater.* **2023,** *6* (4), 3095–3110.

21. Abinaya, N., Robert, M. C., Srinivasan, N., and Saravanakumar, S., Electron density mapping and bonding in Mn doped $CoFe_2O_4$ using XRD, and its correlation with room temperature optical and magnetic properties. *J. Magn. Magn. Mater.* **2023,** *580*, 170938.

22. McCusker, L., Von Dreele, R., Cox, D., Louër, D., et al., Rietveld refinement guidelines. *J. Appl. Crystallogr.* **1999,** *32* (1), 36–50.

23. Khandpur, R. S., *Handbook of Analytical Instruments*. McGraw Hill Professional: **2007**.

24. Adhikari, B. and Banerjee, A., Facile synthesis of water-soluble fluorescent silver nanoclusters and HgII sensing. *Chem. Mater.* **2010,** *22* (15), 4364–4371.

25. Yang, X., Feng, Y., Zhu, S., Luo, Y., et al., One-step synthesis and applications of fluorescent Cu nanoclusters stabilized by L-cysteine in aqueous solution. *Anal. Chim. Acta* **2014,** *847*, 49–54.

26. Wang, J., Yao, J., Sun, N., and Deng, C., Facile synthesis of thiol-polyethylene glycol functionalized magnetic titania nanomaterials for highly efficient enrichment of N-linked glycopeptides. *J. Chromatogr. A* **2017,** *1512*, 1–8.

27. He, Z., Que, W., Chen, J., He, Y., et al., Surface chemical analysis on the carbon-doped mesoporous TiO_2 photocatalysts after post-thermal treatment: XPS and FTIR characterization. *J. Phys. Chem. Solids* **2013,** *74* (7), 924–928.

28. Doolaanea, A., Harun, A. F., and Mohamed, F., Quantification of Nigella sativa oil (NSO) from biodegradable PLGA nanoparticles using FTIR spectroscopy. *Int. J. Pharm. Pharm. Sci.* **2014,** *6* (10), 228–232.

29. Xiong, G., Pal, U., Serrano, J., Ucer, K., et al., Photoluminesence and FTIR study of ZnO nanoparticles: The impurity and defect perspective. *Phys. Status Solidi C* **2006,** *3* (10), 3577–3581.

30. Banwell, C. and Mc Cash, E., *Fundamentals of Molecular Spectroscopy*, Tata McGraw Hill. *New Delhi* **1994**.

31. Skoog, D. A., West, D. M., Holler, F. J., and Crouch, S. R., *Fundamentals of Analytical Chemistry*. Saunders College Pub. Fort Worth: **1996**; Vol. 33.

32. Zhang, P., Haage, T., Habermeier, H. U., Ruf, T., et al., Raman spectra of ultrathin $YBaCuO_{7-\delta}$ films. *J. Appl. Phys.* **1996,** *80* (5), 2935–2938.

33. Li, Y., Shelley, C., Cohen, L., Caplin, A., et al., Raman studies of laser-written patterns in $YBa_2Cu_3O_x$ films. *J. Appl. Phys.* **1996,** *80* (5), 2929–2934.

34. Zhou, X., Cardona, M., Colson, D., and Viallet, V., Plane oxygen vibrations and their temperature dependence in $HgBa_2Ca_2Cu_3O_{8+\delta}$ single crystals. *Phys. Rev. B.* **1997,** *55* (18), 12770.

35. Vetterhöffer, J., Wagner, J., and Weber, J., Isolated hydrogen molecules in GaAs. *Phys. Rev. Lett.* **1996,** *77* (27), 5409.

36. Champagnon, B., Gregora, I., Monin, Y., Duval, E., et al., Raman scattering in silicon nanostructures. *Porous Silicon Science and Technology: Winter School Les Houches, 8 to 12 February 1994* **1995**, 247–253.

37. Townsley, C., Davies, J., Wolverson, D., Boyce, P., et al., Spin-flip Raman-scattering studies of compensating donor centers in nitrogen-doped zinc selenide grown by molecular-beam epitaxy. *Phys. Rev. B.* **1996,** *53* (16), 10983.

38. Bokobza, L. A. and Zhang, J., Raman spectroscopic characterization of multiwall carbon nanotubes and of composites. *Express Polym. Lett.* **2012,** *6* (7).

39. López-Lorente, A., Simonet, B., and Valcárcel, M., Raman spectroscopic characterization of single walled carbon nanotubes: Influence of the sample aggregation state. *Analyst* **2014,** *139* (1), 290–298.

40. Bayle, M., Reckinger, N., Felten, A., Landois, P., et al., Determining the number of layers in few-layer graphene by combining Raman spectroscopy and optical contrast. *J. Raman Spectrosc.* **2018,** *49* (1), 36–45.

41. Silva, D. L., Campos, J. L. E., Fernandes, T. F., Rocha, J. N., et al., Raman spectroscopy analysis of number of layers in mass-produced graphene flakes. *Carbon* **2020,** *161*, 181–189.

42. Mueller, N. S., Heeg, S., Alvarez, M. P., Kusch, P., et al., Evaluating arbitrary strain configurations and doping in graphene with Raman spectroscopy. *2D Mater.* **2017,** *5* (1), 015016.

43. Sampathkumar, K., Diez-Cabanes, V., Kovaricek, P., Del Corro, E., et al., On the suitability of Raman spectroscopy to monitor the degree of

graphene functionalization by diazonium salts. *J. Phys. Chem. C* **2019,** *123* (36), 22397–22402.

44. Belyaeva, L. A., Jiang, L., Soleimani, A., Methorst, J., et al., Liquids relax and unify strain in graphene. *Nat. Commun.* **2020,** *11* (1), 898.

45. Judek, J., Pasternak, I., Dabrowski, P., Strupinski, W., et al., Hydrogen intercalation of CVD graphene on germanium (001)–Strain and doping analysis using Raman spectroscopy. *Appl. Surf. Sci.* **2019,** *473*, 203–208.

46. Zhao, W., Ghorannevis, Z., Amara, K. K., Pang, J. R., et al., Lattice dynamics in mono-and few-layer sheets of WS_2 and WSe_2. *Nanoscale* **2013,** *5* (20), 9677–9683.

47. Yu, Y., Li, C., Liu, Y., Su, L., et al., Controlled scalable synthesis of uniform, high-quality monolayer and few-layer MoS_2 films. *Sci. Rep.* **2013,** *3* (1), 1866.

48. Li, S.-L., Miyazaki, H., Song, H., Kuramochi, H., et al., Quantitative Raman spectrum and reliable thickness identification for atomic layers on insulating substrates. *ACS Nano* **2012,** *6* (8), 7381–7388.

49. Zhang, X., Han, W., Wu, J., Milana, S., et al., Raman spectroscopy of shear and layer breathing modes in multilayer MoS_2. *Phys. Rev. B Condens. Matter.* **2013,** *87* (11), 115413.

50. Desai, S. B., Seol, G., Kang, J. S., Fang, H., et al., Strain-induced indirect to direct bandgap transition in multilayer WSe_2. *Nano Lett.* **2014,** *14* (8), 4592–4597.

51. Wang, Y., Cong, C., Qiu, C., and Yu, T., Raman spectroscopy study of lattice vibration and crystallographic orientation of monolayer MoS_2 under uniaxial strain. *Small* **2013,** *9* (17), 2857–2861.

52. Zhu, C., Wang, G., Liu, B., Marie, X., et al., Strain tuning of optical emission energy and polarization in monolayer and bilayer MoS 2. *Phys. Rev. B Condens. Matter.* **2013,** *88* (12), 121301.

53. Hui, Y. Y., Liu, X., Jie, W., Chan, N. Y., et al., Exceptional tunability of band energy in a compressively strained trilayer MoS2 sheet. *ACS Nano* **2013,** *7* (8), 7126–7131.

54. Sharma, M. D., Mahala, C., Modak, B., Pande, S., et al., Doping of MoS2 by "Cu" and "V": An efficient strategy for the enhancement of hydrogen evolution activity. *Langmuir* **2021,** *37* (16), 4847–4858.

55. Vandalon, V., Sharma, A., Perrotta, A., Schrode, B., et al., Polarized Raman spectroscopy to elucidate the texture of synthesized MoS_2. *Nanoscale* **2019,** *11* (47), 22860–22870.

56. Froehlicher, G., Lorchat, E., Zill, O., Romeo, M., et al., Rigid-layer Raman-active modes in N-layer transition metal dichalcogenides: interlayer force constants and hyperspectral Raman imaging. *J. Raman Spectrosc.* **2018,** *49* (1), 91–99.

57. (a) Kim, H., Ko, H., Kim, S. M., and Rho, H., Polarization-dependent anisotropic Raman response of CVD-grown vertically stacked MoS_2 layers. *J. Raman Spectrosc.* **2020,** *51* (5), 774–780. (b) Soundiraraju, B. and George, B. K., Two-dimensional titanium nitride (Ti_2N) MXene: Synthesis, characterization, and potential application as surface-enhanced Raman scattering substrate. *ACS Nano,* **2017,** *11* (9), 8892–8900.

58. Aakyiir, M., Yu, H., Araby, S., Ruoyu, W., et al., Electrically and thermally conductive elastomer by using MXene nanosheets with interface modification. *Chem. Eng. J.* **2020,** *397,* 125439.

59. Hu, M., Li, Z., Hu, T., Zhu, S., et al., High-capacitance mechanism for $Ti_3C_2T_x$ MXene by in situ electrochemical Raman spectroscopy investigation. *ACS Nano* **2016,** *10* (12), 11344–11350.

60. Stevie, F. A. and Donley, C. L., Introduction to X-ray photoelectron spectroscopy. *J. Vac. Sci. Technol. A* **2020,** *38* (6).

61. Shang, Y., Xu, X., Gao, B., Wang, S., et al., Single-atom catalysis in advanced oxidation processes for environmental remediation. *Chem. Soc. Rev.* **2021,** *50* (8), 5281–5322.

62. Stevie, F. A., Garcia, R., Shallenberger, J., Newman, J. G., et al., Sample handling, preparation and mounting for XPS and other surface analytical techniques. *J. Vac. Sci. Technol. A* **2020,** *38* (6).

63. Isaacs, M. A., Davies-Jones, J., Davies, P. R., Guan, S., et al., Advanced XPS characterization: XPS-based multi-technique analyses for comprehensive understanding of functional materials. *Mater. Chem. Front.* **2021,** *5* (22), 7931–7963.

64. Wang, H. and Linford, M. R., X-ray photoelectron spectroscopy and auger electron spectroscopy. *Vacuum Technology & Coating* **2015,** 2–7.

65. Bagus, P. S., Ilton, E., and Nelin, C. J., Extracting chemical information from XPS spectra: A perspective. *Catal. Lett.* **2018,** *148,* 1785–1802.

66. Di Castro, V. and Polzonetti, G., XPS study of MnO oxidation. *J. Electron Spectros. Relat. Phenomena* **1989,** *48* (1), 117–123.

67. Zuo, Y., Liu, Y., Li, J., Du, R., et al., In situ electrochemical oxidation of Cu_2S into CuO nanowires as a durable and efficient electrocatalyst for oxygen evolution reaction. *Chem. Mater.* **2019,** *31* (18), 7732–7743.

68. Bagus, P. S., Nelin, C. J., Al-Salik, Y., Ilton, E. S., et al., Multiplet splitting for the XPS of heavy elements: Dependence on oxidation state. *Surf. Sci.* **2016,** *643*, 142–149.

69. De Groot, F., Multiplet effects in X-ray spectroscopy. *Coord. Chem. Rev.* **2005,** *249* (1–2), 31–63.

70. Mansour, A., Characterization of NiO by XPS. *Surf. Sci. Spectra* **1994,** *3* (3), 231–238.

71. Gupta, A., Ghosh, S., Bhalothia, D., Thangarasu, S., et al., Stabilizing a Ru single atom catalyst through electronic metal–support interaction with a $NiCo_2O_4$ support for overall water splitting and urea electrolysis. *J. Mater. Chem. A* **2024,** *12* (35), 23819–23836.

72. Fageria, P., Uppala, S., Nazir, R., Gangopadhyay, S., et al., Synthesis of monometallic (Au and Pd) and bimetallic (AuPd) nanoparticles using carbon nitride (C_3N_4) quantum dots via the photochemical route for nitrophenol reduction. *Langmuir* **2016,** *32* (39), 10054–10064.

73. Zhai, P., Xia, M., Wu, Y., Zhang, G., et al., Engineering single-atomic ruthenium catalytic sites on defective nickel-iron layered double hydroxide for overall water splitting. *Nat. Commun.* **2021,** *12* (1), 4587.

74. Lin, C., Zhao, Y., Zhang, H., Xie, S., et al., Accelerated active phase transformation of NiO powered by Pt single atoms for enhanced oxygen evolution reaction. *Chem. Sci.* **2018,** *9* (33), 6803–6812.

75. Li, L., Yang, H., Miao, J., Zhang, L., et al., Unraveling oxygen evolution reaction on carbon-based electrocatalysts: effect of oxygen doping on adsorption of oxygenated intermediates. *ACS Energy Lett.* **2017,** *2* (2), 294–300.

76. Li, D., Chen, Y., Zhou, C., Shi, C., et al., XPS depth profiling of functional materials: Applications of ion beam etching techniques. *Mater. Chem. Front.* **2024**.

77. Yano, J. and Yachandra, V. K., X-ray absorption spectroscopy. *Photosynth. Res.* **2009,** *102*, 241–254.

78. Zimmermann, P., Peredkov, S., Abdala, P. M., DeBeer, S., et al., Modern X-ray spectroscopy: XAS and XES in the laboratory. *Coord. Chem. Rev.* **2020,** *423*, 213466.

79. Newville, M., Fundamentals of XAFS. *Rev. Mineral. Geochem.* **2014,** *78* (1), 33–74.

80. Kuo, C.-H., Mosa, I. M., Thanneeru, S., Sharma, V., et al., Facet-dependent catalytic activity of MnO electrocatalysts for oxygen reduction and oxygen evolution reactions. *Chem. Commun.* **2015,** *51* (27), 5951–5954.

81. Gaur, A., Shrivastava, B., and Joshi, S. Copper K-edge XANES of Cu (I) and Cu (II) oxide mixtures. *J. Phys. Conf. Ser.* **2009,** *190*, 012084.

82. Krishnan, V., Feth, M. P., Wendel, E., Chen, Y., et al., EXAFS spectroscopy: Fundamentals, measurement techniques, data evaluation and applications in the field of phthalocyanines. *Z. Phys. Chem.* **2004,** *218* (1), 1–16.

83. Ravel, B., Quantitative EXAFS analysis. *X-Ray Absorp. X-Ray Emiss. Spectrosc: Theory Appl.* **2016,** 281–302.

84. Newton, M. A., Zimmermann, P., and van Bokhoven, J. A., X-ray absorption spectroscopy (XAS): XANES and EXAFS. In *Springer Handbook of Advanced Catalyst Characterization*, Springer: **2023,** pp. 565–600.

85. Teo, B. K., *EXAFS: Basic Principles and Data Analysis.* Springer Science & Business Media: **2012,** Vol. 9.

86. Kuang, P., Wang, Y., Zhu, B., Xia, F., et al., Pt single atoms supported on N-doped mesoporous hollow carbon spheres with enhanced electrocatalytic H_2-evolution activity. *Adv. Mater.* **2021,** *33* (18), 2008599.

87. Shard, A. G., Schofield, R. C., and Minelli, C., Ultraviolet–visible spectrophotometry. In *Characterization of Nanoparticles*, Elsevier: **2020,** pp. 185–196.

88. Akash, M. S. H. and Rehman, K., *Essentials of Pharmaceutical Analysis.* Springer: **2020**.

89. Njoki, P. N., Lim, I.-I. S., Mott, D., Park, H.-Y., et al., Size correlation of optical and spectroscopic properties for gold nanoparticles. *J. Phys. Chem. C* **2007,** *111* (40), 14664–14669.

90. Haiss, W., Thanh, N. T., Aveyard, J., and Fernig, D. G., Determination of size and concentration of gold nanoparticles from UV–Vis spectra. *Anal. Chem.* **2007,** *79* (11), 4215–4221.

91. Huang, X. and El-Sayed, M. A., Gold nanoparticles: Optical properties and implementations in cancer diagnosis and photothermal therapy. *J. Adv. Res.* **2010,** *1* (1), 13–28.

92. He, R., Qian, X., Yin, J., and Zhu, Z., Preparation of polychrome silver nanoparticles in different solvents. *J. Mater. Chem.* **2002,** *12* (12), 3783–3786.

93. Wei, C. and Liu, Q., Shape-, size-, and density-tunable synthesis and optical properties of copper nanoparticles. *Cryst. Eng. Comm.* **2017,** *19* (24), 3254–3262.

94. Soltani, N., Gharibshahi, E., and Saion, E., Band gap of cubic and hexagonal CdS quantum dots. Experimental and theoretical studies. *Chalcogenide Lett.* **2012,** *9* (7), 321–328.

95. Sakthivel, P., Muthukumaran, S., and Ashokkumar, M., Structural, band gap and photoluminescence behaviour of Mn-doped ZnS quantum dots annealed under Ar atmosphere. *J. Mater. Sci.: Mater. Electron.* **2015,** *26*, 1533–1542.

96. Blitz, J. P., Diffuse Reflectance Spectroscopy. John Wiley & Sons New York: **1998,** Vol. 14, pp. 185–219.

97. Aydın, C., Abd El-sadek, M., Zheng, K., Yahia, I., et al., Synthesis, diffused reflectance and electrical properties of nanocrystalline Fe-doped ZnO via sol–gel calcination technique. *Opt. Laser Technol.* **2013,** *48*, 447–452.

98. Singh, J. P., Sharma, Y. K., Nag, A., Pal, S., et al., Optical and electronic characterization of CdS: Nd^{3+} nanoparticles using diffuse reflectance spectroscopy techniques. *J. Opt.* **2024**, 1–7.

99. Shrestha, K. M., Sorensen, C. M., and Klabunde, K. J., Synthesis of CuO nanorods, reduction of CuO into Cu nanorods, and diffuse reflectance measurements of CuO and Cu nanomaterials in the near infrared region. *J. Phys. Chem. C* **2010,** *114* (34), 14368–14376.

100. Biju, C., Raja, D. H., and Padiyan, D. P., Glycine assisted hydrothermal synthesis of α-Fe_2O_3 nanoparticles and its size dependent properties. *Chem. Phys. Lett.* **2014,** *610*, 103–107.

Chapter 4

Characterization of Materials Using Microscopic Techniques

Sucheta Sengupta[a] and Manoj Raula[b]
[a]*Amity Institute of Advanced Research and Studies (Materials and Devices), Amity University, Sector-125, Noida 201303, India*
[b]*Department of Chemistry, Amity Institute of Applied Sciences, Amity University, Sector-125, Noida 201303, India*
mraula@amity.edu

Microscopic methods are crucial in materials science for understanding how a material's internal features, such as grains, defects, and phases, influence its overall behavior and performance. Different types of microscopic techniques are used depending on the specific information required. The choice of technique depends on the scale of features being examined, the material type, and the information sought. This chapter includes an introduction to the microscopic techniques in general, followed by a detailed discussion about the fundamental principles of each microscopic technique, along with the state-of-the-art research activities in the respective fields.

Materials Design for Electrocatalysis
Edited by Sagarika Bhattacharya and Biswarup Chakraborty

ISBN 978-981-5352-32-0 (Hardcover), 978-1-042-02558-9 (eBook)
www.jennystanford.com

DOI: 10.1201/9781042025589-4

4.1 Introduction

Advanced microscopic techniques play a pivotal role in exploring and understanding the properties of materials at the nanoscale level. These microscopic methods enable scientists to visualize, manipulate, and analyze materials at the atomic and molecular levels, leading to significant advancements in technology, engineering and fundamental science. Different types of microscopic techniques are used depending on the specific information required. Grain boundaries, dislocations, and other imperfections that influence the material strength and conductivity are critical in the fields, such as metallurgy and composite materials and can be visualized with techniques like transmission electron microscopy (TEM). Microscopic techniques like atomic force microscopy (AFM) and scanning electron microscopy (SEM) are essential for analyzing the surface topography, roughness and mechanical properties of thin films, coatings and biomaterials. Real-time observation of thin film growth, using in situ TEM or reflection high-energy electron diffraction (RHEED), enables the optimization of processes like sputtering, epitaxy, and chemical vapor deposition (CVD), leading to high-quality films in photovoltaics and electronic devices. Confocal laser scanning microscopy (CLSM) allows for the 3D visualization of soft materials like polymers and biomolecules and their molecular structures crucial for understanding material properties, such as elasticity, biodegradability, and drug delivery capabilities.

Thus, microscopic characterization is essential for materials science, because it provides unparalleled insights into the structure, composition, and behavior of materials at scales that directly influence their performance. Thus, an in-depth understanding is critical for advancements in industries, such as electronics, energy, aerospace, nanotechnology, and biomedicine. The underlying principle for majority of the microscopic techniques is the interaction of some form of radiation (light, electrons, or mechanical probes) with the materials being studied. The different ways in which these interactions occur form the basis of various microscopy methods. Based on the source of radiation, microscopic technique can be broadly subdivided into two broad categories: optical microscopy (where light acts as a radiation source) and electron microscopy (where electrons act as a radiation source). The different types of microscopic techniques most popularly used are listed in the flowchart described in Fig. 4.1.

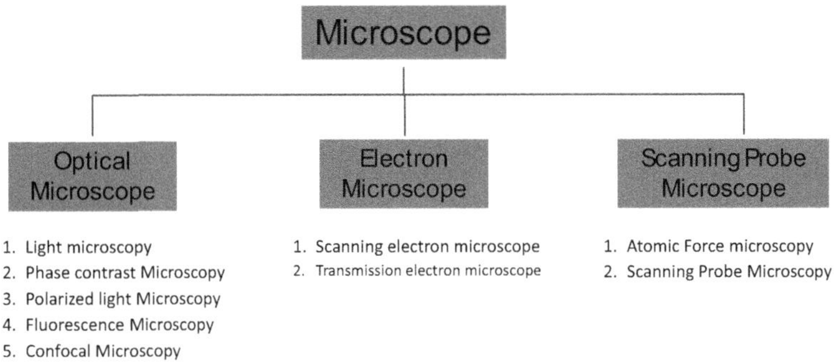

Figure 4.1 Flowchart depicting different types of microscopes.

In this chapter, we will delve into the fundamental principles of microscopic techniques and provide a broad overview of the various methods used for material characterization.

4.2 Light Microscopy: Basic Principles and Applications

Optical microscope, also known as light microscope, is one of the most used microscopes specially for the biological and medical researchers. It uses the visible wavelengths of the light and a system of lenses to magnify samples. Optical microscopy is widely used in biology, materials science, medicine, and environmental science to analyze samples at the cellular or structural level. Since its evolution in the 17th century, the optical microscope has gradually improved its design, optics, and imaging technology, leading to increasingly powerful tools over the centuries. Optical microscopy continues to evolve, with advancements like fluorescence, confocal, and super-resolution microscopy expanding its applications and enabling more detailed analysis at the cellular and molecular levels. The key applications include identifying pathogens, aiding in the diagnosis of infections, and examining material structure, composition, and defects at a microscopic scale. It is essential for studying cells, bacteria, viruses, and tissues to observe cell structure, function, and division, as well as interactions between cells and microorganisms.

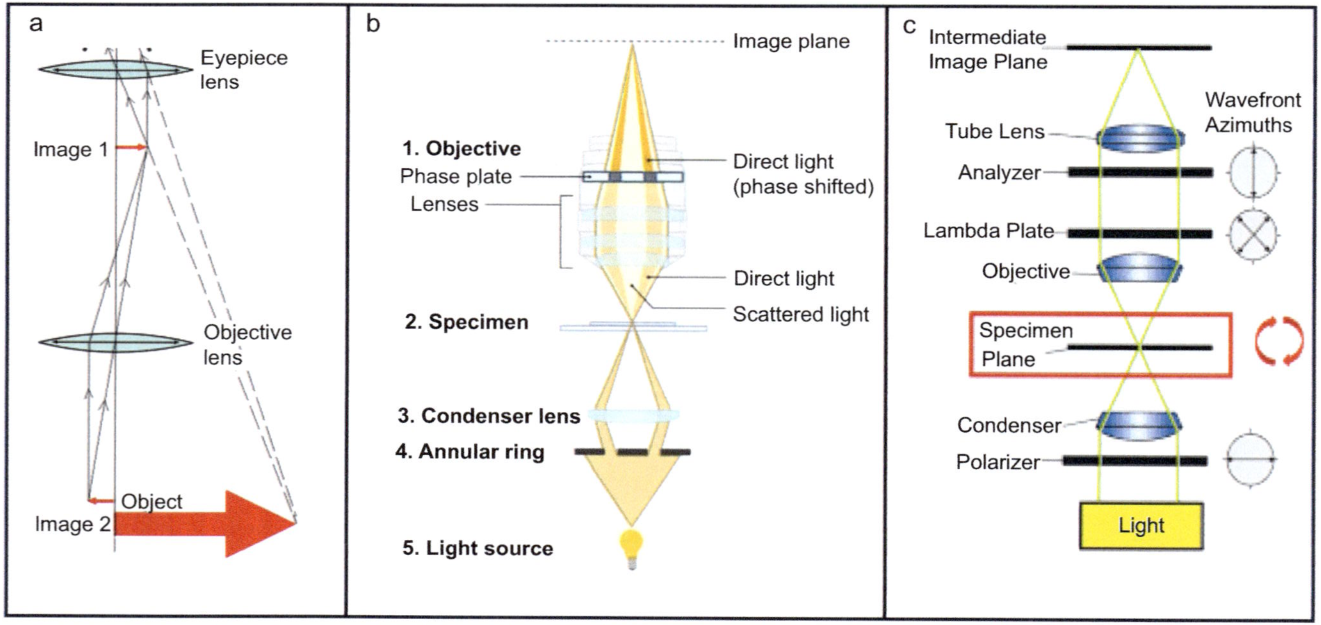

Figure 4.2 Schematic diagram of (a) compound microscope, (b) phase-contrast microscope, and (c) polarized microscope.

Basic Principle: The basic principle of optical microscopy is to use visible light and one/multiple lenses (simple or compound microscope) to magnify and visualize small objects. The general principle of the optical microscope is schematically described in Fig. 4.2. A light microscope focuses a beam of light through a small and transparent specimen to produce an image, which is further magnified by passing through one or two lenses. The strength of the lens is directly proportional to the focal length, with the maximum magnification obtained for the short focal length. The resolution of a light microscope is determined by the numerical aperture of its lens system and by the wavelength of the light it employs. A few of the modern-day microscopes, along with their brief descriptions, are discussed below:

4.2.1 Bright-Field Light Microscope

A bright-field light microscope is one of the most common types of microscopes used in biology and materials sciences. Brightfield microscopy, also known as compound light microscopy, uses light to illuminate a sample and create a dark image against a bright background. A typical bright field light microscope is composed of the following basic components:

(i) A light source to illuminate the sample
(ii) An objective lens to magnify the image
(iii) An eyepiece or camera to view the image
(iv) A stage to hold the sample
(v) A diaphragm to control the amount of light entering the microscope
(vi) Focusing knobs to adjust the focus of the image

Brightfield microscopy is one of the most widely used imaging techniques because of its simplicity, availability, and affordability. Brightfield microscopy can be used to study a wide range of samples, from cells and tissues to minerals and fibers, making it a versatile tool for many fields of research and industry. Limitations of brightfield microscope include low contrast for weakly absorbing samples and low resolution for out-of-focus material. Also, to improve the contrast and visibility, most organisms need to be stained before observation, which is toxic.

4.2.2 Phase-Contrast Microscope

A phase-contrast light microscope is a specialized type of microscope that enhances the visualization of transparent specimens, such as living cells or thin tissue sections, without the need for staining. It is particularly useful in biology and medical research, because it allows the observation of live samples in their natural state. Phase-contrast microscopes modify the path of light so that the sample modifies part of the beam, but not all of it. This allows the microscope to use phase shifts to make structures visible. A phase-contrast microscope is sequentially built up of (Fig. 4.2b):

i. Light source (mercury arc lamp)
ii. Collective lens
iii. Aperture
iv. Condenser
v. Condenser annular
vi. Specimen
vii. Objective
viii. Phase plate
ix. Deflected light
x. Phase ring

The condenser of the phase-contrast microscope has an opaque disk that is known as an annular ring, with a transparent ring that produces a cone of light that passes through a specimen. Due to light variations, some light bends at the specimen, caused by variations in light density, forming an image at the objective lens. The undeviated light will strike the phase ring on the phase plate, and the deviated light will miss the phase ring, passing through the phase plate directly, this forms an image. Since no staining is required for this type of microscopy, hence the sample integrity is preserved in this case, making it ideal for studying dynamic processes, such as cell division, motility, and intracellular changes. However, in this case, slight halo artifacts may occur around the specimen due to the phase-contrast technique. Also, the phase-contrast microscope is suitable only for thin samples, thick specimen images may appear distorted.

4.2.3 Polarized Light Microscope

A polarized light microscope is a powerful optical tool, which visualizes specimens that interact with polarized light. It is specifically designed to evaluate the three-dimensional structure and composition of anisotropic samples with better image quality. It typically consists of (Fig. 4.2c):

i. Light source
ii. Polarizer
iii. Stage
iv. Analyzer
v. Optical components

In polarized light microscopy, light waves are polarized by a polarizer, forcing the waves to move in just one direction, and hence only those light waves that vibrate in parallel pass through, while light waves in all the other directions are blocked. The two filters can be adjusted to allow light at different angles, so that the sample can be visualized from different perspectives. Since this technique does not require the addition of a contrast agent or dye, it is known as one of the most reliable and sensitive noninvasive methods of imaging variety of samples from different scientific disciplines. In geology, it is used to identify minerals and study rock textures, while in biology it can be used to examine structures like muscle fibers, spindles in mitosis. Double-refractive structures like polymers, fibers, crystal structures, and other birefringent materials can be successfully investigated by polarized light microscopy.

4.2.4 Fluorescent Microscope

A fluorescent microscope is a powerful optical microscopy that utilizes fluorescence or phosphorescence instead of, or in addition to, reflection and absorption to study the properties of materials. In comparison to the conventional light microscope, fluorescence microscope uses a more intense light source and filters to produce a magnified image with higher resolution. The principle of fluorescent microscope is to use fluorescent materials (commonly known as fluorophores) to label specific parts of the sample which can be later visualized under microscope since the fluorophores are expected to emit light when the samples are illuminated by absorbing the same.

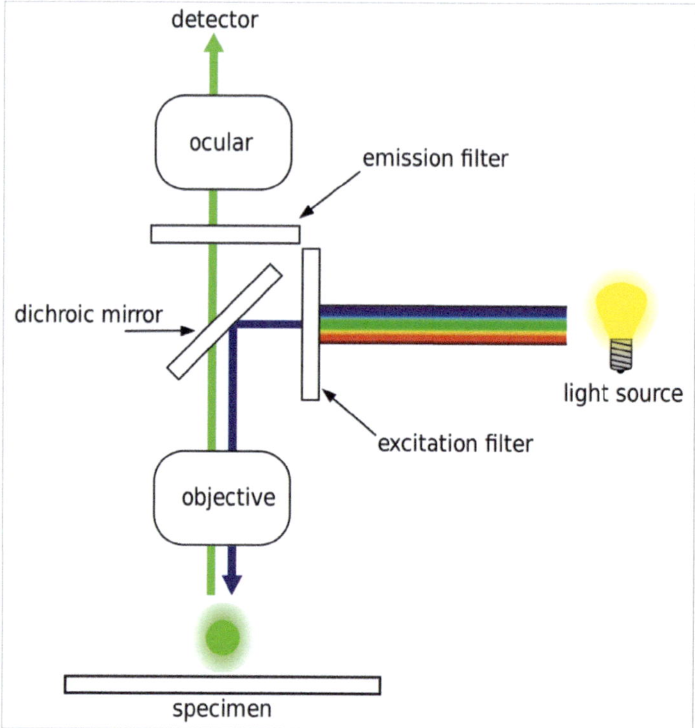

Figure 4.3 Schematic diagram of fluorescent microscope.

4.2.5 Confocal Microscope

In light microscopy, light is directed through the sample as evenly as possible across the viewing area. For thicker specimens, where the objective lens lacks the necessary depth of focus, light from planes above and below the focal plane will also be captured. This out-of-focus light introduces blur to the image, consequently lowering the resolution. In fluorescence microscopy, any dye molecules present in the field of view will be excited, including those in planes that are out of focus. Confocal microscopy offers a solution to eliminate the out-of-focus light from the detector, preventing it from contributing to the blurriness of the collected images. This method enables high-resolution imaging in dense tissues.

In a confocal microscope, both the illumination and detection optics are focused on the same spot in the sample that is limited by diffraction, which is the only area captured by the detector during a confocal scanning process. To create a complete image, the spot must be moved across the sample while data is collected incrementally. One major benefit of the confocal microscope is its capacity for optical sectioning, which enables the 3D reconstruction of a sample using high-resolution image stacks. Various types of confocal microscopes have been designed for this purpose, each offering unique advantages and drawbacks.

4.2.5.1 Components of confocal microscopy

A variable dichromatic mirror or an Acousto-Optic Beam Splitter (AOBS) is at the heart of a confocal microscope, which expertly harnesses light from a laser source through precisely designed collimating optics. This light is reflected to the objective lens, where it is focused on a specific point within the sample, enabling sharp and detailed imaging. The innovative use of scanning mirrors allows the excitation beam to traverse the sample point by point, culminating in high-resolution images. The photomultiplier tube (PMT) plays a crucial role by detecting the fluorescence emitted as it returns through the objective lens and dichromatic mirror or AOBS. Importantly, a pinhole strategically located at the conjugate picture plane to the sample's focal point effectively rejects out-of-focus light, ensuring that only the detector captures the essential data. This epifluorescence setup streamlines the process, requiring only the detector-side pinhole due to shared pathways for illumination and emission light. Adjusting the pinhole size directly influences the amount of light collected and the optical section thickness, allowing for customization according to experimental needs. Furthermore, spectral imaging capabilities can be enhanced using an array of PMTs paired with a diffraction grating or prism in the emission light path, enabling even greater analytical precision. Additionally, a schematic illustrates the scanning mirrors in action within confocal microscopes, demonstrating how they seamlessly sweep the excitation light across the sample to generate striking and informative images.

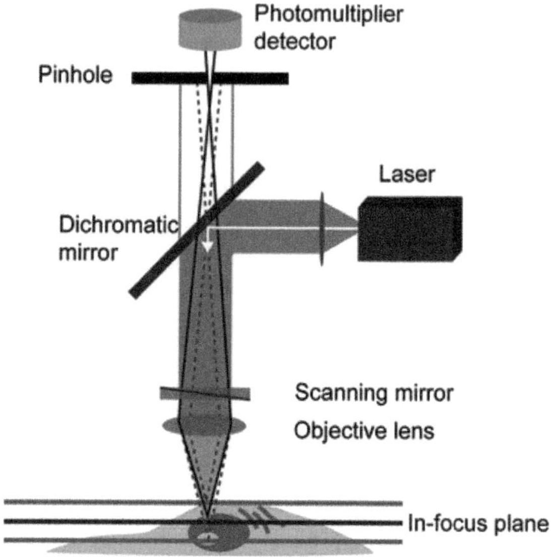

Figure 4.4 Schematic representation of confocal microscope.

4.2.5.2 Types of confocal microscopes

(a) Laser scanning confocal microscopes
(b) Spinning disk confocal microscopes
(c) Hybrid scanning confocal microscopes

4.2.5.3 Applications of confocal microscopy

Confocal microscopy is a powerful imaging technology that has transformed biomedical and materials science research. This specialized type of microscopy has various advantages over regular optical microscopy, including higher spatial resolution, greater contrast, and the capacity to capture high-quality, three-dimensional images of samples. One of the principal uses of confocal microscopy is the investigation of plant–microbe interactions. The capacity to visualize the spatial distribution and localization of microorganisms within their natural microhabitats has provided essential insights into these intricate ecological connections. Confocal laser scanning microscope (CLSM), in particular, has become a widely used tool for analysing the composition and organization of microbial communities associated with plant tissues and surfaces.

Confocal microscopy is also widely used in neuroscience, where it has helped to understand the nanoscale architecture and dynamics of brain systems. Confocal systems' high spatial resolution and fluorescence-based imaging capabilities have allowed researchers to monitor the movement and interactions of synaptic proteins in real time, as well as identify nanoscale brain activity. Furthermore, confocal microscopy has been used to investigate bacterial biofilms, which are complex, surface-associated microbial communities with higher antibiotic resistance. This technology enables three-dimensional visualization of biofilm structures, which provides important information on the spatial organization and composition of these microbial communities.

4.3 Analytical Electron Microscopy: Basic Principles, Constituents, and Operations

In nanomaterials and microstructural evolution, electron microscopy has had a significant impact on materials exploration. TEM, SEM, energy-dispersive X-ray spectroscopy (EDS), electron diffraction, operando electron microscopy, and aberration-corrected electron microscopy are tools for studying nanoscale material properties and structure. TEM is excellent for studying atomic arrangements and defects in materials, whereas SEM provides micro- to nanoscale topographical, morphological, and compositional information. EDS, often used in conjunction with SEM or TEM, analyzes elements and compounds to determine material compositions. Operando electron microscopy enables scientists to monitor and evaluate materials during catalytic processes and battery charge/discharge cycles. This technique is critical for understanding how dynamic processes affect nanoscale material properties and behavior. Another sophisticated approach, aberration-corrected electron microscopy, corrects lens aberrations that impair electron microscope resolution. This modification allows imaging at sample-limited resolutions, allowing for additional investigation of atomic structures and defects. These electron microscopy approaches considerably advance our understanding of nanomaterial microstructural evolution. These have permitted considerable research into defects, phase transitions, and formation methods, all of which are necessary for generating ideal materials.

Interaction between Electrons and Materials

The interaction between electrons and materials is a core concept in materials science and solid-state physics, forming the basis for the various physical, chemical, and electronic properties seen in different materials. Grasping these interactions is essential for creating advanced technologies like electronics, optoelectronics, and energy solutions. Materials are mainly categorised into three types based on electronic properties: insulators, semiconductors, and metals. These categories differ primarily due to their electronic band structures, which arise from interactions among the elements and their electrons.

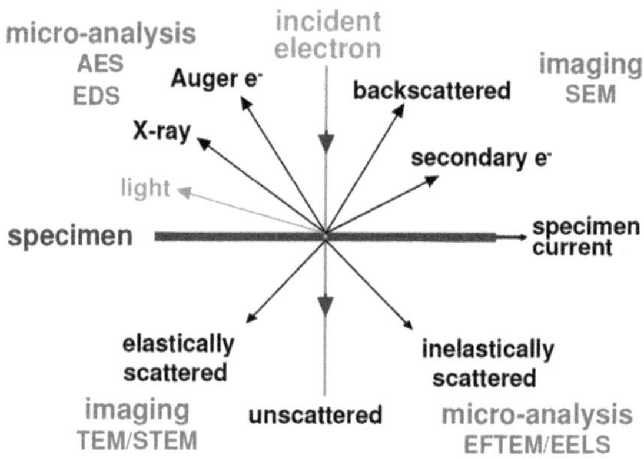

Figure 4.5 Products of the interactions between the electron beam and the target material.

Scattering of Electrons

The behavior of electrons within a material is governed by its electronic band structure, which delineates the permitted energy levels and the distribution of electrons among these levels. Insulators are characterized by a substantial energy gap between the valence and conduction bands, thereby preventing electron flow and rendering them poor conductors. Semiconductors have a smaller energy gap, enabling their electronic properties to be controlled through doping or external stimuli. Metals, however, exhibit a continuous band structure with partially filled bands, which facilitates the free movement of electrons and results in high electrical conductivity. When interacting with high-energy electrons, materials produce various radiations such as elastically scattered electrons, inelastically scattered electrons, back-scattered electrons, Auger electrons, secondary electrons, and X-rays, as illustrated in Fig. 4.5. These different energetic electrons are useful for characterizing the materials.

4.3.1 Basic Components of Analytical Electron Microscope

Analytical electron microscopy is a powerful technique that combines the high-resolution imaging capabilities of an electron microscope with the ability to perform detailed chemical and structural analysis of materials at the nanoscale. The key components of an analytical electron microscope include the electron source, the electron optical column, the sample stage, and the various detection and analysis systems. The electron source, typically a tungsten filament or a field emission gun (FEG), produces a beam of accelerated electrons that are focused onto the sample surface by a series of electromagnetic lenses. The electron beam interacts with the sample, generating a variety of signals that can be detected and analyzed. One of the most common types of analytical electron microscope is the scanning electron microscope (SEM). In an SEM, the electron beam is scanned across the sample surface, and the resulting signals are used to generate a high-resolution image of the sample's topography and composition. In addition to imaging, analytical electron microscopes can also be equipped with a range of analytical tools, such as EDS and electron energy-loss spectroscopy. These techniques allow for the elemental and chemical analysis of the sample, providing valuable information about its structure and properties. The versatility and high-resolution capabilities of analytical electron microscopes have made them indispensable tools in a wide range of scientific and engineering fields, from materials science and nanotechnology to biology and medicine.

4.3.1.1 Electron gun

Electron microscopes, such as TEMs and SEMs, rely on the generation and application of electron beams to produce high-resolution images of specimens. The key component responsible for generating these electron beams is the electron gun. The electron gun is a device that produces a focused beam of electrons, which is then accelerated and focused onto the sample being imaged. In a typical electron gun, a heated filament or other source of electrons generates a stream of electrons. These electrons are then accelerated by a high voltage (typically 1–30 kV) and focused into a narrow beam using a series of electromagnetic lenses.

In SEM, the electron gun is responsible for generating the primary electron beam that interacts with the sample surface. This primary beam is focused to a fine spot, typically around 10 to 20 nanometers in diameter, which is then scanned across the sample to build up an image. The interactions between the primary electrons and the sample surface produce secondary electrons (SEs) and backscattered electrons, which are then detected to create the final SEM image.

4.3.1.2 Lens systems and deflectors

The glass lenses in a visible light microscope (VLM) are comparable to electron lenses in the TEM, and we can make many similarities between the two. For instance, the action of a convex lens can be roughly compared to the behavior of all the lenses in a typical TEM. In essence, the lens serves two purposes. (1) Create an image by taking all the rays that come from a spot in an item. (2) Direct parallel rays toward a location in the lens's focal plane. The lens cannot collect all the rays from the object and we often deliberately limit the collection angle with an aperture. We can draw ray diagrams showing how electron lenses control beams of electrons (Fig. 4.6). These diagrams correspond directly to the ray diagrams used in physical optics.

As in a VLM, the lenses in a TEM control all the basic operational functions of the instrument. As you are well aware, we have to physically move glass lenses up and down in a VLM to control the intensity of the illumination and to focus the image. The focal length of a glass lens is fixed so we have to change lenses to change the magnification. We choose stronger lenses for higher magnification. By contrast, in a TEM, the positions of the lenses are fixed, but we can change the strength of the lens at will.

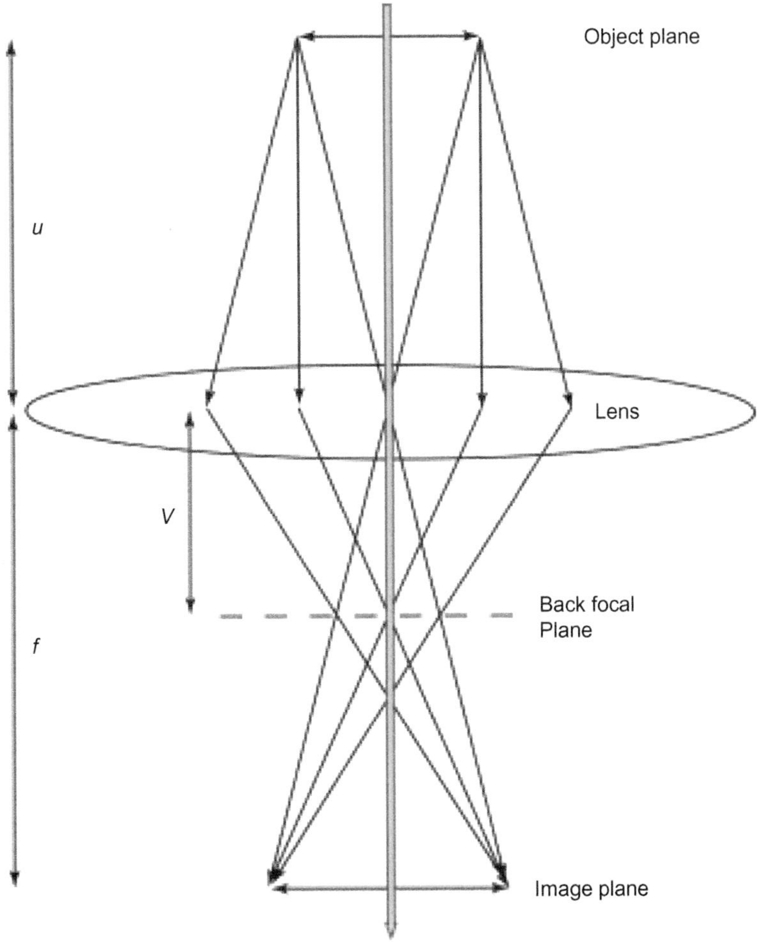

Figure 4.6 A complete ray diagram for a finite object, symmetrically positioned around the optic axis. All rays emerging from a point in the object (distance do from the lens) that are gathered by the lens converge to a point in the image (distance di from the lens) and all parallel rays coming from the object are focused in the focal plane (distance f from the lens).

The lenses we use are electromagnetic, so we change their strength by changing the current through a coil around a soft-iron core which changes the strength of the resultant magnetic field. Almost any operation we carry out on the TEM involves changing magnification or focus; we use electron lenses to magnify and focus the electron beam, the images, and the DPs. These factors are critical in the principal functions of a TEM: imaging, diffraction, and analysis. An aperture is used to select different electron beams to form different images, thus manipulating the image contrast.

4.4 Scanning Electron Microscope

An SEM is a powerful imaging tool used to obtain high-resolution images of surfaces at the nanoscale. Unlike traditional optical microscopes, which use light waves, SEM employs a focused beam of electrons to scan the sample, producing detailed images with excellent depth of field. The electron beam interacts with the surface atoms once it penetrates to a depth of the few microns, generating many types of signals of which SEs, back scattered electrons (BSEs), and characteristic X-rays are mainly utilized for SEM. SEs are emitted because of the inelastic scattering and are primarily used for topographical images. Emission of backscattered electrons occurs due to elastic scattering at high angles and are responsible for the contrast providing sample composition. Characteristic X-ray peaks are observed for EDS data for the precise identification of elements present in the sample. The maximum resolution obtained in SEM depends on various factors, including the electron beam spot size and the interaction volume of the electron beam with the sample. The different signals generated by the electron beam and the sample surface interaction is schematically described in Fig. 4.7(A), and the schematic diagram of the different components of a typical SEM is depicted in Fig. 4.7(B).

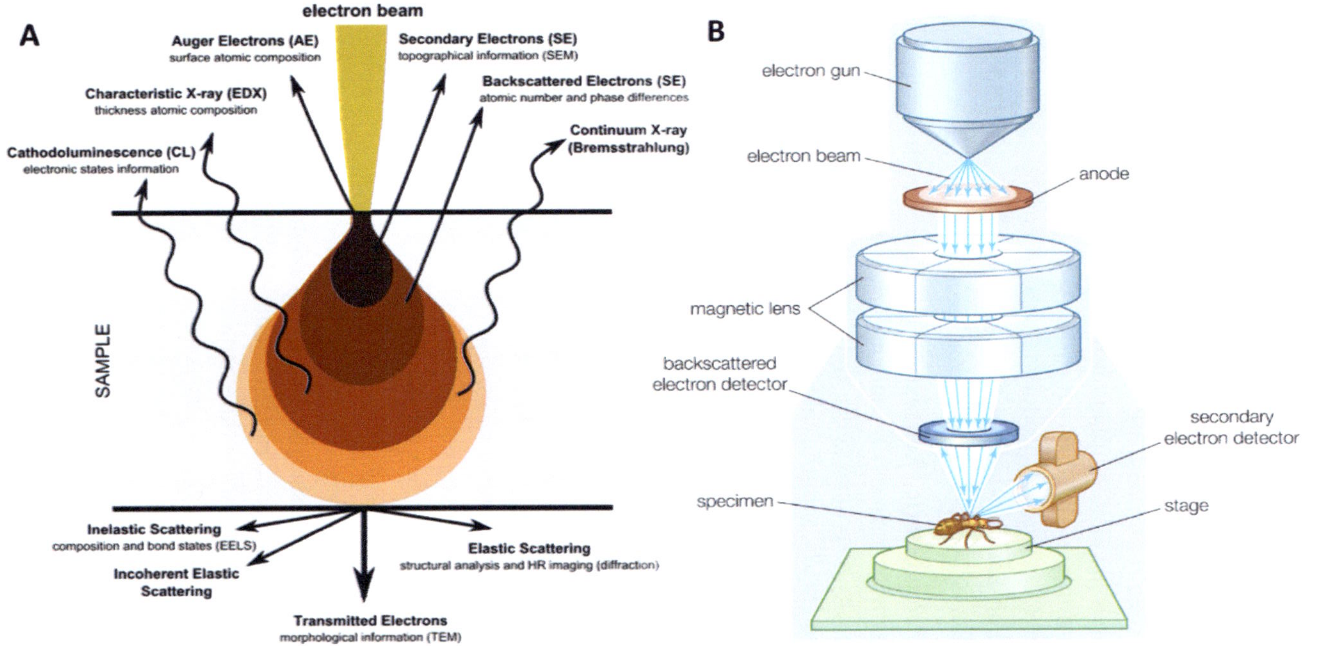

Figure 4.7 (A) Onion-shaped interaction of electron beams with the materials. (B) Schematic representation of all the components of SEM. Reproduced from [12].

4.4.1 Instruments

The different components of the SEM instrument are described as follows:

(a) Electron Gun: The electron gun is the source of beam of the electrons required for imaging. The commonly used electron sources are tungsten filament, lanthanum hexaboride (LaB_6) filament and field emission gun (FEG) of which the latter provides the highest resolution. Electrons are produced at the top of the column in the electron gun and accelerated through the column at a specified accelerating voltage (1 keV–30 keV).

(b) Electromagnetic Lenses: SEM contains two types of electromagnetic lenses: (i) condenser lenses for controlling the beam intensity and convergence and (ii) objective lenses for focusing the electron beam onto the sample with high precision. A series of condenser lenses is used to focus the electron beam and to exclude off-axis electrons, thereby narrowing the spot upon the sample surface. The objective lens is responsible for optimizing the working distance and probe size. Longer working distance correspond to a larger depth of field and lower spatial resolution.

(c) Scanning System: After the beam is focused onto the sample, deflector coils guide the electron beam in X and Y axes in a raster pattern over the sample surface. These scanning coils are synced with the scan generator software that controls the beam movement and interacts with the display system, allowing real-time images to be viewed.

(d) Sample Chamber and Sample Stage: The sample is placed in a vacuum chamber to ensure the free movement of electrons. In the sample chamber, the specimen is usually mounted on a stage that can be tilted, rotated, and moved in multiple directions in 3D space (X, Y, and Z axes) to position the sample under the electron beam. For cryogenic applications, it can also contain temperature controller for the heating or cooling of the stage and a variety of other accessories for specific uses.

(e) Detectors: Several types of detectors are available to collect the different types of signals produced upon interaction of the electron beam with the sample surface. The mostly used ones are described below:

i. **Secondary Electron Detector:** SED images allow for a visualization of the inelastically scattered electrons generated close to the sample surface and provide topographical information with the best possible resolution. SEs are detected with an Everhart-Thornley (E-T) detector. The ET detector is composed of a scintillator inside a Faraday cage and is placed above the sample to one side. A positive bias is applied to the Faraday cage to attract the low-energy electrons where a scintillator converts the electrons into light (photons). The photon signals is then amplified using a PMT to yield the final signal.

ii. **BSE Detector:** This detects high-energy electrons that are reflected or backscattered from the sample. BSE images provide information about the composition and topography of the sample. Backscattered electrons vary in their amount and direction due to the composition and topography of the specimen. The contrast of the backscattered electron image depends on multiple factors, including the atomic number (Z) of the sample material, the acceleration voltage of the primary beam, and the specimen angle (tilt) in relation to the primary beam. Materials with elements composed of higher Z elements yield more backscattered electrons than lower Z elements. The most common type of BSD is composed of a solid-state sensor.

iii. **Energy-Dispersive Spectroscopy (EDS):** It detects characteristic X-rays emitted from the sample, allowing elemental analysis of the sample's composition. It is a fast, accurate, and non-destructive method for identifying elemental composition on the micron scale. EDS data can be obtained at a point, along with a line, or mapped over an area. A silicon drift detector (SDD) is the most common type of EDS detectors used in an SEM instrument. The SDD is composed of a solid-state sensor that generates electron-hole pairs by absorbing X-rays.

4.4.2 Applications

SEM is used for variety of fields including Industrial uses, nanoscience, biological studies, and geological studies.

(a) Material Science: SEM is extensively used in materials science for characterizing microstructures, analyzing surface properties, and understanding failure mechanisms. SEM helps visualize grain size, shape, porosity, and active sites for analyzing nanomaterials like nanotubes, nanofibers, and quantum dots. SEM is used to detect cracks, voids, and delamination in microelectronic devices. SEM with electron backscatter diffraction (EBSD) provides information on crystal orientation and texture. It is also essential for studying film thickness, uniformity, and surface roughness. SEM examines corrosion layers and their composition using EDS/EDX.

(b) Biological Science: SEM is widely used in biological research due to its ability to provide high-resolution, three-dimensional images of biological specimens. SEM reveals the morphology of bacteria, fungi, and viruses, including their surface structures, biofilms, and interactions. SEM can be used in entomology, archaeology, plant science, cell research, and taxonomy among other topics. SEM is also widely used in medical research for identify structural abnormalities in tissues, useful in diagnosing diseases and examining surface properties of medical implants, such as artificial joints, stents, and dental implants.

(c) Geological Science: SEM is a powerful tool in geology, mineralogy, and petrology for analyzing surface textures, mineral compositions, and rock structures at a microscopic scale. It plays a crucial role in understanding Earth's history, natural resource exploration, and environmental science. SEM, often combined with energy dispersive X-ray spectroscopy (EDS/EDX), helps identify mineral phases and chemical compositions. SEM is used with EBSD to study crystal orientation and structural defects.

(d) Forensic Science: SEM is widely used in forensic science due to its ability to provide high-resolution images and detailed surface analysis. SEM with EDS helps detect and analyze characteristic elements like lead (Pb), barium (Ba), and antimony (Sb) in Gun shot residues (GSR) particles.

4.5 Transmission Electron Microscopy

TEM is a strong analytical tool that has changed the field of materials science and nanoscale research. The size, shape, lattice spacing, etc., can be determined using TEM. The basic idea behind TEM is the interaction of a thin specimen with a high-energy electron beam, which enables the visualization and examination of the atomic-scale internal structure, composition, and characteristics of the sample.

A TEM instrument's electron optical column, which is made up of various essential parts, is its central component. The high-energy electron beam is produced by the electron cannon, which is usually a FEG or a tungsten filament. The objective lens creates the first image of the sample, and the condenser lenses concentrate the electron beam onto the specimen.

The electron beam interacts with the atoms in the material as it travels through the thin specimen, experiencing both elastic and inelastic scattering. The several imaging and analytical methods used in TEM, including diffraction-based methods, bright-field imaging, and dark-field imaging, are based on these interactions. Since the specimen must be thin enough (usually fewer than 100 nanometers) to permit the electron beam to pass through it, sample preparation is an essential stage in TEM investigation. Ion milling, focused ion beam milling, and ultramicrotomy are common techniques for sample preparation.

4.5.1 Instrumentation

Figure 4.8 depicts the components of a TEM which consists of an electron source, electromagnetic lenses, and detectors. The electron source emits electrons that travel through a vacuum chamber. These electrons are driven through a sample, and the transmitted electrons are recorded by a detector. The image is then amplified on a fluorescent screen. There are several magnification modes available, including low, high, convergent beam electron diffraction, and chosen region diffraction. The TEM can obtain magnifications of up to 0.1 nm. TEM can reveal the atomic distribution and chemical components of a material. TEM requires thin samples that can conduct electron beams through them. Materials with small dimensions can provide detailed information about the sample. The instrumentation of a TEM is complex, encompassing several key components that work in concert to generate high-resolution images, as shown in Fig. 4.8. The components of TEM are discussed in brief as follows:

(a) **Electron Gun:** The heart of the TEM, the electron gun, generates a high-energy beam of electrons. Conventional TEM, high-resolution TEM, and atomic emission microscope (AEM) can be equipped with two types of electron sources: thermionic and field emission. Thermionic sources, such as W hairpin filaments, zirconium oxide-coated W needle, and refractory crystals, such as LaB_6, function at high temperatures, emitting electrons that are then accelerated by the anode potential. The thermionic sources are heated either by current flowing through the emitting material (for the W hairpin cathode filament) or by thermal contact with a low-work function emitter material, such as LaB_6, and resistive heating of a W wire supporting material. The field-emission gun (FEG) source uses electron tunnelling from the tip to vacuum with a strong electric field (10^9 V/m) to create a significant electric field gradient at the cathode tip. This high field creates a very small potential barrier, allowing electrons to tunnel from a metallic tip with a low work function into vacuum. This emission originates from a very small region (on the scale of 10 nm or less) of a single crystal tip, resulting in a high current density. This gun configuration has multiple versions. Cold FEGs emit at room temperature, but require ultrahigh vacuum (10^{-8}–10^{-9} PA) to prevent gas adsorption on the tip's surface. To lower adsorption sensitivity, thermally assisted field emission and Schottky emission sources have been introduced. Thermally aided FEGs use a strong electric field to heat W single crystal tips to around 1600 K. The emission also happens via tunnelling from a tiny area of the tip, with characteristics comparable to cold field emission.

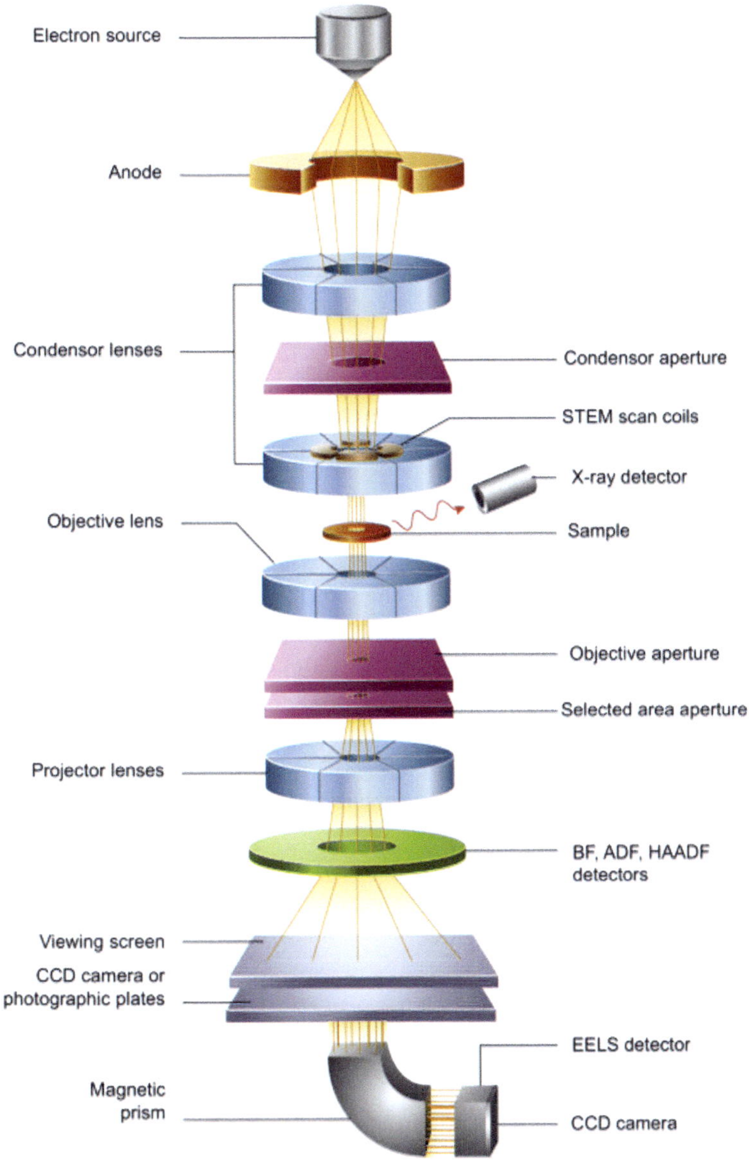

Figure 4.8 A schematic diagram showing different parts of a TEM. Reproduced from [12].

(b) Condenser Lens System: Within the column, electromagnetic lenses shape the electron beam. Because the electron beam is negatively charged, an electromagnetic force can function as a lens. Wire coils surround the beam, generating a field that deflects the electrons. Each lens is made of a coil of copper wire through which a current flows. There is a hole in the center through which the beam passes. Magnetic lenses allow the user to adjust the magnification and focus of the beam. A magnetic electron lens consists of two parts: The pole-piece consists of a soft iron core with a drilled hole (bore) and is surrounded by a copper wire coil as shown in Fig. 4.9. When we pass current through the coil, a magnetic field forms in the bore. The strength of the field in a magnetic lens determines the ray trajectories, bringing off-axis rays back into focus. When comparing the operation of an electromagnetic lens to that of an optical lens, we can see that the picture is rotated to a degree determined by the lens's strength. The focal length can be changed by adjusting the current strength. The coil's resistive heating necessitates cooling the lenses; thus, a water cooling system is a crucial component of TEM lenses.

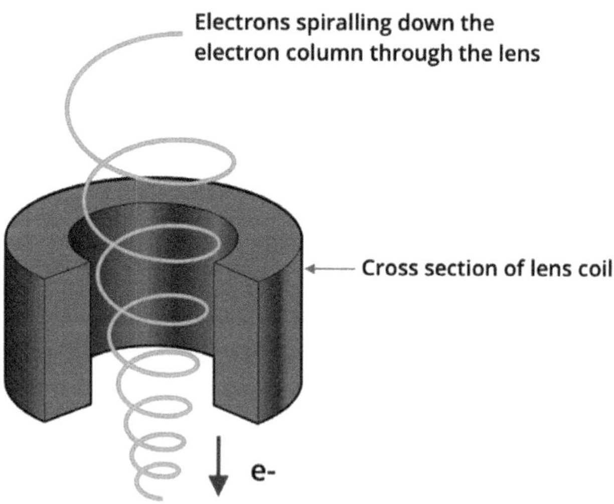

Figure 4.9 Schematic representation of electromagnetic lenses and movement of high energy electrons through the lenses.

The anode stack accelerates electrons to the TEM's high-voltage potential, after which the condensing optics demagnify the source picture in the sample plane. Electromagnetic lenses contain positive spherical aberration coefficients (Cs = 0.5–1.0 mm for high-resolution microscopes with restricted sample tilt and a short polepiece gap; Cs = 1.5–3 mm for significant tilt and large polepiece gaps). Electromagnetic lenses likewise have chromatic aberrations, Cc of the order of C_c = 1–2 mm, and the wave structure of electrons, like in light optics, imposes a diffraction limit to resolution, as enforced by Rayleigh scattering.

(c) **Specimen Stage:** TEM specimens are always round, having a diameter of 3 mm. Perpendicular to this disc, the specimen must be thin enough (at least in certain areas) to let electrons transmit and generate the magnified image. The specimen stage is intended to keep the specimen as steady as possible, as any drift or vibration would be amplified in the final image, reducing spatial resolution (particularly if the image is captured by a camera over many seconds). However, in order to observe all possible regions of the specimen, the specimen must be moved horizontally for up to 3 mm, if necessary. The stage must also be designed so that the specimen may be put into the TEM column's vacuum without introducing any air. This is accomplished by passing the specimen through an airlock, a tiny compartment into which the specimen is originally inserted and can be emptied before entering the TEM column. Not unexpectedly, the specimen stage and airlock are the TEM's most mechanically difficult and precisely produced components. The specimen stage has two fundamental designs: side entry and top entry. The sample is inserted into the well of the specimen holder using a flange or ring that screws into the well to keep it secure. It is critical that the grid is securely fastened so that it does not fall out of the specimen holder. The holder is then placed in the column. During this process, the sample airlock is evacuated, which may take several minutes. The O-ring on the holder must be free of lint and dust, and well oiled, or it will interfere with the hoover. When not in use, the holder is kept clean by placing it in a covering sleeve.

The stage allows the precise movement and positioning of the specimen. This also allows tilting, rotation, in situ cooling, and heating of the samples.

(d) **Other Lenses:** Other than the condenser lenses, the TEM also contains objective lenses, intermediate lenses and projector lenses. Objective lenses are the most critical lens in the TEM; They form the primary magnified image of the specimen. It forms the initial magnified image of the specimen and determines resolution and depth of field. These lenses also deal with any aberrations that will directly affect image quality. Intermediate lenses magnify the image from the objective lens before final projection. They also help to obtain diffraction patterns of the materials. The projector lens helps to magnify and project the final image onto the viewing screen or detector. The magnifications will also be determined by projector lenses. These lenses also help in correcting astigmatism in the electron beam. In scanning transmission electron microscopy (STEM), lenses also control beam alignment and scanning of the specimen.

4.5.2 Applications

TEM is a powerful analytical technique that has found widespread applications in various scientific and technological fields, from materials science to biology and nanotechnology. One of the most significant advantages of analytical TEM is its capacity to give high-resolution imaging and characterization of specimens at the nanoscale level, allowing for the investigation of complicated structures and compositions that other microscopy techniques cannot.

In materials research, analytical TEM has become an essential technique for studying the microstructure and composition of a wide range of materials, including metals, ceramics, polymers, and composites. By combining high-resolution imaging with techniques like electron diffraction and EDS, researchers can gain detailed insights into materials' atomic-scale structure, defects, and chemical composition, which is critical for understanding their mechanical, electrical, and optical properties.

Analytical TEM has made major contributions to the study of electronic materials and devices. TEM techniques can be used to determine the crystal structure, interface morphology, and elemental distribution of semiconductor devices, which is critical for understanding and optimizing their performance. Furthermore, TEM has contributed to the development and characterization of nanomaterials, such as carbon nanotubes, graphene, and quantum dots, which have transformed industries such as electronics, energy, and catalysis.

A TEM's analytical capabilities extend beyond imaging to include spectroscopy and diffraction. These capabilities enable thorough material analysis at the atomic and molecular levels. EDS is an analytical technique used in TEM and SEM to assess materials' elemental makeup. A high-energy electron beam stimulates atoms in the sample, causing them to generate recognizable X-rays. X-rays are detected and analyzed to establish the material's elemental composition. EDS's major features are element identification, element mapping in a sample, and quantitative analysis.

Electron energy loss spectroscopy (EELS) is an analytical technique used in TEM to investigate the elemental composition, chemical bonding, and electrical characteristics of nanomaterials. EELS typically uses a high-energy electron beam (100–300 keV) to interact with samples. Inelastic scattering causes some electrons to lose their energy. The energy loss spectrum is recorded, providing elemental and chemical information. EELS can identify light elements, such as Li, B, C, N, and O (better than EDS). It can also be used to determine the chemical bonding and oxidation states, bandgap, and dielectric characteristics of semiconductors. It boasts a high spatial resolution of 0.1 nm and an energy resolution of 0.1 eV.

Selected area electron diffraction (SAED) is a TEM technique that analyzes the crystalline structure of materials by examining their electron diffraction patterns. A parallel electron beam interacts with a crystalline area of the sample. The diffracted electrons create a pattern of brilliant spots on a detector. The spacing and arrangement of these spots reveal information about the crystal structure, phase, and orientation of the material. This approach will help determine the crystal structure and phase. It identifies polymorphs of materials, calculates interplanar distances using Bragg's Law, and detects grain orientation, twinning, stacking defects, and strain.

In the biological sciences, analytical TEM has emerged as a critical tool for studying cellular and subcellular structures, allowing for the nanoscale visualization of organelles, membranes, and macromolecular complexes. TEM techniques can be used to explore the ultrastructure of biological samples, revealing information about the organization and function of numerous cellular components. Furthermore, the capacity to perform elemental analysis with TEM has proven critical in understanding the distribution and localization of certain elements within biological tissues, as well as their significance in cellular processes. Overall, analytical TEM applications span a wide range of disciplines, including materials science, nanotechnology, biology, and medicine. As developments in electron microscopy push the boundaries of what is feasible, the value of this powerful analytical tool is projected to expand, resulting in new discoveries and innovations in a variety of disciplines of science and industry.

4.6 Conclusion

Microscopic material characterization is an important component of materials science and engineering, because it provides significant insights into the structure, characteristics, content, and behaviors of materials at different sizes. Researchers can delve into the nanoscale features of materials by using advanced techniques such as SEM, TEM, etc. Microscopic techniques are important in this characterization, because they provide precise information about the material's surface and interior structure. Optical microscopy magnifies samples using visible light and optical lenses. This is ideal for analyzing the microstructure of metals, polymers, and biological substances. Additionally, this process is easy, non-destructive, and produces colour photographs. SEM uses a concentrated electron beam to scan the surface of a sample. SEM is a technique for imaging surface topography and composition at high resolution. SEM has a high depth of field and resolution, as well as the ability to provide elemental analysis via energy-dispersive X-ray spectra. TEM creates an image by sending high-energy electrons through a thin material. It is utilized to conduct extensive internal structure investigation at the atomic level. It has numerous applications in a variety of fields, including materials science, nanotechnology, biology, and medicine. These microscopic techniques are critical tools in materials science, allowing researchers to examine and comprehend materials at both the microscopic and atomic levels. Each technique has distinct advantages and is selected based on the study's specific requirements.

References

1. J. Girkin. *A Practical Guide to Optical Microscopy*; 2020, CRC press, Taylor and Francis Gr. USA.
2. D. B. Murphy and M. W. Davidson. *Fundamentals of Light Microscopy and Electronic Imaging*; 2013, J. Wiley and Sons; USA.
3. J. B. Sanderson. *Phase Contrast Microscopy,* Encyclopedia of Life Sciences; 2002, 1–4.
4. B. Heit. *Fluorescent Microscopy*; 2022, Humana New York, NY.
5. Y. Wang, J. Lin, Q. Zhang, X. Chen, et al. Fluorescence Nanoscopy in Neuroscience; *Engineering*; 2022, 16, 29–38.
6. C. M. St. Croix, S. H. Shand, S. C. Watkins. Confocal microscopy: Comparisons, applications, and problems; *BioTechniques;* 2005, 39, 2–5.
7. A. D. Elliott. *Confocal Microscopy: Principles and Modern Practices; Current Protocols in Cytometry*; 2019, 92, 1.
8. S. Amelinckx, D. van Dyck, J. V. Landuyt, G. V. Tendeloo. *Electron Microscopy: Principles and Fundamentals*; 1997, Wiley VCH, USA.
9. A. Ul-Hamid. *A Beginners' Guide to Scanning Electron Microscopy*; 2018, Springer Nature Switzerland AG.
10. N. Erdman, D. C. Bell, R. Reichelt. *Scanning Electron Microscopy*; 2019, Springer Handbooks. Springer.
11. D. B. Williams and C. B. Carter. *Transmission Electron Microscopy*, 2009, Springer New York, USA.
12. B. J. Inkson. Scanning electron microscopy (SEM) and transmission electron microscopy (TEM) for materials characterization. *Materials Characterization Using Nondestructive Evaluation (NDE) Methods.* G. Hübschen, I. Altpeter, R. Tschuncky, H. G. Herrmann (eds.); 2016, 17–43, Elsevier.

Chapter 5

Fabrication of Electrodes Using Different Materials and Cell Setup

Yashwant Pratap Kharwar,[a] Rathindranath Biswas,[a] Chandan Das,[a] and Arnab Dutta[a,b,c]
[a]*Chemistry Department, Indian Institute of Technology Bombay, Powai, Maharashtra 400076*
[b]*National Center of Excellence in Carbon Capture and Utilization (NCoE-CCU), Indian Institute of Technology Bombay, Powai, Maharashtra 400076*
[c]*Centre for Climate Studies, Indian Institute of Technology Bombay, Powai, Maharashtra 400076*
arnab.dutta@iitb.ac.in

5.1 Preparation of Catalyst Ink or Slurry

The preparation of catalyst ink or slurry is a crucial step in fabricating electrodes for various electrochemical applications, such as water splitting, nitrate reduction, N_2 reduction, CO_2 reduction, fuel cells, and electrolyzers. There are several methods used to prepare catalyst inks or slurries, and the choice of method depends on the application and the materials involved. In the following segment, the typical methods for catalyst ink preparation have been discussed.

Materials Design for Electrocatalysis
Edited by Sagarika Bhattacharya and Biswarup Chakraborty

ISBN 978-981-5352-32-0 (Hardcover), 978-1-042-02558-9 (eBook)
www.jennystanford.com

DOI: 10.1201/9781042025589-5

5.1.1 Simple Mixing

This method involves direct mixing of a certain amount of catalyst powder with a solvent and binder. The mixture is sonicated or stirred until a uniform slurry is obtained. Most commonly, Nafion™ is used as a binder, and water, ethanol, isopropanol, or a mixture of solvents are used to enhance the dispersibility of the catalyst. The typical ratio of catalyst, solvent, and Nafion binder may vary depending on the application and the substrate on which the ink is to be cast. Conventionally, the common composition of 80% catalyst, 10% binder, and 10% solvent by mass is utilized to prepare a thick ink for electrodes for applications in supercapacitors and batteries. On the other hand, 4 to 6 mg of catalyst, 400 to 700 μL of solvent, and 20 to 30 μL of Nafion binder are the usual ratio of catalyst, solvent, and binder to prepare the slurry for electrode preparation for water splitting, nitrate reduction, N_2 reduction, and CO_2 reduction reactions in small scale. This method is very simple and easy to scale up; however, it can result in poor dispersion if not optimized properly, resulting in poor performance. Hence, we have to follow other factors during the preparation of catalyst ink or slurry as discussed in the following subsections.

5.1.2 Sonication

Once the catalyst has been mixed in, use an ultrasonic bath or a probe sonicator to ensure that the catalyst particles are uniformly dispersed in the solution. Sonication for 30 minutes to an hour is required, depending on the viscosity of the ink and the size of the particles. This step helps to break up any agglomerates to make a uniform dispersion, but excessive sonication can degrade some catalysts, so the duration and power should be carefully optimized. In addition, long-term sonication can generate heat, which might affect sensitive catalysts.

5.1.3 Adjusting Viscosity

Depending on the intended coating method (e.g., drop casting, spraying, painting, etc.), the viscosity of the ink may need adjustment.

The addition of small amounts of solvent to dilute or letting some solvent evaporate to thicken the slurry is commonly deployed for adjusting the desired viscosity of the ink material.

5.1.4 Resting and Defoaming

It is better to keep the slurry for a few minutes to release any trapped air bubbles formed during mixing or sonication. A vacuum desiccator can also be used to degas the ink and remove bubbles.

5.1.5 Storage

If the catalyst ink does not need to be used immediately, it is better to store it in a sealed container to prevent solvent evaporation. However, it is best to use freshly prepared ink for optimal results.

5.1.6 Binder Concentration

Binder concentration affects the electrode's mechanical properties and ionic conductivity. It is also essential to adjust the amount of binder depending on the application. An excessive amount of binder can shield or block the electrochemical active centres of the catalyst.

5.1.7 Ball Milling

This is a mechanical grinding process where catalyst materials, binders, and solvents are mixed in a ball mill, which can be operated at a predefined speed and duration to grind and disperse the materials. The fine and uniform particle size and good mixing can be achieved to prepare the slurry, but this process damages the catalyst structure if excessive energy is applied.

5.1.8 Electrodeposition

In this process, the catalyst is deposited directly onto the electrode surface from a solution by applying a current. The electrode is immersed in an electrolyte that contains dissolved catalyst precursors, that is, the desired metal ion or its chemical composition.

Then a controlled potential needs to be applied to deposit the catalyst onto the electrode. Electrodeposition leads to direct and uniform deposition of the catalyst on the electrode, and the major advantage is that a binder or solvent is not required, but precise deposition conditions are needed.

5.2 Common Methods for Coating Catalyst Ink or Slurry on Electrodes

The prepared catalyst ink or slurry can then be applied to electrode substrates in different ways, such as drop-casting, spray-coating, brush painting, doctor-blading, or spin-coating.

5.2.1 Drop-Casting

A specific volume of catalyst ink is pipetted directly onto the flat surface of the electrode and allowed to spread naturally or with minor manipulation. Another way one can spread the ink evenly is by tilting or spinning the electrode. Then, allow the drop cast electrode to dry at room temperature or under controlled conditions (e.g., heating or vacuum). In this process, there is no requirement for special equipment, but it may result in nonuniform coatings and uneven catalyst distribution on the electrode surface.

5.2.2 Brush Coating

Brush coating is another technique for electrode preparation; it involves applying a slurry or a liquid mixture of active materials onto a substrate (typically a conductive material like metal foil or carbon paper) using a painting-type brush. In this method, we can adjust the coating process by controlling the brush pressure, angle, stroke speed, and slurry concentration to achieve a uniform and desired thickness of the catalyst. Low heat (40–60°C) is sometimes applied to dry the ink-loaded electrode. Drop casting and brush coating are useful on a laboratory scale but may not be ideal for large-scale or highly automated production.

5.2.3 Spray Coating

Spray coating is another technique that can be used for electrode preparation; this method involves atomizing a liquid suspension of active material and spraying it onto a substrate to form a thin, uniform coating. The active material (e.g., catalyst particles, conductive additives) is suspended in a solvent with binders or surfactants to stabilize the dispersion, and this mixture must be homogeneous with the right viscosity for atomization. Then, the suspension is loaded into a spray gun or nozzle, which atomizes the liquid into fine droplets and sprays it onto the substrate (e.g., metal foil, carbon cloth/paper). The thickness and uniformity of the coating of active material can be controlled by tuning the parameters such as spray pressure, nozzle type, distance from the substrate, and spraying time. One of the major advantages of spray coating is that we can get a more uniform and controlled layer compared to manual techniques like brush coating or drop-casting. This method is highly scalable and can be adapted for large-scale industrial production, but there may be overspray or wastage of the suspension material. Notably, the equipment (spray guns, atomizers, and drying systems) can be more expensive than simpler techniques like brush coating, but a uniformly loaded electrode surface can be achieved, which is much more beneficial for reproducible electrochemical performance.

5.2.4 Spin Coating

In this method, a small volume of catalyst ink is deposited on the electrode, which is then spun at high speed to spread the ink evenly across the surface by centrifugal force. A high-speed spinning (typically between 1000 and 4500 rpm) of the electrode is required for a few seconds to a minute. The centrifugal force spreads the ink uniformly across the surface. The thickness of the coating depends on the speed and duration of spinning, as well as the viscosity of the solution. Spin coating provides excellent control over film thickness and uniformity, especially for thin coatings.

The selection of the appropriate coating method depends on the application requirements, desired catalyst loading, and geometrical structure of the electrode.

5.3 Common Electrode Supports for Electrochemical Cells

In the fabrication of electrochemical cells, the choice of electrode support is crucial to achieving optimal performance. Various supports, such as carbon paper, carbon cloth, nickel foam, and copper foam, are commonly used in applications like water splitting, nitrate reduction, N_2 reduction, CO_2 reduction, fuel cells, electrolyzers, supercapacitors, and batteries. These supports serve as substrates onto which the catalyst is deposited, playing a critical role in ensuring electrical conductivity, mechanical stability, and mass transport of reactants and products. Some key factors in selecting electrode supports are discussed below.

5.3.1 Electrical Conductivity

First of all, the electrode supports must have high electrical conductivity to promote effective electron transfer between the active material and the external circuit during electrochemical performance. Materials such as carbon-based supports (carbon paper, carbon cloth, graphite) or metal-based supports (stainless steel, nickel foam, copper foam, titanium mesh) are commonly used due to their excellent conductivity.

5.3.2 Mechanical Stability

The second one is the mechanical stability, that is, the support must maintain structural integrity under operating conditions, such as high pressure, temperature, or mechanical stress. For example, nickel foam and titanium mesh are mostly used in electrolyzers because of their robustness, while carbon fiber papers are used for their flexibility and strength in batteries.

5.3.3 Chemical Stability

The chemical stability of the electrode support is another important parameter. The support material must be resistant to corrosion or degradation in highly acidic and alkaline conditions. Titanium and stainless steel are often chosen for their corrosion resistance in electrolyzers, whereas carbon-based materials are favored for their resistance to chemical attack in neutral and alkaline conditions.

5.3.4 Porosity and Surface Area

The porosity and surface area of support materials impact the distribution of active material and the accessibility of the electrolyte to the electrode surface. It is well documented that high surface area and porosity enhance catalytic activity and increase the reaction sites. Carbon cloth, carbon paper, and metal foams offer a porous structure, enhancing the performance of electrodes in applications like metal-air batteries where oxygen diffusion is critical.

Table 5.1 Properties of different support materials for electrodes

Properties	Carbon Paper	Carbon Cloth	Nickel Foam	Copper Foam
Conductivity	High	High	Very high	Very high
Porosity	Moderate	High	Very high	Very high
Mechanical flexibility	Low	High	Moderate	Moderate
Surface area	Moderate	Moderate-high	Very high	Very high
Chemical stability	High	High	Good in alkaline	Good in neutral/ alkaline
Corrosion resistance	Excellent	Excellent	Good (alkaline)	Moderate (acidic)
Cost	Moderate	Moderate-high	High	High

5.3.5 Carbon Paper and Carbon Cloth

Carbon paper is commercially available in various thicknesses but is more brittle and prone to cracking, whereas carbon cloth is more flexible, durable, less brittle, and has better mechanical properties than carbon paper. Carbon paper is mostly used in proton exchange membrane (PEM) fuel cells and polymer electrolyte membrane electrolyzers, while carbon cloth is used in metal-air batteries and flow batteries.

5.3.6 Nickel Foam and Copper Foam

Nickel foam and copper foam have high porosity (typically around 95%), which provides a large surface area for catalyst deposition and is followed by electrochemical performance. The excellent electrical conductivity of nickel foam and copper foam is due to the metallic nature of nickel and copper. Ni foam has good corrosion resistance in alkaline media but not in acidic media. Carbon paper has excellent corrosion resistance in acidic as well as alkaline media.

5.3.7 Stainless Steel (Solid or Mesh)

Stainless steel support can be used in both acidic and alkaline electrolyzers. It is durable and relatively inexpensive compared to nickel foam and copper foam. Electrochemical alkaline water electrolysis has been studied using stainless steel electrodes in a highly alkaline medium (8 M potassium hydroxide, i.e., KOH).[1]

5.3.8 Titanium Mesh or Foil

It is highly resistant to corrosion and ideal for use in electrolyzers, operating in harsh, acidic, or alkaline conditions. Though it is more expensive, it is often used in high-performance applications where long-term stability is critical.

Most recently, Yadav et al. used four different support materials (viz., nickel foam, flexible graphite sheet, copper foil, and stainless-steel foil) to deposit $NiCo_2O_4$ (NCO) as the active electrode material. The electrochemical supercapacitor performance was performed with and without active material loaded on different electrodes. It has been observed that Ni foam possesses its specific capacity of 128 Fg^{-1} as an electrode, and NCO–Ni possesses the highest specific capacity of 960 Fg^{-1} at a current density of 1 Ag^{-1}, compared with NCO-FG (237 Fg^{-1}), NCO–Cu (169 Fg^{-1}), and NCO-SS (64 Fg^{-1}) electrodes.[2] Each support material has specific advantages and disadvantages, and the selection depends on the requirements and conditions of electrochemical studies.

5.4 Electrolytes and Membranes

In H-cells and electrolyzers, different types of membranes are used to separate the compartments and facilitate ion transport. Here is a rundown of the main types:

5.4.1 Cation Exchange Membranes

Cation exchange membranes (CEMs) impede anions (negatively charged ions) while selectively allowing cations (positively charged ions) to pass through. Electrolyzers and fuel cells are two examples of electrochemical cells in which CEMs are essential parts. They selectively let through positively charged ions, known as cations, and obstruct negatively charged ions, known as anions.

Perfluorosulfonic acid (PFSA) polymers used as CEMs include:

- Nafion: A widely used PFSA-based membrane.
- Ultrex: Another PFSA-based membrane.
- Flemion: A long-lasting polymer with high proton conductivity, based on PFSA, used in fuel cells and electrolyzers.
- Dow PFSA polymers: Commonly utilized in PEM fuel cells and electrolyzers due to excellent ionic conductivity and chemical resistance.

Cost, chemical resistance, conductivity, durability, and durability are some of the characteristics that influence the choice of CEM.

5.4.2 Anion Exchange Membranes

Anion exchange membranes (AEMs) allow anions (negatively charged ions) to pass through while blocking cations (positively charged ions). They are designed to selectively conduct anions (negatively charged ions) while blocking cations (positively charged ions) and are used in a variety of electrochemical applications, including alkaline electrolyzers, fuel cells, and redox flow batteries.

Common types and examples of AEMs include:

- Quaternary ammonium-based membranes: Contain quaternary ammonium groups that exchange anions such as hydroxide (OH^-), chloride (Cl^-), or carbonate (CO_3^{2-}).

- Fumapem and Selemion membranes: Contain polyethylene-based membranes modified with anion-exchange groups.
- Polybenzimidazole (PBI)-based membranes: Contain functionalized PBI polymers for anion exchange, offering high temperature and chemical stability, useful in some fuel cell and electrolyzer systems.

For a variety of electrochemical processes where the movement of anions is critical, AEMs are indispensable. Ion conductivity, chemical resistance, mechanical strength, and particular application requirements are some of the parameters that influence the choice of AEM. New materials and composites are constantly being developed, which improves AEM performance and applications across a range of technical domains.

5.4.3 Proton Exchange Membranes

Membranes are specially made to be impermeable to gasses and other ions but capable of conducting protons (H^+ ions) from the anode to the cathode. Solid polymer electrolytes, or PEMs, are crucial parts of electrolyzers and fuel cells. Protons (H^+ ions) can conduct more easily through them, while gasses and other substances are blocked.

Notable examples of PEMs and their applications include:

- Nafion: Well-known for strong mechanical stability, chemical resistance, and excellent proton conductivity; extensively utilized in electrolyzers, PEM fuel cells, and other electrochemical devices.
- Aciplex: Provides high proton conductivity and durability; used in PEM fuel cells and electrolyzers.
- Flemion: Used in PEM fuel cells and electrolyzers; offers high proton conductivity and stability.
- Celtec: Renowned for excellent PEM fuel cell and other electrochemical system performance.
- PBI (polybenzimidazole): Used in high-temperature PEM fuel cells due to chemical stability and temperature resistance.

PEMs are essential for effective proton conduction in electrochemical systems. The choice of membrane depends on the specific application requirements, including conductivity, chemical resistance, durability, and operating circumstances. Nafion remains the most popular and extensively utilized PEM, while PBI and SPEEK are important in certain applications. The applications of PEM are expanding, with continuous improvements in performance driven by advancements in membrane technology.

5.4.4 Solid Oxide Electrolyte Membranes

Solid oxide fuel cells (SOFCs) and solid oxide electrolysis cells (SOECs) are two examples of high-temperature electrochemical devices that rely heavily on solid oxide electrolyte membranes (SOEMs). For chemical energy to be efficiently converted to electrical energy, or vice versa, they are essential.

Common types of SOEMs include:

- Zirconia-based electrolytes
- Lanthanum-based electrolytes

5.4.5 Polymer Electrolyte Membranes (PEMs)

Another term for proton exchange membranes, while it can also sometimes be used to describe a wider range of membranes made of polymers that are employed in fuel cells and electrolysis.

5.4.6 Electrolyte (Solid and Liquid)

Depending on the electrochemical process and the particular design of the electrolyzer, many types of electrolytes are used in electrolyzers. The electrolyzer's performance, longevity, and efficiency are all impacted by the electrolyte selection. The primary electrolyte types utilized in various electrolyzer types are listed below.

- Alkaline electrolytes: KOH, sodium hydroxide (NaOH)
- Acidic electrolytes: Sulfuric acid (H_2SO_4), phosphoric acid (H_3PO_4)
- Perfluorosulfonic acid (PFSA) electrolyte: Nafion.
- Solid oxide electrolyte: Barium zirconate (BZ), barium cerate (BC); utilized in SOFCs and electrolyzers to produce hydrogen at high temperatures.

- Organic electrolytes: Tetraalkyl quaternary ammonium salts; tetrabutyl ammonium- perchlorate ($TBAClO_4$), tetrafluoroborate ($TBABF_4$), hexafluorophosphate ($TBAPF_6$); occasionally added in non-aqueous solvents .

5.5 Electrochemical Cell Setup

There are two types of electrochemical cells: one which generates electrical energy from chemical reactions (galvanic cells) and the other which uses electricity to conduct chemical reactions to produce chemicals like H_2 and O_2 (i.e., an example of water electrolysis). This cell is called an electrolytic cell. Electrolysis of water has been a green process to produce green hydrogen from water. As we know, in the case of water electrolysis, two reactions occur, that is, at anode oxidation of water (OER-oxygen evolution reaction) and at cathode reduction of hydrogen (HER-hydrogen reduction reaction). As seen in the following Eqs. 5.1 and 5.2,

Anode: $$H_2O_{(l)} \rightarrow 2H^+ + 2e^- + O_{2(g)}\uparrow \quad (5.1)$$

Cathode: $$2H^+ + 2e^- \rightarrow H_{2(g)}\uparrow \quad (5.2)$$

In electrochemical CO_2 reduction, the cathodic reaction in general is as follows:

$$xCO_2 + nH^+ + ne^- \rightarrow \text{product} + yH_2O \quad (5.3)$$

Since water is the only renewable source of electrons and protons, protons (H^+) can be produced by splitting water using electrical energy from renewable sources such as wind and solar, as shown in Eq. 5.1.

5.5.1 Three Electrodes Cell

The three-electrode system is used to set up the study of electrochemical properties of materials and reactions. The cell contains three electrodes: a working electrode (WE), which is the material to be studied, and a counter electrode (CE), and a reference electrode (RE) (Fig. 5.1).

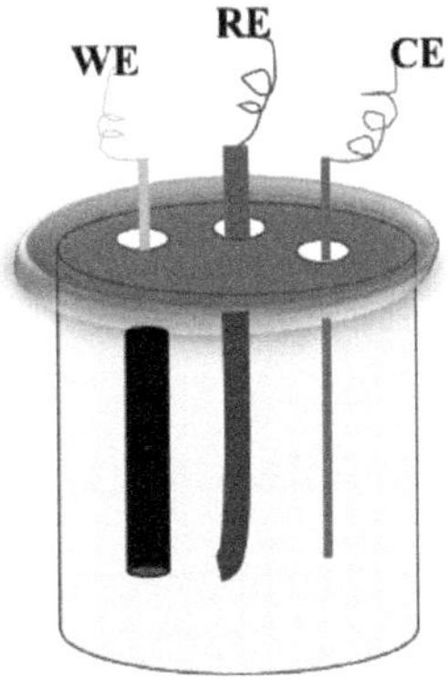

Figure 5.1 A typical three-electrode system for electrochemical experiments.

5.5.2 Two Electrodes Cell (H-cell)

This is a divided electrochemical cell shaped like the English alphabet H. It is mainly designed to study the selectivity, crossover, resistance, and conductivity of a membrane or separator in a liquid medium, keeping two different concentrations in two separate compartments. Figure 5.2 illustrates the diagram of a conventional H-cell, where two compartments, A and B, are separated by a glass frit/membrane. In the case of water electrolysis, if we consider compartment A as the anode, then the H_2O molecules oxidize and produce H^+ ions, and those ions flow through the diaphragm of distance *l*, reach to compartment B, and are reduced at the electrode to H_2.

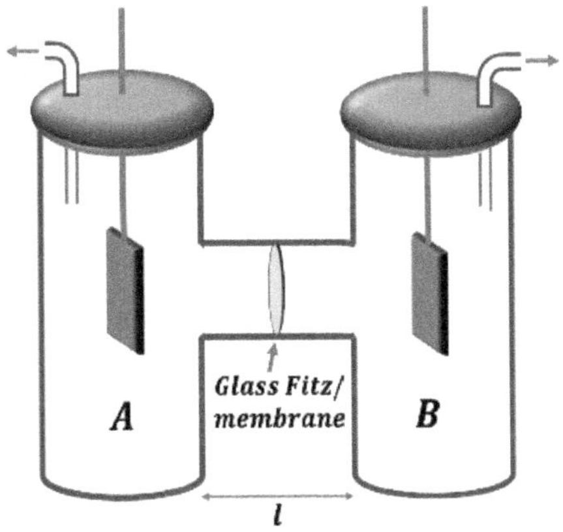

Figure 5.2 Diagram of the state-of-art conventional H-cell setup.

The electrolytic cell, which splits water into H_2 and O_2, is known as a water-electrolyzer. Although there are many electrolyzers present in the market, many of them produce a mixture of H_2 and O_2 after water electrolysis. To separate these again, we require extra energy. However, a few electrolyzers or cells are there which produce H_2 and O_2 separately from water.

5.5.3 Two-Electrode Cells (Electrolyzers)

An electrolyzer is a zero-gap two-electrode cell that uses electricity to drive a non-spontaneous chemical reaction to a spontaneous reaction with some valuable products through electrolysis. This can also be used as carbon dioxide (CO_2) reduction into useful chemicals and fuels.

5.6 Components of Electrolyzers

The electrode supports for catalyst materials for electrolyzers can be any porous conducting substrate such as gas diffusion layers (GDLs): carbon papers/cloths/felts, Ni-foam, etc. Two-catalyst electrodes bonded on either side of the solid polymer electrolyte membrane are called a membrane electrode assembly (MEA). Other components that are used to assemble the MEA into fuel cells/electrolyzers are gaskets, bipolar plates, current collectors, and end plates (Fig. 5.3). Each of these components of the fuel cell is explained as follows:[3,4]

Catalyst layer: It is made of Pt or non-Pt-based materials. These materials with a Nafion ionomer binder are made into an ink, which is loaded on the microporous carbon-coated carbon paper.

GDL: This acts as a support layer for the catalytic layer. It is an electrically conducting layer with micropores through which reactant gases diffuse to reach the reaction sites in the catalytic layer.

Ion-conducting polymer membrane: It is an ion-conducting polymer that separates reactions at both electrodes and prevents the reactants' mix-up.

Gaskets: It is an electrically insulating material to prevent the leakage of reactant gases.

***Bipolar plate*:** It is a graphite plate with gas flow channels on both sides. This plate is required when assembling a fuel cell stack. One side of this plate is in touch with the anode of the MEA, and the other side is in contact with the cathode of another MEA.

***Current collectors*:** It is a highly electrical conducting material such as gold-plated copper, which is in close contact with monopolar graphitic plates.

***End plates*:** The outer parts that compact all the above components together.

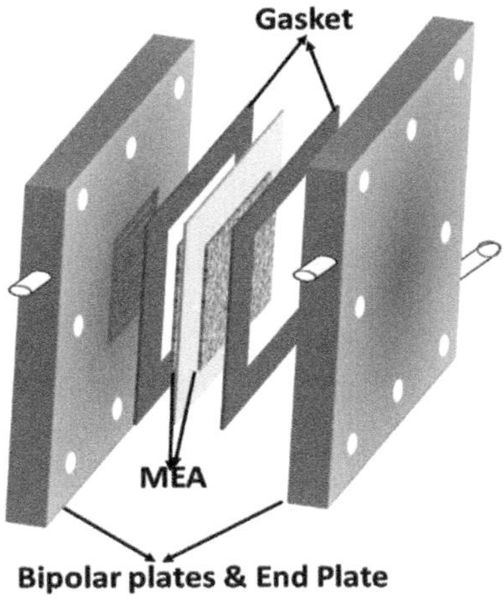

Figure 5.3 Components of a single-stack electrolyzer.

5.6.1 Types of Electrolyzers

Alkaline Electrolyzers: These electrolyzers use liquid alkaline electrolytes (e.g., KOH). Alkaline electrolyzers are known for their durability and lower cost but generally operate at lower efficiencies compared to PEM electrolyzers.

Polymer Electrolyte Membrane (PEM) Electrolyzers: These electrolyzers use a solid polymer electrolyte as the electrolyte. PEM electrolyzers are compact, operate at higher current densities, and can start up quickly, but they tend to be more expensive due to the materials used.

Solid Oxide Electrolyzers (SOECs): These electrolyzers operate at high temperatures (typically 500–850°C) and use a solid oxide electrolyte. SOECs are highly efficient and can also produce syngas (a mixture of hydrogen and carbon monoxide, i.e., CO) when used with CO_2, but they require more robust materials and are still under development for widespread use.

5.7 Application of Electrolyzers

Electrolyzers have wide applications such as the production of hydrogen, which is going to be next-generation fuel, CO_2 reduction to various value-added chemicals such as CO, methane (CH_4), ethylene (C_2H_4), ethane (C_2H_5OH), and other hydrocarbons, which can be used as fuels or chemical feedstocks.

5.7.1 Electrochemical Nitrogen Reduction Reaction (ENRR)

As the world's second-largest chemical producer, NH_3 has an annual supply of more than 180 million metric tonnes, valued at US$67 billion.[5] Nearly 80% of the NH_3 supply is used to make nitrogen-based fertilizers, which are critical to ensure food security in many nations. As the world's population grows, so will the demand for NH_3.[6] Beyond agriculture, NH_3 is gaining popularity as a possible energy carrier in sustainable energy applications. As a carbon-free fuel, NH_3 has a wide range of applications in fuel cells and direct-combustion gas turbines as a single or co-fuel. These applications are predicted to grow in popularity as emission control measures improve to restrict nitrogen oxides in exhaust flows. More crucially, with a high-energy density, NH_3 beats several renewable energy carriers, including hydrogen (H_2), when it comes to long-term storage and transit. As a result, NH_3 is particularly competitive in terms of integrating renewables while minimizing their intermittent nature.[7]

For more than a century, the Haber–Bosch process has met the growing demand for NH_3, which is becoming increasingly unsustainable due to high energy and environmental costs. The Haber–Bosch process (Eq. 5.4) catalytically converts nitrogen (N_2)

and hydrogen (H_2) into NH_3 on a heterogeneous surface at around 700 K and 100 bar pressures.[8] Hydrocarbons such as natural gas, heavy oil, and coal are frequently consumed during the steam reforming process to produce H_2, producing CO_2 as a by-product. As a result, the Haber–Bosch process consumes around 2% of the world energy and emits more than 300 million metric tons of greenhouse gases every year.[9]

$$N_2(g) + 3H_2(g) \leftrightarrow 2NH_3(g)\ \Delta G°298\ K = -16.5\ kJ\ molNH_3^{-1} \quad (5.4)$$

Owing to the Haber–Bosch process's high infrastructural and environmental costs, it is a large source of CO_2 emissions. Researchers are creating sustainable and modular alternatives that have the potential to revolutionize the NH_3 manufacturing sector. Consequently, there is an urgent need for a more sustainable production process, which might be accomplished by directly electrochemically reducing N_2 (ENRR) to NH_3. In both aqueous and non-aqueous conditions, this research creates electrolyzers appropriate for NH_3 electrosynthesis. Nitrogenous impurities obscure the identification of real products from N_2 under aqueous conditions, despite the literature's numerous reports of this phenomenon.

Gen 1 NH_3 vs. Gen 2 NH_3 vs. Gen 3 NH_3

Particularly at low NH_3 concentrations, the widely used separator Nafion cannot stop NH_3 or ammonium ion (NH_4^+) crossover or adsorption, which raises the risk of false positives. Alternatively, by employing a gas-diffusing electrode made with an inexpensive carbon reaction layer, non-aquatic NH_3 electrosynthesis was shown to be more dependable than the aqueous scenario. The performance and durability of NH_3 electrosynthesis are improved by changing the carbon material, binder, electrolyte, and electrolysis settings.

In many nations, the idea of using green H_2 as the feedstock for the Haber–Bosch process (also known as Gen 2 NH_3[10] has been conceptualized and commercialized as a means of bridging the gap between the electrochemical NH_3 production (Gen 3 NH_3) and the Haber–Bosch process (Gen 1 NH_3). With this intermediate method, the electrochemically produced H2 is obtained either by solid-oxide electrolysis at high temperatures using the waste heat discharged from the exothermic Haber–Bosch reactor or by water electrolysis at (near) ambient temperature and pressure using renewable electricity.[5,11] In the short term, by approximately 2030, Gen 2 has the potential to complement the Haber–Bosch process with significantly reduced carbon emissions, providing the commercially available mega-watt-scale water electrolyzers that are capable of achieving high current densities and high energy efficiencies. Nevertheless, high capital investment, availability of renewable energy, and centralized production modes remain to be addressed for the Gen 2 NH_3 production, which in turn is making the ENRR a more attractive alternative to the Haber–Bosch process in the long term. With considerably lower carbon emissions, Gen 2 has the potential to supplement the Haber–Bosch process by 2030. This would allow for the production of megawatt-scale water electrolyzers that are commercially available and capable of achieving high current densities and high energy efficiencies. On the other hand, the manufacture of Gen 2 NH_3 still requires significant capital investment, access to renewable energy sources, and centralized production processes. For these reasons, the ENRR is becoming a more alluring long-term substitute for the Haber–Bosch process.

First, the cost of producing Gen 2 NH_3 is increased by around USD$800 per tonne of NH_3 from USD 1000 to 1500 per tonne of NH_3, depending on the size of the plant, due to the requirement for costly electrolyzer modules on top of the prices of Haber–Bosch plants. On the other hand, the only capital investment needed for Gen 3 NH_3 manufacturing is the ENRR electrolyzers. Additionally, the catalyst in the Haber–Bosch reactor is not able to withstand O_2 impurities in the N_2 stream at ppm levels. Nevertheless, O_2 can be electrochemically reduced to H_2O on the cathode, which is unlikely to be detrimental to the electrode's electrocatalytic stability. Therefore, compared to the manufacture of Gen 1 and Gen 2 NH_3, which is still connected to the Haber–Bosch process, the ease with which N_2 may be separated from air could result in further capital and operational cost savings.

Activation of N≡N

Mechanistic insights into the bottlenecks in NH_3 electrosynthesis activity and efficiency promotion have been obtained through theoretical models.[12–15] First of all, N_2 has a substantial triple-bond energy of 941 kJ mol^{-1}, making it chemically stable. As a result, it is very difficult to fully dissociate the N–N bond at ambient settings. Only breaking the first bond requires an enormous 410 kJ mol^{-1} energy input. Furthermore, N_2 has a very low solubility in aqueous solutions; for example, at 25°C and 1 atm N_2 partial pressure, the mole fraction of N_2 in water is about 1.17×10^{-5}. This affects the electro-reduction of N_2 in aqueous solutions in two ways. Second, the high ionization energy (1525 kJ mol^{-1}) and wide energy gap (1044 kJ mol^{-1}) between the highest occupied and lowest empty molecular orbitals make N_2 electrically stable. Consequently, the electron transfer to N_2 molecules has a very negative equilibrium potential (Eq. 5.5) and is kinetically challenging. The activation of N_2 still needs a rather significant electric potential (–3.2 V vs. RHE) to overcome the intrinsic kinetic barrier, even when coupled with the proton transfer (Eq. 5.6).[16]

Thirdly, enhancing the selectivity toward N_2 reduction to ammonia over hydrogen evolution is not helped by a significantly lower concentration of N_2 in the electrolyte relative to that of proton donors (such as water molecules and protons). Second, the mass transfer of N_2 molecules from the bulk solution to the electrode surface may be the limit on the ammonia generation rate, provided that a very promising N_2 reduction electro-catalyst can be produced. N_2 electro-reduction at a triple-phase boundary appears to offer a possible solution to this problem. The triple-phase boundary is a zone of contact between three separate phases, including reactant gas, electrolyte, and electrode. It is a particularly important concept in the field of fuel cells. Fourthly, N_2 has a lower proton affinity than ENRR intermediates such as diazene (N_2H_2) and hydrazine (N_2H_4), hence making the initial N_2 protonation a more rate-determining step than the subsequent ones (Eqs. 5.7–5.10). And last, because most ENRR tests were performed in aqueous circumstances, surplus protons from H_2O or H_3O^+ can be electrochemically reduced to H_2 with quicker kinetics (Eq. 5.11), resulting in a significant reduction in ENRR FE. Furthermore, H atoms bind to the heterogeneous surface more strongly than N atoms at the same overpotential for transition metals (like Mo and Fe), which are predicted to protonate N_2 most actively. This results in fewer catalytic sites being available to bind to and activate N_2, ultimately decreasing the efficiency of the ENRR.

$$N_2 + e- \leftrightarrow N2- \; E°_{298\,K} = -4.16 \text{ V vs. RHE} \tag{5.5}$$

$$N_2 + H+ + e- \leftrightarrow N_2H \; E°_{298\,K} = -3.20 \text{ V vs. RHE} \tag{5.6}$$

$$N_2 + 2H+ + 2e- \leftrightarrow N_2H_2 \; E°_{298\,K} = -1.19 \text{ V vs. RHE} \tag{5.7}$$

$$N_2H_2 + 2H+ + 2e- \leftrightarrow N_2H_4 \; E°_{298\,K} = 0.529 \text{ V vs. RHE} \tag{5.8}$$

$$N_2H_4 + 2H+ + 2e- \leftrightarrow 2NH_3 \; E°_{298\,K} = 0.939 \text{ V vs. RHE} \tag{5.9}$$

$$2H^+ \text{ (aq.)} + 2e- \leftrightarrow H_2(g) \; E°_{298\,K} = 0 \text{ V vs. SHE at pH} = 0 \tag{5.10}$$

$$2H_2O(l) + 2e- \leftrightarrow 2OH^- \text{ (aq.)} + H_2(g) \; E°_{298\,K} = -0.828 \text{ V vs. SHE at pH} = 14 \tag{5.11}$$

5.7.2 Electrochemical Nitrogen Reduction Reaction (ENRR)-Based Electrolyzer

NH_3 electrosynthesis via the ENRR is a useful technique where the $6e^-$ cathodic reduction of N_2 leads to NH_3. Equations 5.12 and 5.13 represent electrochemical reduction of nitrogen in acidic and neutral pH :

$$N_2(g) + 6H^+ \text{ (aq.)} + 6e^- \leftrightarrow 2NH_3(aq.) \; E°_{298\,K} = 0.092 \; V_{SHE} \text{ at pH} = 0 \tag{5.12}$$

$$N_2(g) + 6H_2O(l) + 6e^- \leftrightarrow 2NH_3(aq.) + 6OH^- \text{ (aq.)} \; E°_{298\,K} = -0.736 \; V_{SHE} \text{ at pH} = 14 \tag{5.13}$$

a. As NH_3 electrosynthesis is thermodynamically feasible at room temperature and pressure and is powered by electricity in the form of electrochemical potentials, it offers operational flexibility and affordability without the need for expensive capital equipment or strict operating conditions, which are common in Haber–Bosch plants.
b. Protons (H^+) can be used directly by *ENRR* to hydrogenate N_2, negating the need to convert hydrocarbons into H_2 as a feedstock. In particular, H^+ from water (H_2O) or the hydronium ion (H_3O^+) can convert N_2 to NH_3 via proton-coupled electron transfer using an aqueous liquid electrolyzer without the need for H_2 directly.
c. To reduce the process's environmental impact, NH_3 electrosynthesis can be directly combined with renewable

electricity produced by solar, wind, or hydro energy. While the utilization of renewable energy sources in water electrolysis can produce hydrogen feedstock for the Haber–Bosch process, potentially mitigating the related carbon footprint, supplementary infrastructure for processing and storing hydrogen is necessary to meet the substantial and ongoing demand for hydrogen. Furthermore, the ENRR process exhibits a significant cost decrease in capital investment and operation and maintenance costs compared to scenarios utilizing the Haber–Bosch process at a lower production scale (e.g., 30 kg day^{-1}). This is illustrated in Fig. 5.4. To eliminate dependency on economies of scale and ensure ecologically benign operation, direct electrosynthesis of NH_3 is therefore appealing.

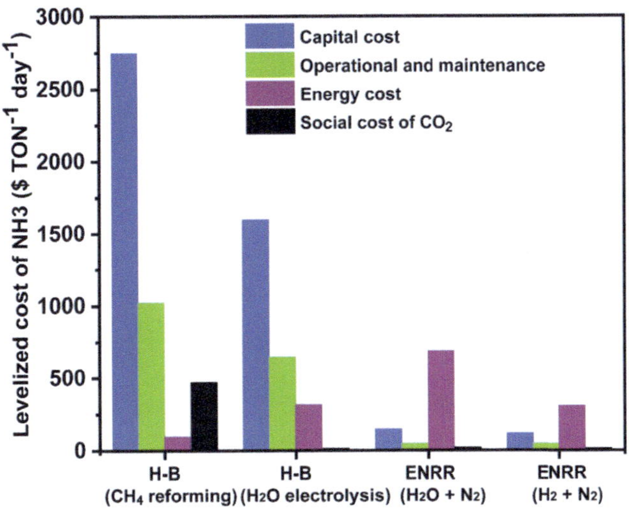

Figure 5.4 Cost comparison to produce NH_3 with the thermochemical (Haber–Bosch) or electrochemical (ENRR) process coupled with/without H_2O electrolysis for the H_2 feedstock. Reproduced from Fernandez and Hatzell.[17]

5.7.3 Electrolyzer Setup

Guidelines and a thorough grasp of how various cell components work together to enhance or degrade ENRR performance are still lacking from the standpoint of electrolyzer design. The design of the flow cell, where gaseous N_2 may be transported using GDEs, provides more chances to efficiently reduce the mass transfer limitations of N_2, which are brought on by its weak diffusivity and marginal solubility under ambient settings.[14,18] A separator is typically employed for both types of cell layouts, and its purpose is to separate the catholyte and anolyte so that any potential NH_3 (or ammonium, NH_4^+, in neutral or acidic aqueous electrolytes) created by the ENRR cathode can be contained (Fig. 5.5). For these purposes, the Nafion membrane is often used as the separator and solid electrolyte.

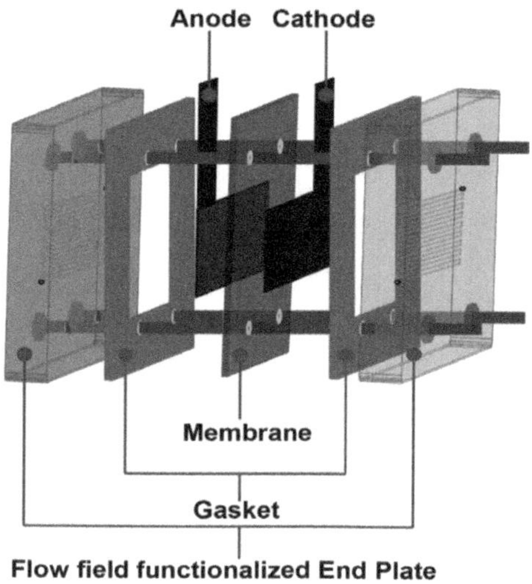

Figure 5.5 An overall three-dimensional pictorial diagram of an electrolyzer.

5.7.4 Potential Measurement

Reports of experimental outcomes are often based on the potential that is applied between the anode and cathode. This makes it difficult to accurately determine the overpotential values for N_2 reduction processes. It should be mentioned that typical REs cannot be included in thin electrolyte membranes (like Nafion membranes) due to their size.

5.7.5 Choice of Electrolyte

Membrane electrode assemblies for N_2 electroreduction have been widely constructed using solid polymer electrolytes, such as Nafion membranes. A class of PFSA ionomer membranes known as Nafion is made up of pendant side chains that are terminated with hydrophilic sulfonic groups and a hydrophobic backbone resembling poly(tetrafluoroethylene). Since water molecules must participate in proton conduction in Nafion membranes, they are often not suitable for temperatures beyond the boiling point of water. Because of their exceptional proton conductivity and superior chemical and electrochemical stabilities, Nafion membranes have been employed extensively in PEM fuel cells and have set the standard for membrane performance.

Nafion membranes have been used as electrolytes to separate the anode and cathode compartments of the electrolytic cell in numerous experimental research related to N_2 electroreduction. However, solid acids are a class of inorganic acid compounds that have been thoroughly researched for application as electrolyte materials in hydrogen fuel cells due to their strong proton conductivity and stability at intermediate temperatures (100–300°C).[19] Solid acids such as $CsHSO_4$, CsH_2PO_4, and $CsH_5(PO_4)_2$ are common. Of these, $CsHSO_4$ and CsH_2PO_4 experience phase changes at 141 and 230°C, respectively. This change is followed by an order of magnitude increase in proton conductivity, with values as high as 10^{-2} Scm^{-1}.[20,21] Notably, the interfacial chemical reaction between CsH_2PO_4 and SiP_2O_7 at 220°C leads to the formation of $CsH_5(PO_4)_2$. As a result, the resulting composite is highly conductive at temperatures below the phase transition temperature of CsH_2PO_4. These two materials are mixed and heated to 220°C. The most significant finding is that, in the absence of adequate water vapor (such as 30 vol%), CsH_2PO_4 and $CsH_5(PO_4)_2$ readily decompose, or become dehydrated, at high temperatures. Poly(acrylic acid) homopolymer was combined with a 6 M KOH aqueous solution to create the gel polymer electrolyte, which was then supported on a porous polypropylene separator. The following is the suggested concept. Using a hydroxide-conducting electrolyte, N_2 and water react at the cathode to produce ammonia and hydroxide ions, and H_2 and hydroxide ions react at the anode to produce water in the electrochemical ammonia synthesis process. Water molecules diffuse in the opposite direction as a result, and hydroxide ions are transferred from the cathode to the anode. Consequently, by restricting water movement within the cell, the use of a gel polymer electrolyte may be able to decrease the availability of water molecules at the cathode, thereby inhibiting hydrogen evolution and enhancing ammonia selectivity. However, ammonia synthesis using Pt/C and Ir/C catalysts showed that the ammonia formation rate and Faradaic efficiency (FE) were still low.

5.7.6 Catalysts Used

Many types of N_2 reduction catalysts have been used, such as metal complexes in the H-B process for ammonia synthesis; transition metals (Ru, Pt, Ir, Pd, Au, Fe, and Ni); and transition-metal oxides (Fe_2O_3 and $Sm_{0.5}Sr_{0.5}CoO_{3-\delta}$). In a further investigation, Kordali et al.[22] employed Ru electrochemically deposited on carbon felt as a catalyst for N_2 reduction when it came into contact with a Nafion membrane and an N_2 flow. A 2 M KOH aqueous solution was placed within the anode chamber, which held the reference and CEs. Electrocatalytic conversion of N_2 to ammonia using carbon paper coated with a transition-metal complex ([Cp_2TiCl_2] or [$(Cp_2TiCl)_2$]) liquid (Pyr_4FAP or Pyr_4OTf) was reported by Katayama et al.[23] as well as using Ni and a combination of Ni and $Ce_{0.8}Sm_{0.2}O_{2-\delta}$ as the anode material (Fe_2O_3/CNTs). Ni/$Ce_{0.8}Sm_{0.2}O_{2-\delta}$; $Sm_{0.5}Sr_{0.5}CoO_{3-\delta}$; $Sm_{1.5}Sr_{0.5}MO_4$ (M = Ni, Co, and Fe); $SmFe_{0.7}Cu_{0.3-x}Ni_xO_3$ (x = 0–0.3);[24] and $SmBaCuMO_{5+\delta}$ (M = Fe, Co, and Ni) are among the cathode materials that have been tested.[25]

5.7.7 Effect of Any Ion as a Mediator

To investigate the impact of lithium ions (Li^+) on the N_2 reduction activity of the Pt/C catalyst using a mixed H+/Li^+ conducting Nafion membrane, researchers worked on it. This membrane was created by exchanging the protons in the membrane with Li^+ ions by immersing it in a 0.1 M Li_2SO_4 aqueous solution. However, it was found that the addition of Li^+ ions to the Nafion membrane had a negative effect on the rate at which ammonia was produced. Nash et al.[26] further analyzed five noble-metal catalysts (Pt/C, Ir/C, Pd/C, Ru/C, and Au/C) for N_2 electro-reduction using PEM and hydroxide exchange membrane (HEM) as two types of electrolyte materials.

5.7.8 Separation of NH_4^+/NH_3

It is interesting to note that Pt/C's ammonia formation rates are many orders of magnitude lower than the numbers Lan et al.[27] previously reported. The usage of Nafion membranes in the H^+ form by Nash et al.[26] and NH_4^+-exchanged Nafion membranes by Lan et al. is a significant distinction between the two investigations. One important reason for the disparity in reported ammonia creation rates was thought to be the leaching of NH_4^+ from the membrane during ammonia synthesis tests. Actually, quaternary ammonium leached from the membrane during ammonia synthesis in HEM electrolyzers and was detected in both the cathode and anode exhaust. Nessler's reagent method of quantifying ammonia is hampered by the leached quaternary ammonium. In order to address this issue, a control experiment was carried out in which H_2 was supplied at 0.2 V to both the cathode and the anode until no quaternary ammonium was seen in the exhaust.

Recently, Wang et al.[28] reported a high NH_4^+ yield using a polyoxometalate as the redox-mediator in a flow cell to efficiently activate N_2, and the synthesized cNH_4^+ accumulated substantially up to 61.0 ppm. Nafion N115, used in the study as the separator, was found to allow NH_4^+ crossover and impair product collection, which demonstrates the importance of selecting appropriate separators for the accurate performance assessment toward the ENRR and effective product collection. However, the Nafion 112 membrane is used as an inert and ineffective barrier against NH_4^+ crossover in electrolyzers with a flow cell configuration, even at the 1-ppm level. Furthermore, the N112 membrane's inability to retain NH_4^+ within the electrolyte recirculation loop results from different factors, including crossover, volatilization, and adsorption. Thus, discrete relocation sites—such as the opposite electrolyte compartment, the acid trap, and the membrane—are revealed in various electrolyte solutions. In particular, uneven NH_4^+ crossover in a neutral electrolyte under varying cell operating circumstances is mostly attributed to the transient behavior of the electrolyte pH. Finally, although it was demonstrated that ion exchange with Na^+ was a successful cleaning method for eliminating adsorbed NH_4^+ from the Nafion membrane and/or ionomer, electrolysis conditioning could also be employed to guarantee that membrane-bound nitrogenous contaminants and

H^+ are adequately eliminated to prevent interference with NH_4^+ measurement and NH_4^+ crossover interpretation.

5.7.9 N_2 Electroreduction in Nonaqueous Solvents

Utilizing non-aqueous solvents is a crucial tactic to inhibit hydrogen evolution and boost the FE of ammonia synthesis since it can drastically lower the amount of protons available in electrochemical cells. By adding an aprotic, hydrophobic layer to the electrode surface, it is also possible to further restrict the proton transfer rate. These techniques have made it possible to produce ammonia electrochemically with extremely high Faradaic efficiencies.

5.7.10 Nitrate Reduction

Due to the high water solubility of nitrate ions and the low bond dissociation energy (204 kJ mol^{-1}) of the N=O bond in nitrate, which is 4.6 times lower than that of the N=N dissociation energy (941 kJ mol^{-1}), the electrochemical nitrate reduction reaction (NO_3RR/NO_2RR) is a promising alternative strategy for greener ammonia synthesis.[29,30] Remarkably, NO_3^- is a common contaminant found in industrial wastewater and agricultural runoff. Therefore, the process of converting nitrate to ammonia presents a "one stone–two birds" approach, addressing pollution while simultaneously producing a product with added value. The complexity of the reaction necessitates a multistep transfer of electrons and protons, with the risk of undesirable side product production and competing HER. As a result, the current status of the NO_3RR is still not a completely simple solution.[31,32] Therefore, a key component of ammonia generation by NO_3RR is selecting an appropriate catalyst to meet the challenge of achieving high selectivity and FE. Thus, to increase selectivity and efficiency toward NO_3RR to NH_3, using a well-designed catalyst with control over the diffusion and interactions with both protons and nitrate ions may be a wise option. Anode and cathode side reactions must both function in full-cell reactions, such as those in an electrolyzer or H-cell. Therefore, in order to fully understand NO_3RR in the complete cell (NO_3RR-OER) electrolysis, it is imperative to take into account the corresponding OER at the anode for the successful synthesis of NH_3.[33–35]

The slow kinetics of OER resulting from the high energy need for O–O bond formation (4e–transfer) acts as a bottleneck. By replacing the OER at the anode with other oxidation reactions that are thermodynamically more feasible, such as the glucose oxidation reaction (GOR), one can reduce the energy input required for ammonia production and the value-added products at the anode (NO_3RR-GOR). This approach lowers the overall potential and increases the economic value of the product by skillfully assimilating two industrially significant areas: the anodic oxidation of glucose to glucaric/gluconic acid.[36–38]

References

1. Amaral, L., Minkiewicz, J., Šljukić, B., Santos, D. M. F., et al. Toward Tailoring of Electrolyte Additives for Efficient Alkaline Water Electrolysis: Salicylate-Based Ionic Liquids. *ACS Appl. Energy Mater.* **2018**, *1* (9), 4731–4742. https://doi.org/10.1021/acsaem.8b00858.
2. Yadav, S., Ghrera, A. S., Rana, A., and Devi, A. Evaluating the Role of Current Collectors in Supercapacitor Electrodes with $NiCo_2O_4$ Nanospheres. *J. Phys. Chem. Solids* **2023**, *178*, 111347. https://doi.org/10.1016/j.jpcs.2023.111347.
3. Nafil, R. Q. Fuel Cells as a Source of Green Energy; Vizureanu, M. S. M. E.-P., Ed.; IntechOpen: Rijeka, **2020**; Ch. 11. https://doi.org/10.5772/intechopen.89736.
4. Hogarth, M. P. and Ralph, T. R. Catalysis for Low Temperature Fuel Cells. *Platinum Met. Rev.* **2002**, *46* (4), 146–164.
5. Rajabloo, T. Ammonia, Production, Applications, and the Effect of Its Phase-Out. **2022**.
6. Lim, J., Fernández, C. A., Lee, S. W., and Hatzell, M. C. Ammonia and Nitric Acid Demands for Fertilizer Use in 2050. *ACS Energy Lett.* **2021**, *6* (10), 3676–3685. https://doi.org/10.1021/acsenergylett.1c01614.
7. MacFarlane, D. R., Cherepanov, P. V., Choi, J., Suryanto, B. H. R., et al. A Roadmap to the Ammonia Economy. *Joule* **2020**, *4* (6), 1186–1205. https://doi.org/10.1016/j.joule.2020.04.004.
8. Ono, K. and Erhard, A. Nondestructive Testing, 3. Ultrasonics. In *Ullmann's Encyclopedia of Industrial Chemistry*; John Wiley & Sons, Ltd, **2011**. https://doi.org/10.1002/14356007.o17_o02.
9. *Beyond Fossil Fuel–driven Nitrogen Transformations.* https://doi.org/10.1126/science.aar6611.

10. Li, S., Zhou, Y., Li, K., Saccoccio, M., et al. Electrosynthesis of Ammonia with High Selectivity and High Rates via Engineering of the Solid-Electrolyte Interphase. *Joule* **2022**, *6* (9), 2083–2101. https://doi.org/10.1016/j.joule.2022.07.009.

11. Salmon, N. and Bañares-Alcántara, R. Impact of Grid Connectivity on Cost and Location of Green Ammonia Production: Australia as a Case Study. *Energy Environ. Sci.* **2021**, *14* (12), 6655–6671. https://doi.org/10.1039/D1EE02582A.

12. Qing, G., Ghazfar, R., Jackowski, S. T., Habibzadeh, F., et al. Recent Advances and Challenges of Electrocatalytic N_2 Reduction to Ammonia. *Chem. Rev.* **2020**, *120* (12), 5437–5516. https://doi.org/10.1021/acs.chemrev.9b00659.

13. Skúlason, E., Bligaard, T., Gudmundsdóttir, S., Studt, F., et al. A Theoretical Evaluation of Possible Transition Metal Electro-Catalysts for N 2 Reduction. *Phys. Chem. Chem. Phys.* **2012**, *14* (3), 1235–1245. https://doi.org/10.1039/C1CP22271F.

14. Singh, A. R., Rohr, B. A., Schwalbe, J. A., Cargnello, M., et al. Electrochemical Ammonia Synthesis—The Selectivity Challenge. *ACS Catal.* **2017**, *7* (1), 706–709. https://doi.org/10.1021/acscatal.6b03035.

15. Chen, G.-F., Ren, S., Zhang, L., Cheng, H., et al. Advances in Electrocatalytic N_2 Reduction—Strategies to Tackle the Selectivity Challenge. *Small Methods* **2019**, *3* (6), 1800337. https://doi.org/10.1002/smtd.201800337.

16. Guo, X., Gu, J., Lin, S., Zhang, S., et al. Tackling the Activity and Selectivity Challenges of Electrocatalysts toward the Nitrogen Reduction Reaction via Atomically Dispersed Biatom Catalysts. *J. Am. Chem. Soc.* **2020**, *142* (12), 5709–5721. https://doi.org/10.1021/jacs.9b13349.

17. Fernandez, C. A. and Hatzell, M. C. Editors' Choice—Economic Considerations for Low-Temperature Electrochemical Ammonia Production: Achieving Haber-Bosch Parity. *J. Electrochem. Soc.* **2020**, *167* (14), 143504. https://doi.org/10.1149/1945-7111/abc35b.

18. Grayer, S. and Halmann, M. Electrochemical and Photoelectrochemical Reduction of Molecular Nitrogen to Ammonia. *J. Electroanalytic. Chem. Interfacial Electrochem.* **1984**, *170* (1), 363–368. https://doi.org/10.1016/0022-0728(84)80059-5.

19. Haile, S. M., Chisholm, C. R. I., Sasaki, K., Boysen, D. A., et al. Solid Acid Proton Conductors: From Laboratory Curiosities to Fuel Cell Electrolytes. *Faraday Discuss.* **2006**, *134* (0), 17–39. https://doi.org/10.1039/B604311A.

20. Haile, S. M., Boysen, D. A., Chisholm, C. R. I., and Merle, R. B. Solid Acids as Fuel Cell Electrolytes. *Nature* **2001**, *410* (6831), 910–913. https://doi.org/10.1038/35073536.

21. Boysen, D. A., Uda, T., Chisholm, C. R. I., and Haile, S. M. High-Performance Solid Acid Fuel Cells through Humidity Stabilization. *Science* **2004**, *303* (5654), 68–70. https://doi.org/10.1126/science.1090920.

22. Kordali, V., Kyriacou, G., and Lambrou, C. Electrochemical Synthesis of Ammonia at Atmospheric Pressure and Low Temperature in a Solid Polymer Electrolyte Cell. *Chem. Commun.* **2000**, No. 17, 1673–1674. https://doi.org/10.1039/B004885M.

23. Katayama, A., Inomata, T., Ozawa, T., and Masuda, H. Electrochemical Evaluation of Titanocenes in Ionic Liquids with Non-Coordinating and Coordinating Anions and Application for NH_3 Synthesis. *ChemElectroChem* **2017**, *4* (12), 3053–3060. https://doi.org/10.1002/celc.201700557.

24. Xu, G., Liu, R., and Wang, J. Electrochemical Synthesis of Ammonia Using a Cell with a Nafion Membrane and $SmFe_{0.7}Cu_{0.3-x}Ni_xO_3$ (x = 0–0.3) Cathode at Atmospheric Pressure and Lower Temperature. *Sci. China Ser. B-Chem.* **2009**, *52* (8), 1171–1175. https://doi.org/10.1007/s11426-009-0135-7.

25. Zhang, Z., Zhong, Z., and Liu, R. Cathode Catalysis Performance of $SmBaCuMO_{5+\delta}$ (M=Fe, Co, Ni) in Ammonia Synthesis. *J. Rare Earths* **2010**, *28* (4), 556–559. https://doi.org/10.1016/S1002-0721(09)60153-8.

26. Nash, J., Yang, X., Anibal, J., Wang, J., et al. Electrochemical Nitrogen Reduction Reaction on Noble Metal Catalysts in Proton and Hydroxide Exchange Membrane Electrolyzers. *J. Electrochem. Soc.* **2017**, *164* (14), F1712. https://doi.org/10.1149/2.0071802jes.

27. Lan, R., Irvine, J. T. S., and Tao, S. Synthesis of Ammonia Directly from Air and Water at Ambient Temperature and Pressure. *Sci. Rep.* **2013**, *3* (1), 1145. https://doi.org/10.1038/srep01145.

28. Lin, Y. and W. Connell, J. Advances in 2D Boron Nitride Nanostructures: Nanosheets, Nanoribbons, Nanomeshes, and Hybrids with Graphene. *Nanoscale* **2012**, *4* (22), 6908–6939. https://doi.org/10.1039/C2NR32201C.

29. Tang, C. and Qiao, S.-Z. How to Explore Ambient Electrocatalytic Nitrogen Reduction Reliably and Insightfully. *Chem. Soc. Rev.* **2019**, *48* (12), 3166–3180. https://doi.org/10.1039/C9CS00280D.

30. Chauhan, R. and Srivastava, V. C. Electrochemical Denitrification of Highly Contaminated Actual Nitrate Wastewater by Ti/RuO_2 Anode and Iron Cathode. *Chem. Eng. J.* **2020**, *386*, 122065. https://doi.org/10.1016/j.cej.2019.122065.

31. Yang, G., Zhou, P., Liang, J., Li, H., et al. Opportunities and Challenges in Aqueous Nitrate and Nitrite Reduction beyond Electrocatalysis. *Inorg. Chem. Front.* **2023**, *10* (16), 4610–4631. https://doi.org/10.1039/D3QI00148B.

32. Wang, Z., Richards, D., and Singh, N. Recent Discoveries in the Reaction Mechanism of Heterogeneous Electrocatalytic Nitrate Reduction. *Catal. Sci. Technol.* **2021**, *11* (3), 705–725. https://doi.org/10.1039/D0CY02025G.

33. Lu, Z., Wang, J., Huang, S., Hou, Y., et al. N,B-Codoped Defect-Rich Graphitic Carbon Nanocages as High Performance Multifunctional Electrocatalysts. *Nano Energy* **2017**, *42*, 334–340. https://doi.org/10.1016/j.nanoen.2017.11.004.

34. Thakur, N., Mehta, D., Chaturvedi, A., Mandal, D., et al. Glucose Oxidation Assisted Hydrogen and Gluconic/Glucaric Acid Production Using NiVP/Pi Bifunctional Electrocatalyst. *J. Mater. Chem. A* **2023**, *11* (29), 15868–15877. https://doi.org/10.1039/D3TA02133E.

35. Chaturvedi, A., Gupta, D., Kaur, S., Garg, K., et al. Glucose Oxidation Assisted Ammonia Production via Electrochemical Dinitrogen Reduction over $CoWO_4$. *J. Mater. Chem. A* **2023**, *11* (34), 18280–18290. https://doi.org/10.1039/D3TA03302C.

36. Chen, G., Li, X., and Feng, X. Upgrading Organic Compounds through the Coupling of Electrooxidation with Hydrogen Evolution. *Angew. Chem. Int. Ed.* **2022**, *61* (42), e202209014. https://doi.org/10.1002/anie.202209014.

37. Liu, W.-J., Xu, Z., Zhao, D., Pan, X.-Q., et al. Efficient Electrochemical Production of Glucaric Acid and H_2 via Glucose Electrolysis. *Nat. Commun.* **2020**, *11* (1), 265. https://doi.org/10.1038/s41467-019-14157-3.

38. Zhang, Y., Zhou, B., Wei, Z., Zhou, W., et al. Coupling Glucose-Assisted Cu(I)/Cu(II) Redox with Electrochemical Hydrogen Production. *Adv. Mater.* **2021**, *33* (48), 2104791. https://doi.org/10.1002/adma.202104791.

Chapter 6

Materials for Anodic Oxidation Reaction: Water Oxidation, Organic Oxidation, Ammonia Oxidation

R. Bhavana and Biswajit Mondal
Department of Chemistry, Indian Institute of Technology Gandhinagar, Palaj, 382355, Gujarat, India
mondal.biswajit@iitgn.ac.in

Electrocatalytic water splitting is a sustainable pathway to large-scale hydrogen production and aligns with global renewable energy goals by providing clean, zero-carbon fuel. Hybrid water electrolysis addresses the limitations of the oxygen evolution reaction (OER), including sluggish kinetics, large overpotential, and explosive risk, thereby allowing for continuous hydrogen generation and the conversion of organic compounds into valuable chemicals. Feedstock for this process can be small alcohols such as methanol and ethanol, biomass-derived alcohols, and plastic-waste-derived alcohols, and electrocatalytic treatment also offers an effective solution for organic wastewater pollutants. Ammonia from fertilizer, industry, and water

Materials Design for Electrocatalysis
Edited by Sagarika Bhattacharya and Biswarup Chakraborty

ISBN 978-981-5352-32-0 (Hardcover), 978-1-042-02558-9 (eBook)
www.jennystanford.com

DOI: 10.1201/9781042025589-6

treatment must be oxidized to prevent environmental harm through water pollution and ecological system damage. Electrochemical ammonia oxidation in fuel cells enables ammonia to be a hydrogen carrier, in which clean energy is obtained by producing byproducts such as nitrogen and water. Apart from that, it reduces pollutants and the nitrogen recovered helps in utilizing fertilizers thus promoting eco-friendly agriculture, and reducing the carbon footprint in chemical industries. The performance and stability in chemical synthesis, energy conversion, and environmental remediation through anodic oxidation reactions require highly engineered materials that make up the electrode. This chapter explores recent advancements in anodic material design, highlighting challenges and future directions for developing efficient, cost-effective, and sustainable electrodes tailored to specific oxidation processes.

6.1 Introduction

Electrocatalytic water splitting for the mass production of clean hydrogen fuel is a promising technology and a sustainable alternative to fossil fuels while offering zero carbon emissions that align with global renewable energy goals [1]. Water splitting is improved by using catalysts to overcome energy barriers, making hydrogen generation possible at an economy of scale. Such hydrogen molecules, produced electrocatalytically, hold a promise for industrial use to fulfill energy demand and progress toward a carbon-neutral economy. Cathodic hydrogen generation by electrocatalytic water splitting is accompanied by an anodic process of the OER. The intrinsic limitation of sluggish kinetics, chances of an explosive reaction by mixing hydrogen and oxygen, and large overpotential hamper the overall cell efficiency. Replacement of conventional water electrolysis by assisted or hybrid water electrolysis is another avenue to improve continuous hydrogen production through the electro-oxidative conversion of organic compounds into value-added chemicals [2]. As oxidation of organic molecules requires less potential to get oxidized as compared to OER, the kinetic efficiency of this process permits yielding a pure stream of hydrogen simultaneously valorizing organic molecules. Various small aliphatic alcohols such as methanol, ethanol, isopropanol, and butanol;

aromatic alcohols such as benzyl alcohol; biomass-derived alcohols such as 5-hydroxymethylfurfural, lignin derivatives, and glycerol; plastic waste–derived alcohols like ethylene glycol; and aromatic amines such as benzylamine and aniline among others, have been commonly used for this. Wastewater having organic environmental pollutants that comprise a wide range of carbon-based chemical compounds and functionalities that are harmful both to the environment and human health finds a solution by electro-oxidative treatment [3].

Ammonia sourced from agricultural fertilizers, industrial processes, and wastewater from livestock and food processing needs to be oxidized primarily to mitigate its environmental impact, as excessive ammonia can lead to water pollution, eutrophication, and harmful effects on aquatic ecosystems. The electrocatalytic oxidation process of ammonia and its derivatives finds application in energy generation, where it acts as a hydrogen carrier in fuel cells to produce electricity with just nitrogen and water as byproducts, making it a clean energy source [4]. By efficiently lowering contaminants and allowing the recovery of precious nitrogen resources for fertilizers, it also plays a critical role in environmental management by treating wastewater that is rich in ammonia. Moreover, the chemical sector relies on ammonia oxidation to synthesize nitrogen-based fertilizers and other substances, promoting sustainable farming methods and reducing the carbon footprint associated with ammonia manufacturing.

An essential class of electrochemical reactions, anodic oxidation reactions, are crucial for chemical synthesis, energy conversion, and environmental remediation. Since each type of reaction has unique material requirements and mechanistic pathways, specific tailoring of the electrode design must be performed to optimize performance, efficiency, and durability. The material design for anodic processes has to address the intricacy of the substrate and the complexity of the reaction pathways. This chapter gives a comprehensive overview of the latest developments in material science for anodic oxidation reactions, concerning the interplay between material properties, electrochemical performance, and mechanisms of reactions as shown in Scheme 6.1. The insights by comparative analysis present the challenges and directions of next-generation tailored anodic materials toward specific oxidation reactions to be used in highly efficient, cost-effective, and durable electrodes in sustainable chemical and energy applications.

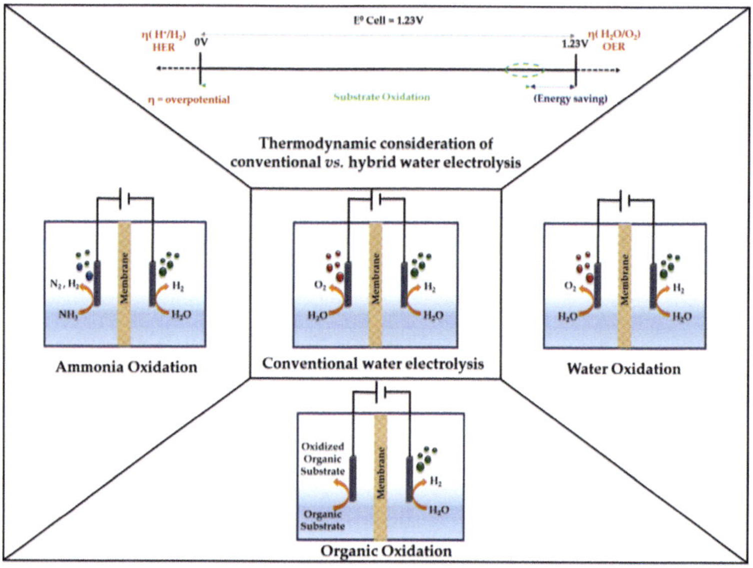

Scheme 6.1 Schematic representation of overall anodic oxidation reactions (water oxidation, organic oxidation, and ammonia oxidation) along with thermodynamic consideration.

6.2 Materials for Water Oxidation Reaction

Water oxidation reaction (WOR) or OER, is an important process in many electrochemical energy conversion technologies where water splitting based on electrochemical energy conversion is used to generate hydrogen from water. WOR is significant even though its sluggish kinetics and overpotential requirements pose a significant challenge; hence, highly efficient and durable anodic materials are the need of the hour. There have been several research on developing anodic material as electrocatalysts for WOR. Multiple catalysts have been reported for this reaction both in acidic and alkaline medium, which can be found in detail in several review papers as provided in the references [5, 6]. In the following section, a brief overview along with the mechanism and selected catalysts for WOR will be discussed as a prelude to the hybrid water electrolysis.

6.2.1 Overview of WOR

The anodic half-cell reaction involves the extraction of electrons from water molecules, resulting in the formation of oxygen gas, protons, and electrons. The reactions are highly endergonic. They are very hefty, requiring an input of huge energies, with complex multi-electron transfer steps, thus introducing a major kinetic barrier into the reaction mechanism. So, these need to be overcome critically with the design and selection of the anodic materials, such that the reaction could better be facilitated with reduced overpotential, long-term stability, and better performance [7].

Good efficiency and performance heavily depend on the intrinsic properties of the anodic materials, mainly on their electrical conductivity, surface area, catalytic activity, and corrosion resistance [8]. Several types of materials have been elaborately studied in the process of looking for highly efficient and stable OER catalysts. Noble metal oxides, especially iridium oxide (IrO_2) and ruthenium oxide (RuO_2), were for many years considered the gold standard for this reaction due to their outstanding catalytic features and stability. However, the complexity of these catalysts leads to high costs and scarcity, making it indispensable to replace them with other materials. Conversely, within the past decade, there have been impressive research efforts toward the development of catalysts based on non-noble metals, such as transition metal oxides, perovskites, and layered double hydroxides, toward lowering the costs without compromising performance.

Today, more than 95% of hydrogen produced worldwide is by means of steam reforming of natural gas or similar fossil fuels [9]. The industrial synthesis produces low-purity hydrogen with large amounts of carbon-based species such as carbon monoxide. Water electrolysis, in principle, produces high-purity hydrogen through the electrochemical conversion of water into hydrogen and oxygen in an efficient process [10, 11]. This approach can make use of a fresh and renewable source of hydrogen, derived from abundant water supplies. The main reactions involved in the process of electrochemical splitting of water are hydrogen evolution reaction (HER) at the cathode and OER at the anode.

Water splitting occurs in two different pathways, in which the reactions are expressed as:

In acidic solution,

Cathode:

$$2H^+ + 2e^- \rightarrow H_2 \quad (6.1)$$

Anode:

$$H_2O \rightarrow 2H^+ + 1/2O_2 + 2e^- \quad (6.2)$$

In neutral and alkaline solutions,

Cathode:

$$2H_2O + 2e^- \rightarrow H_2 + 2OH^- \quad (6.3)$$

Anode:

$$2OH^- \rightarrow H_2O + 1/2O_2 + 2e^- \quad (6.4)$$

The thermodynamic voltage needed for splitting water is 1.23 V at 25°C and 1 atm regardless of the medium in which water existed; however, it drops with increases in temperature. Nevertheless, the applied voltage in electrochemical water splitting is always greater than its value of thermodynamics. This supplemental voltage, or overpotential (η), is mostly to counteract inherent activation barriers at both the anode side (η_a) and the cathode side (η_c), as well as other resistances, including the Ohmic drop (η_{other}). The practical operating voltage for water splitting is described as [12]:

$$E_{OP} = 1.23\ V + \eta_a + \eta_c + \eta_{other} \quad (6.5)$$

Highly active oxygen and hydrogen evolution catalysts can be used to reduce the overpotentials at the cathode as well as at the anode so that η_{other} is reduced by a well-designed electrolytic cell. Reducing these overpotentials will help water splitting in more energy-efficient and cost-effective ways.

6.2.2 Mechanism of WOR

The OER consists of four sequential proton and electron transfer steps. Amongst the most well-known mechanisms responsible for this process is the adsorbate evolution mechanism, or AEM. The mechanism of OER can be described for a single metal site of a metal oxide, M*, as:

In acidic solution,

$$H_2O + M^* \rightarrow M\text{-}OH^* + H^+ + e^- \quad (6.6)$$

$$\text{M-OH}^* \rightarrow \text{M} - \text{O}^* + \text{H}^+ + \text{e}^- \tag{6.7}$$

$$\text{M-O}^* + \text{H}_2\text{O} \rightarrow \text{M} - \text{OOH}^* + \text{H}^+ + \text{e}^- \tag{6.8}$$

$$\text{M-OOH}^* \rightarrow \text{M} + \text{O}_2 + \text{H}^+ + \text{e}^- \tag{6.9}$$

In alkaline solution,

$$\text{OH}^- + \text{M}^* \rightarrow \text{M} - \text{OH}^* + \text{e}^- \tag{6.10}$$

$$\text{M-OH}^* + \text{OH}^- \rightarrow \text{M} - \text{O}^* + \text{H}_2\text{O} + \text{e}^- \tag{6.11}$$

$$\text{M-O}^* + \text{OH}^- \rightarrow \text{M} - \text{OOH}^* + \text{e}^- \tag{6.12}$$

$$\text{M-OOH}^* + \text{OH}^- \rightarrow \text{M}^* + \text{O}_2 + \text{H}_2\text{O} + \text{e}^- \tag{6.13}$$

Although the details of the origin of OER activity are still unclear, depending on the reported results it is widely accepted that the OER activity of the electrocatalyst critically depends on the bonding strength of adsorbents at the metal active site (M), such as M–OH*, M–O*, and M–OOH*, and linearly proportional to their respective adsorption energies [13, 14]. In addition to this, another OER mechanism, the lattice oxygen-mediated mechanism (LOM), has also been confirmed. In this mechanism, O–O bonds are formed between adjacent metal ions. Two HO* groups on neighboring metal ions undergo deprotonation and form two M–O* species at the first step. These adjacent M–O* species subsequently undergo direct coupling to form an O–O bond. At the end, O_2 is released, and the two vacant metal sites are replenished by OH^-. In contrast to AEM, LOM does not incorporate HOO*.

6.2.3 Catalysts for WOR

The splitting of water is one promising approach toward storing sustainable but intermittent resources. Its biggest challenge will be the OER whose overpotentials are not low enough for optimum performance in most cases. A primary area of interest within WOR is the development of catalytic materials with low overpotential, fast kinetics, and excellent electrochemical stability. A few selected, major classes of materials that are widely exploited for enhanced WOR are discussed here.

6.2.3.1 Precious metal-based catalysts

Recent developments in anodic materials for the OER during water electrolysis have centered on activity augmentation, stability, and overall performance. With the focus on materials for anodic transformation associated with OER, balancing catalytic activity and durability under acidic conditions remains an important challenge. In this regard, a 'macro-nano' fcc Ru_3Ir nanocrystal synthesized via two-dimensional self-assembly has recently been reported by Fan et al. as a new approach [15]. This catalyst reaches a low overpotential of 190 mV at 10 mA cm^{-2} and could be stable up to a period of 400 h or 10,000 cycles. Improved activity is observed because the self-reconstructed Ru_3IrOx layer would facilitate the reduction in the barrier energy for crucial reaction steps, enhancing the corrosion as well as the durability of the material. Bimetallic AuRu alloy nanofibers developed by Kwon et al. also exhibit promising potential as bifunctional electrocatalysts for both OER and HERs [16]. These nanofibers, constituted by a RuO_2 framework decorated with Au nanoparticles, outperform traditional catalysts, such as Ru, Au, and commercial Pt/C and Ir/C. They show the lowest overpotential at 10 mA/cm^2 with excellent durability, thereby demonstrating superior stability in acidic conditions and promising electroactivity due to improved nanofibrous structure. Ir, Ru, and Rh doped high-valent noble metals into PdCu nanoparticles exhibit remarkable OER and HER activity [17]. Another interesting synthesis by Liu et al. is that of ternary RuIrTe nanotubes, which work as bifunctional electrocatalysts for the overall splitting of water[18]. These nanotubes derive the electronic coupling between Ru, Ir, and Te, along with a very rough one-dimensional tubular surface that enhances catalytic activity as shown in SEM image (Fig. 6.1 A). They show fantastic performance with an overpotential of only 29 mV for hydrogen evolution and 205 mV for oxygen evolution, resulting in a low cell voltage of 1.511 V in 0.5 M H_2SO_4 (Fig. 6.1 B–C). Qin and coworkers developed Ir_{16}–PdCu/C which exhibits mass activities 50.5-fold higher than that of the standard PdCu/C for OER, with optimized binding energies of the reaction intermediates [17]. Other new materials by Galani et al. comprise RuO_2-loaded CeO_2, which augments the surface area and Ce^{3+} content to increase the catalytic activity, and unsupported RuIr alloys, with low overpotentials over a wide pH range [19]. The alloys exhibit excellent stability in terms of performance for more than 400 hours in an acidic medium, which demonstrates stability and effectiveness for splitting water. Seow et al. reported a wet-chemical synthesis with a core-shell structure, Ru@IrOx, that shows outstanding performance in the OER at high specific activity in acidic and basic media, holding promise for versatile water electrolyzers [20]. These developments herald a trend toward developing anodes with high performance and durability for the OER and are strategically important for the optimization of water electrolysis and clean hydrogen production.

6.2.3.2 Non-precious metal-based catalysts

For practical reasons, attention has been shifted toward the development of OER catalysts based on 3d metals. Recent examples include a Co–P foam, a novel electrocatalyst with a distinct porous structure developed by Li et al., which demonstrates outstanding electrocatalytic activity toward HER and OER, allowing for an overall water electrolysis cell voltage as low as 1.98 V at high current densities [21]. A nickel foam decorated electrode with iron-doped Ni–Fe-hydroxide nanoparticles is an inexpensive electrode that demonstrates excellent reversibility in both OER and HER, with low overpotentials and stable performance that exceeds that of platinum-based electrodes as studied by Shrestha and coworkers [22]. In the work by Adamson et al., a CoO/Fe_3O_4 biphase metal oxide composite is a bifunctional catalyst that shows excellent performance in HER and OER, with high stability for extended periods [23]. XRD pattern of bare CFP, CoO, Fe_3O_4, and Co-Fe NPs along with OER LSV curves for CoO, Fe_3O_4, Co-Fe NPs, bare CFP and Ir/C electrode in 1M KOH are shown in Fig. 6.1 D–E. Chronopotentiometry stability testing at a constant current of 50mA cm^{-2} for OER studies is also shown (Fig. 6.1 F). Studies by Qian et al. produced catalysts by the anion exchange route to afford transition metal oxide hollow nanostructures that demonstrate enhanced OER activity and stability, thus enabling the development of noble metal-free electrocatalysts [24]. Wang et al. developed a single-atom Mo-decorated Co_9S_8@C catalyst that exhibits outstanding activity at different pH values and demonstrates extraordinary stability, far exceeding that of most electrocatalysts based on noble metals [25]. Pan et al. reported $Mn_{7.5}O_{10}Br_3$, a non-noble metal catalyst with a low overpotential value of 295 ± 5 mV at an applied current density of 10 mA cm^{-2}, and the electrocatalytic water splitting exhibits very promising results with such overpotential for over than 500 h [26]. As described in the report, an enhanced electronic conductivity occurring with its self-oxidized surface layer leads to higher enhanced performance for the catalyst. Further, Retuerto et al. observed in sodium-doped $SrRuO_3$ that sodium doping increases the activity and stability of $SrRuO_3$ under acidic OER conditions [27]. DFT calculations reveal that sodium doping optimizes the binding of reaction intermediates, thus enhancing overall performance. Xu et al. reported electrochemical route for the preparation of Sn-doped NiFe oxyhydroxides leads to better OER performance; 100 mA cm^{-2} at 260 mV, and stable performance at 500 mA cm^{-2} for 5 days with small activity loss [28]. Mesoporous Ni_2P@FePOxHy pre-electrocatalyst demonstrates excellent current density performance and has good durability so that it can be considered for use in commercialized applications in alkaline environments reported by Meena et al. [29]. Three-phase heterojunction of Ni/MoO_2@CN architecture developed by Qian et al. maximized both the performance of HER and OER at low overpotentials with excellent stability during long-term testing [30]. Liu et al. reported dendritic V-doped Ni_3S_2@NiO nanoarrays offering mass and charge transport channels that facilitate improved kinetics for the water electrolysis process [31]. Tsai et al. developed a CFeCoNiP-based bifunctional electrocatalyst that has outstanding activity for HER and OER with good charge transfer characteristics, promising to be highly efficient for water splitting [32].

6.2.3.3 Molecular catalysts

The molecular approach of water oxidation catalysts has been significantly advanced in the past few decades by the creation of ruthenium (Ru) polypyridine complexes, especially by a study on the [Ru(bds)$(pic)_2$] complex, which goes by the name Ru-bds. This catalyst comes with an equatorial tetradentate dianionic sulfonate ligand and axial 4-picoline ligands as developed by Yang et al. [33]. The Ru-bds catalyst shows excellent electrochemical activity for water oxidation, reaching a high turnover frequency of 160 s^{-1} at acidic conditions and an astonishing 12,900 s^{-1} at near-neutral conditions, the best-reported value for any Ru-bda catalyst. Water oxidation has also been catalyzed by manganese (Mn) complexes. Li and coworkers studied electrocatalytic water oxidation experiments with Mn(III) tris(pentafluorophenyl)corrole revealing that hydroxide ions are required for the production of oxygen at the $Mn^{V/IV}$ potential [34]. The synthetic $Mn^{V}(O)$ complex exhibits profound differences in reactivity toward hydroxide ($k_2(OH^-)$ = $7.4 \times 10^3\ M^{-1}\ s^{-1}$) relative to water ($k_2(H_2O) = 4.4 \times 10^{-3}\ M^{-1}\ s^{-1}$), indicative of fundamentally different electrocatalytic behaviors. The most promising avenue for electrocatalysts of OER is carboxylate-based metal-organic frameworks (MOFs) [35]. They exhibit unique structural advantages, and recent developments studied by Yang et al. suggest that minor structural changes make a significant difference in the catalytic activity. Recent breakthroughs include anchoring oligomeric Ru complexes onto graphitic materials for water oxidation at neutral pH with good current density through CH-π interactions studied by Hoque et al. [36]. This generalized approach can be applied in many transition metals and diversifies its scope in future sustainable energy technologies. Recently, new cobalt-based metal-organic frameworks (Co-MOFs) have been synthesized by Tarrino and coworkers with strong intermolecular bonding and interesting "memory effects" during delamination and reassembly [37]. The modified Co-MOFs have excellent catalytic activity toward the OER with TOF values of $0.034\ s^{-1}$ at an overpotential of 400 mV in neutral pH conditions due to the unique structural arrangement that prevents direct coupling of active sites. The structures of a few molecular catalysts for water oxidation are shown in Fig. 6.1 G [33, 38]. By investigation of anodic materials like Ru, Mn, and Co-based molecular catalysts that are used to facilitate catalytic efficiency as well as stability for water oxidation, a thorough understanding of the mechanism in catalysis is made that further helps in improving efficiencies as well as stability and structural considerations for optimizing the performance.

6.2.3.4 Carbon-based catalysts

An alternative electrochemical process for the production of hydrogen peroxide (H_2O_2) with carbon-based anodes, that is, an alternate route compared to the established anthraquinone route was established by Pangotra et al. [39]. Optimization of crucial parameters, including current density, electrolyte concentration, and pH conditions significantly improves the efficiency of water oxidation in this process. In particular, sodium metasilicate (Na_2SiO_3) acting as a chemical stabilizer enhances H_2O_2 yield at higher pH. It also established a relation between the carbonate ion activity and the production of H_2O_2, hence reaching the concentration of 33 mmol L^{-1} at a current density of 100 mA cm^{-2}. This method exhibits sustainability and economic viability for the production of H_2O_2 from carbon fiber paper, which is less expensive. Another innovative approach by Gu et al. is through the utilization of a metal-free composite made of polyaniline, and carbon dots, which happen to be bifunctional photo-assisted catalysts for efficient water splitting [40]. In neutral electrolytes under visible light, it strongly reduces overpotentials such as 150 mV for the OER and 70 mV for the HER. At current densities of 30 and 20 mA cm^{-2}, overpotentials decrease to 65 mV, while visible light increases the current density to 13.27 mA cm^{-2} at 2.0 V, a 62.8% increase over the dark condition. An acidified, metal-free electrocatalyst is also prepared from carbon by an "amino-assisted polymerization and carbonization" process by Zhao and coworkers [41]. Such an amino-rich hierarchical-network carbon, namely, amino-HNC framework, displays superior OER performance under acidic conditions, with an overpotential of only 281 mV and a Tafel slope of 96 mV dec^{-1}, and retains about 98% activity after 100 h. Gao et al. studied another metal-free electrocatalyst, with C_{60} as the electron acceptor deposited on the surface of SWCNTs [42]. It has been demonstrated to possess multifunctionality. With its excellent ability to catalyze ORR, OER, and HER at an extremely wide pH range, it outperformed Pt and RuO in electrocatalytic performance and stability. Nitrogen-doped carbon nanostructures, CN_x, promise bifunctional catalysis for both ORR and OER in acidic media [43]. While CN_x catalysts are less active than their commercial Pt/C and Ir/C counterparts, they have a lower combined overpotential for both reactions than the single-function commercial catalysts. Higher densities of pyridinic nitrogen sites in CN_x samples result in higher catalytic activities, and hence nitrogen functionalities are critical to optimize the performance. A nitrogen-doped graphite layer core-shell carbon catalyst developed by Zhu et al. has been able to show outstanding OER performance at 1.70 V at 10 mA cm^{-2} in an acidic environment with special reference toward the protection of the shell layer being useful for its durability and efficiency [44]. Together, these studies highlight the significant role that novel anodic material developments are playing toward achieving efficient WORs and point toward metal-free and carbon-based catalysts as promising alternatives toward the creation of sustainable energy technologies.

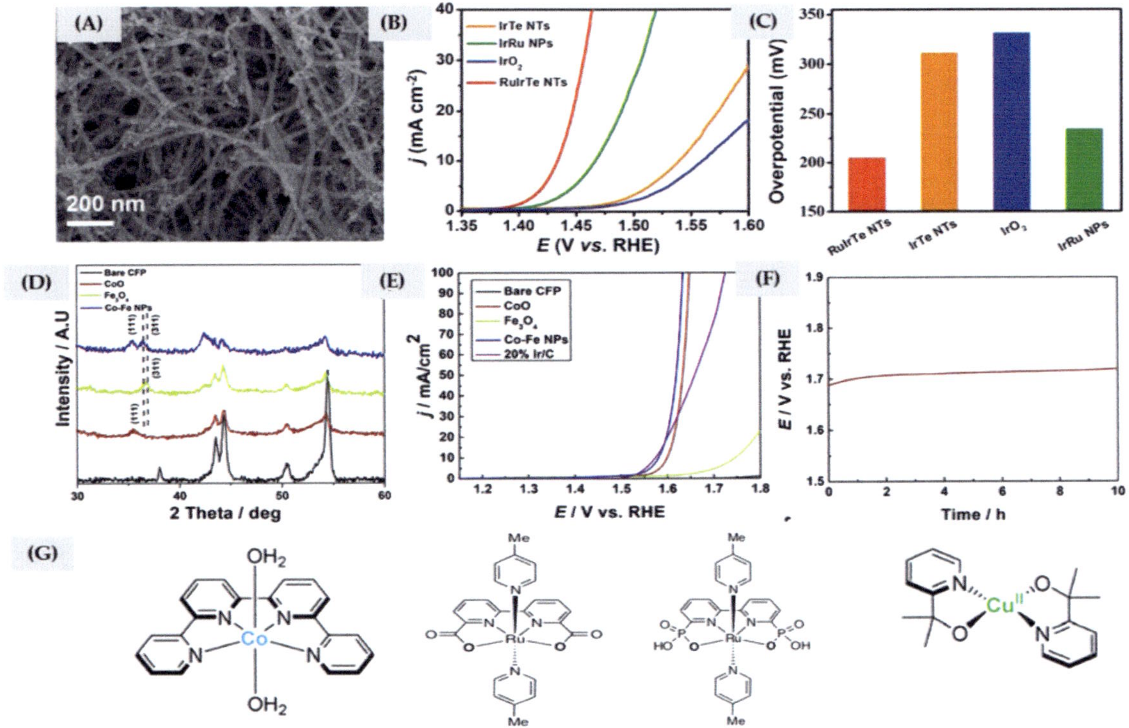

Figure 6.1 (A) SEM image of RuIrTe; (B) OER polarization curves of various catalysts in 0.5 M H_2SO_4; (C) Comparison of the overpotentials at 10 mA cm^{-2}; (D) XRD pattern of bare CFP, CoO, Fe_3O_4, and Co-Fe NPs (E) OER LSV curves for CoO, Fe_3O_4, Co-Fe NPs, bare CFP and Ir/C electrode in 1M KOH; (F) Chronopotentiometry stability testing at a constant current of 50mA cm^{-2} for OER; (G) Structures of molecular water oxidation catalysts. Reprinted from Refs. 18, 23, 33, and 38 from Royal Society of Chemistry, Elsevier, Springer Nature BV, and Royal Society of Chemistry, Copyright 2013, 2020, 2010, and 1972, respectively.

6.2.4 Challenges and Future Directions

Critical challenges abound in the development of anodic materials for OER when applied in real-world conditions for sustainable production and other energy conversion technologies. The high durability of such materials in harsh environments, such as acidic or alkaline media, coupled with low overpotentials and high electron transfer rates, are critical challenges. Traditional noble metal oxides, such as IrO_2 and RuO_2, have high catalytic activity but are expensive to produce and scarce, hence there is a need for a cheaper substitute without losing their performance. Non-noble metal catalysts, namely transition metal oxides, perovskites, etc., have been deemed promising for low-cost systems but still need stability enhancement and long-term performance, particularly in highly acidic or alkaline environments where they tend to degrade. Challenges in the search for active, metal-free carbon-based catalysts include optimization of surface activity and managing reductions in overpotentials. While the improvements through nitrogen doping and hierarchical structure have significantly improved durability and performance, carbon-based catalysts still need major tailoring of their activities toward extreme pH conditions.

The next generation of anodic materials for water oxidation is likely to emphasize designs with multifunctional material compositions and synergies achieving cost-effectiveness, strong catalytic activity, and stability over a wide pH range. Emerging approaches that may reduce overpotentials, and increase stability, include those in MOFs, heteroatom doping, and single-atom catalysis. For example, for carbon-based composites, namely polyaniline-carbon dot catalysts and the nitrogen-doped graphite structure, there is a belief that these will evolve continuously with better electronic configurations for structural stability to withstand conditions relevant to OER. Then, molecular catalysts in hybrid structures that support all three: oxygen reduction reaction (ORR)/ HER and OER-in materials may open up even more versatile all-in-one anodic materials. It would not only enhance the efficiencies of water-splitting technologies but also add to a broader landscape of clean and renewable energy sources by providing sustainable hydrogen production and oxidation of water.

6.3 Materials for Organic Oxidation Reaction

Hybrid water electrolysis with the replacement of OER with organic substrate oxidation is rewarding by overcoming the intrinsic drawbacks of conventional water splitting reactions by operating at less cell potential, enhanced kinetics, and overall cell thermodynamics with simultaneous production of hydrogen and platform chemicals. The scope of organic materials includes both aliphatic and aromatic alcohols, amines, and organic pollutants which encompass a wide range of functionalities [45]. Anodic materials play an important role in the electrochemical oxidation of organics, a trend gaining much popularity because of their effectiveness in water treatment and environmental decontamination along with the production of value-added products. These anodic materials are designed to endure aggressive environments while permitting effective organic pollutant degradation and targeting challenging industrial and municipal wastewater contaminants [46]. Dopants, composite structure, and sustainable material integration all reflect the state of the art in achieving effectiveness balanced with environmental concerns. As research continues, electrochemical oxidation will be an even more versatile and scalable solution in the pursuit of greener water treatment technologies.

6.3.1 Overview of Organic Oxidation Reaction

Hybrid water electrolysis represents promising alternatives for the sustainable production of hydrogen via the electro-oxidation of organic compounds into value-added chemicals [47]. Unlike the OER, the oxidation of organic molecules, such as aliphatic alcohols, for example, methanol and ethanol, aromatic alcohols such as benzyl alcohol, biomass-derived alcohols like HMF, lignin, and aromatic amines such as benzylamine, is more thermodynamically favorable and kinetically faster at lower potentials around 1.0 V vs. RHE [48–50]. In that way, the process allows not only pure hydrogen output but also valorizes organic compounds, generating valuable byproducts. The electrocatalytic oxidation of alcohols is of growing interest due to its applications in direct fuel cells, biomass-derived alcohol upgrading (e.g., methanol, ethanol, glycerol), and the synthesis of valuable chemicals. In hybrid water electrolysis, the lesser oxidation potentials of small aliphatic and aromatic alcohols reduce the total required voltage to produce pure hydrogen at the cathode with concurrent production of valuable byproducts such as aldehydes and carboxylic acids at the anode. Complete oxidation of primary aliphatic alcohols produces CO_2 that has high energy density. Partial oxidation yields valuable chemicals. However, stronger bonding of CO intermediate with the catalyst surface results in the impairment of their performance due to poisoning in the catalyst [51–53]. Weaker-bonding intermediates like formate and acetate minimize deactivation in the catalyst, as well as improve regeneration.

$$HCOOH + 4H^+ + 4e^- \leftarrow CH_3OH \rightarrow CO_2 + 6H^+ + 6e^- \quad (6.14)$$

$$CH_3COOH + 4H^+ + 4e^- \leftarrow CH_3CH_2OH \rightarrow CO_2 + 12H^+ + 12e^- \quad (6.15)$$

Complete and partial methanol oxidation occurs at potential 0.015 V vs. RHE and 0.03 V vs. RHE, and ethanol at –0.49 V vs. RHE and 0.09 V vs. RHE, respectively. This relatively recent development involves small organic alcohols, for example, benzyl alcohol ($PhCH_2OH$), as substrates undergoing anodic oxidation to make value-added benzoic acid at the anode and pure H_2 at the cathode. Because oxidation of benzyl alcohol requires only 0.48 V vs. RHE, it is relatively thermodynamically favorable relative to the OER value of 1.23 V [54].

Organic pollutants comprise thousands of carbon-based chemical compounds that are harmful both to the environment and human health. They are generally synthesized or industrial, agricultural, or domestic byproducts which are stable against natural degradation; hence they last long in the environment. Organic pollutants encompass hydrocarbons, polychlorinated compounds, pesticides, herbicides, phenolic compounds, polycyclic aromatic hydrocarbons, volatile organic compounds, surfactants, pharmaceuticals, personal care products, plasticizers and additives, nitro compounds, amines, persistent organic pollutants, etc., and can be a potential substrate for hybrid water electrolysis, thereby helping in the remediation of environmental contaminants [55].

6.3.2 Mechanism of Organic Oxidation Reaction

Electrocatalytic alcohol oxidation typically involves cleavage of the α-C–H bond in an alcohol molecule, RR′CHOH, where R or R′ can be hydrogen, aromatic, or aliphatic groups. The bond dissociation energy (BDE) of this C–H bond is lower than that of water's O–H bond, so alcohol oxidation is more thermodynamically and kinetically favored than the OER. A lower BDE (α-C–H) has been found to be correlated with a lower catalytic onset potential by Mondal et al., and benzyl alcohol presents the lowest BDE compared to methanol and ethanol, which further results in faster oxidation rates [56].

$$M^{n*} + 3OH^- \rightarrow M^{m+}OOH + H_2O + (m - n)e^-, m > n, \text{fast} \quad (6.16)$$

$$RR'CHOH_{sol} \leftrightarrow RR'CHOH_{ads} \quad (6.17)$$

$$RR'CHOH_{ads} + M^{m+}OOH \rightarrow RR'COH_{ads} + M(OH)_n + (m - n)e^- + H^+ \quad (6.18)$$

$$RCHOH_{ads} + H_2O + 3M^{m+}OOH \rightarrow RCOOH_{sol} + 3M(OH)_n \text{ If}(R'=H) \quad (6.19)$$

The oxidation mechanism follows a number of steps. In some catalysts, the breaking of the C–H bond is associated with the rate-determining step. For example, nanoparticles of $Ni_{12}P_5$ [57] showed step II (Eq. 6.17) as the rate-determining step (r.d.s.), while step III (Eq. 6.18) was shown in the NiFeMOF catalyst during methanol oxidation as the r.d.s. In step III, free radicals may recombine forming unwanted pinacol products. The pinacol formation is typical for secondary free radicals such as those formed from benzyl alcohol may be minimized by doing the reaction at higher potentials and ensuring there is enough metal oxyhydroxide present for complete oxidation. Alcohol-assisted oxidation processes present a potential method for industrial-scale applications since they produce valuable oxidized compounds at the anode and a pure stream of hydrogen.

Anodic oxidation of organic contaminants can be achieved through two primary mechanisms: direct and indirect oxidation [45]. In the mechanism of direct anodic oxidation, electron transfer is the primary process in which the pollutant is broken down directly when adsorbed onto the anode surface. This method functions on the transport of compounds and electron transfer at the electrode-solution interface and typically operates at potentials less than those required for oxygen evolution [58]. However, the most significant drawback of this technique is the electrode fouling through the formation of a polymer layer on the surface of the electrode which reduces decontamination efficiency and the lifespan of electrodes [59, 60]. For instance, oxidation at about 1.67 V causes electrode fouling due to a polymeric film that adheres to its surface, according to studies by Comninellis et al., who carried out experiments on the oxidation of phenol by direct electron transfer using boron-doped diamond (BDD) electrodes [60].

The indirect oxidation processes for mineralization and degradation are quite efficient. Mineralization means destroying organic pollutants; on the other hand, this happens with complete or partial destruction of the pollutant during indirect oxidation as against direct oxidation in which, since no intermediate was used in the reaction process, direct action would lead to only partial or low-effect destruction. Strong in situ generated oxidants applied here include hydrogen peroxide, ozone, persulfate, and active chlorine. "Active" anodes have low oxygen evolution potential (OEP) and offer higher oxides for selective oxidation, while "non-active" anodes are characterized by high OEP and tend to induce full combustion of organics up to CO_2. Most real systems contain both mechanisms in combination, and the anode type controls the pollutant degradation pathways [61, 62]. For example, the Pt anodes caused complete removal of dye color at direct and indirect oxidation conditions, while with BDD anodes, up to 87% of organic load was removed via structural breakdown of the dye.

6.3.3 Catalysts for Organic Oxidation Reaction

Electrochemical oxidation, facilitated by anodes, is an efficient method for breaking down persistent organic pollutants into less harmful compounds and upgrading selected organics into economically valuable compounds. They act as active sites where the degradation occurs through redox reactions by means of reactive oxygen species formed due to the process. A prominent role is played by materials commonly used in anodic construction, which include metal oxides, titanium dioxide, lead dioxide, and the latest BDD anodes, whose potentials for oxidation are highly possible, very stable, and capable of producing reactive species of oxygen

[63]. Developments in mixed metal oxides involving iridium- or ruthenium-based types of anodes offer enhanced catalytic efficiency while increasing their durability for being used on an industrial level. Special emphasis on the oxidation of aliphatic and aromatic alcohols, biomass-derived alcohols, amines, and organic pollutants will be the focus of the following section.

6.3.3.1 Precious metal-based catalysts

In studies by Zaman et al., Pt/C from waste plastic produced greater electrocatalytic activity for methanol and ethanol oxidation compared to commercial Pt/C catalysts, indicating the feasibility of using recycled plastic material as a precursor for catalyst production [64]. Ruthenium nanoparticles supported on nitrogen-doped carbon nanotubes have been excellent for benzyl alcohol oxidation with 100% selectivity and 95% yield of benzaldehyde by virtue of an optimal adsorption configuration that enhances the reaction rates [65]. SEM image of Ru-NPs@NCNTs has been shown in Fig. 6.2 A. LSV curves for the oxidation of benzyl alcohol over Ru-NPs@NCNTs and comparision of Faraday efficiency can be seen in Fig. 6.2 B, C. Puthiyapura et al. reported bimetallic catalysts, PtSn as an efficient catalyst material that accelerates alcohol oxidation efficiency through a bifunctional mechanism and further lowers the activation energy by increasing the oxidation rates associated with n-butanol isomers [66]. $Pd_{87}Cu_{13}/C$ showed excellent performance compared to Pd/C in the electrochemical oxidation of polyalcohols like ethylene glycol and glycerol, where it showed a substantial increase in the power density of alkaline direct liquid fuel cells in research by Munoz et al. [67]. $Pd(DBA)_2$ showed excellent catalytic activity for the electro-oxidation of various alcohols like ethanol and methanol, with good power densities in direct alcohol fuel cells [68]. High entropy alloys- nanoparticles catalyze the electro-oxidation of 5-hydroxymethylfurfural to value-added chemical showing selectively affected by alloy compositions. MnO_x, a promising alternative to anodic OER, converts HMF into the desired oxidized compound, producing higher yields of 53.8% FDCA when dissolved in acidic media [69]. Surface morphology and LSVs of a MnO_x anode were studied (Fig. 6.2 D, E). Conversion of HMF (%) and yield (%) of oxidation products obtained by using MnO_x anodes at 1.6 V versus RHE during the course of electrochemical oxidation of HMF is shown in Fig. 6.2 F.

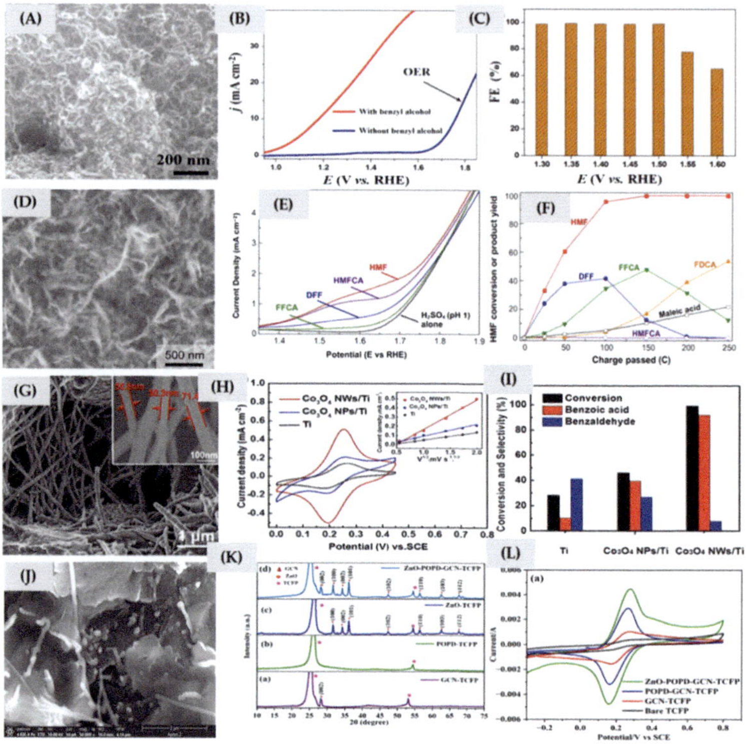

Figure 6.2 (A) SEM image of Ru-NPs@NCNTs; (B) LSV curves for the oxidation of benzyl alcohol over Ru-NPs@NCNTs; (C) Faraday efficiency (FE); (D) SEM image of a MnO_x film after annealing at 400°C for 2 h; (E) LSVs of a MnO_x anode obtained in a pH 1 H_2SO_4 solution without (black) and with 20 mm HMF (red), 20 mm DFF (blue), 20 mm HMFCA (purple), and 20 mm FFCA (green) at room temperature at a scan rate of 5 mVs^{-1}; (F) Conversion of HMF (%) and yield (%) of oxidation products obtained by using MnO_x anodes at 1.6 V versus RHE during the course of electrochemical oxidation of HMF at 60°C in a pH 1 H_2SO_4 solution containing 20 mm HMF at various amounts of charge passed; (G) High-magnification SEM image of the Co_3O_4 NWs/Ti electrocatalytic membrane; (H) CV curves of the Ti membrane and Co_3O_4 NWs/Ti and Co_3O_4 NPs/Ti membranes at a scan rate of 4.0 mV s^{-1}, Inset plot is current density versus $v^{-1/2}$ (0.3–4.0 mV s^{-1}); (I) Effect of different catalysts on conversion of BA and selectivity to benzoic acid and benzaldehyde in the ECMR at 20°C; (J) FESEM image of ZnO-POPD-GCN-TCFP; (K) X-ray diffraction analysis of (a) GCN-TCFP, (b) POPD-TCFP, (c) ZnO- TCFP, and (d) ZnO-POPD-GCN-TCFP electrodes (L) Cyclic Voltammogram of bare TCFP, GCN-TCFP, POPD-GCN-TCFP and ZnO-POPD-GCN-TCFP electrodes measured in 25 mM $K_3[Fe(CN)_6]$ in 1M KCl. Reprinted from Refs. 65, 69, 74, and 77 from the American Chemical Society, John Wiley and Sons - Books, American Chemical Society, and Elsevier, Copyright 2022, 2008, 2017, and 2024, respectively.

PtRu-based anodes developed by Beliaeva et al. were examined with regard to lignin electrolysis where the results were that for hydrogen production, the rates obtained were low due to significant reaction intermediates poisoning the catalyst [70]. The study challenges the oxidation of compounds related to lignin and calls for better designs of the catalyst. Dynamic Potential Cycling Technique, strongly reduces the deactivation of noble metal catalysts in the electro-oxidation of ethylene glycol derived from polyethylene terephthalate, thus allowing higher catalyst life and activity together with efficient removal of the surface oxides during the reaction [71].

Platinum is a widely utilized catalyst in electrochemical amine oxidation reactions, and recent studies by Peng's research group have explored its effectiveness in the electrochemical dehydrogenation of ethylamine (EDH) [72]. They investigated several factors, namely substrate concentration, with an optimum at 2 M, electrolyte pH, with the range from 0 to 3 M KOH, optimum at 1 M, and the temperature of 276 to 303 K, whereby 303 K was efficient. It was an exploratory work involving two types of electrochemical cells: an anion exchange membrane was used for one type, while the other type did not apply an anion exchange membrane. The processed results, at a hydrogen production rate of 309 mmol/h with 97% Faradaic efficiency from the cathode, as well as 100% Faradaic efficiency for the production of acetonitrile at the anode at a low cell voltage of 0.6 V, prove the efficiency of such a process and its use for hydrogen storage applications (Fig. 6.3 A–C).

Electrochemical oxidation is a treatment for wastewater generated in petrochemical processes, which typically contain high COD ranging from 20,450 to 52,300 mg/L. The experiment used a ruthenium-iridium mixed metal oxide (Ru-MMO) anode in a parallel plate reactor in direct anodic oxidation with a COD reduction efficiency of 79% in just 6 minutes, while it consumed 117 kWh/m^3 of energy [73]. Pseudo-second-order kinetics with high coefficients ($R^2 > 98\%$) were demonstrated during the kinetic analysis of these systems, indicating a successful efficiency of this process, especially at acidic pHs less than 5, while showing a much lower effectiveness in terms of COD removal under alkaline conditions.

6.3.3.2 Non-precious metal-based catalysts

Metal oxides are very important solid catalysts in heterogeneous catalysis that have both acid-base as well as redox natures and have been of tremendous importance as active phases and supports. Fine chemical synthesis is going to favor electrochemical oxidation, due to the effectiveness of newer anode materials in obtaining good efficiency and high selectivity. A novel Co_3O_4-based nanowire (NW)-coated titania membrane developed by Yin et al. [74] exhibited over 99% conversion and about 92% selectivity for benzoic acid from benzyl alcohol at a very low reaction temperature of 0°C (Fig. 6.2 G–I). Cox et al. developed $NiCo_2O_4$ anode for selectively oxidizing primary alcohols to carboxylic acids and secondary alcohols to ketones with stable performance under alkaline conditions and with satisfactory current efficiencies [75]. Panizza et al. studied the oxidation of *p*-nitrophenol by Pb/PbO_2 and 2-naphthol through a BDD electrode, with the latter having the fastest oxidation rate and current efficiency [76]. These anode materials do indeed promote selective electrochemical oxidation processes which were conventional. A broad range of metal oxide anode materials are used in the electrochemical oxidation of organic compounds for sustainable production purposes. Recently, electropolymerized deposition of ZnO nanoparticles over the Toray carbon fiber paper coated by graphitic carbon nitride (gCN(H)) has been established by Rodrigues et al. [77] for FA oxidation into furfural (Fig. 6.2 J, K). Regarding its great electrocatalytic activity and stability, the ZnO-POPD-GCN-TCFP electrode promises to serve as an exciting new example for green chemistry methods (Fig. 6.2 L). The alcohol oxidation reaction took place at a low potential of 1.42 V vs. RHE and a current density of 10 mA/cm^2, with Co_2NiO_4 displaying excellent catalytic activity toward various straight-chain primary alcohols [78]. The product distribution was dependent on the chain length of the alcohol, and short chains resulted in aldehydes and long chains in acids. Furthermore, Jud et al. demonstrated a new type of electrochemical oxidation using a heterogeneously dispersed mediator based on NiOOH, which resulted in a maximum of 77% yield for the oxidation of nicotinic acid from nicotinyl alcohol in the flow setup [79]. Apart from that, ferberite $FeWO_4$, has been identified by Adak et al. for potential applications as an aqueous as well as a possible ethanol oxidation anode; even at 10 mA/cm^2, at only 282 mV overpotential [80]. These findings outline the potential of these metal oxide and metal-based anode materials for efficient electrochemical oxidation reactions and renewable energy applications.

Electrocatalysts for HMF oxidation have a large potential in sustainable chemical synthesis. Lu et al. explored Co-based spinel oxides, especially Co_3O_4, which are effective due to mixed valences and specific active sites: Co^{3+} octahedral sites enhance reactivity by forming cobalt oxyhydroxide, while Co^{2+} tetrahedral sites can assist HMF adsorption through Lewis acidity [81]. Iridium single atom doping of Co_3O_4 strengthens the interaction more with the C=C bonds of HMF, while further boosting catalytic performance. The oxidation activity is also four times increased when the Co^{2+} is substituted by Cu^{2+} in $CuCo_2O_4$. Barwe et al. reported high faradaic efficiency and yield for the oxidation of HMF to 2,5-furandicarboxylic acid (FDCA), which are also obtained on nickel foam modified with nickel boride (Ni_xB), a low-cost catalyst offering economic solutions toward green polymers [82].

Honorato et al. synthesized a P-NiFeCo/NC catalyst with a Ni, Fe, and Co nanoalloy core encased in a phosphidated nitrogen-doped carbon shell, enabling lignin oxidation at a low potential of 0.2 V vs. Hg/HgO, resulting in up to fivefold turnover enhancement and significantly higher hydrogen production compared to pure KOH [83]. Meanwhile, Bateni et al. showed that β-PbO_2/MWNTs nanocomposites exhibit low-cost, high-surface-area catalysis for lignin depolymerization and produce valuable byproducts such as vanillin [84]. In addition, Zhou et al. reported that MnCoOOH can effectively break the C(OH)-C bonds in lignin derivatives and transform them into carboxylates with good yields and stable operation, thereby providing a promising avenue for both the petrochemical industry and biomass conversion [85]. It succeeds in depolymerizing ethanol organosolv lignin, which was from different biomass sources, at a unique oxidation product depending upon the lignin. The overall anode materials have potential use in efficiently oxidizing lignin toward renewable energy as well as sustainable production of chemicals.

One of the most used polymers, PET, has slow degradation and some environmental concerns related to the persistence of waste. Behera et al. reported a cobalt-based 1D coordination polymer as an electrocatalyst for oxidizing ethylene glycol and PET upcycling with a low onset potential of 1.27 V vs. RHE. With 100% terephthalic acid (TPA) and about 80% selectivity toward potassium diformate (KDF) as isolated products, along with hydrogen co-production, were achieved [86]. Mechanistic analysis indicates the existence of a side reaction that produces carbonate from the oxidation of formate by circumventing the oxalate pathway. Such studies would be helpful in optimizing the upcycling technologies for PET, increasing the formate and KDF yield. Taghaddosi et al. [87] reported nitrogen-doped hollow carbon nanospheres synthesized by carbothermal treatment of polyaniline-co-polypyrrole (PACP) shells and confirmed by XRD (Fig. 6.3 E). This work of N-doped HCNSs provides a core structure for not one but two species of nano-scaled forms of nickel(II) oxide; in the nanosheet shape named NiO-NS and the NiO-NWs, in which the morphology as well as the nanostructured framework could be defined simply by just hydrothermal plus calcinations (Fig. 6.3 D). The NiO-NWs@N-HCNSs/GCE electrode showed highly enhanced electrochemical properties when acting as electrocatalysts in the oxidation of hydrazine, such as significant increase in electrochemical active surface area (ECSA), catalytic activity, reduced overpotential, and improved stability compared with other electrodes such as GCE, N-HCNSs/GCE, NiO-NPs/GCE, and NiO-NSs@N-HCNSs/GCE (Fig. 6.3 F). The performance of anodes Ti/RuO_2-SnO_2 and Ti/RuO_2-IrO_2 prepared by thermal decomposition during a separate experiment of aniline oxidation were investigated by Zhu et al. in different electrolytes: 0.05 M Na_2SO_4, 0.05 M NaCl, and both with 0.005 M $FeSO_4$ at different current densities [88, 150]. The experiments indicated that aniline is broken down more efficiently using the Ti/RuO_2-SnO_2 as an anode, with up to 64.2% maximum TOC removal efficiency recorded at 40 mA cm^{-2} in electrolyte Na_2SO_4. The average MCEs of 1.6% and 1.51 kWh (g TOC)$^{-1}$ were reported in ECTOC for all the experiments. Comparatively, the Ti/RuO_2-IrO_2 anode exhibited the same MCE but 1.95 kWh (g TOC)$^{-1}$ ECTOC to achieve a TOC removal efficiency of 63.1% in NaCl electrolyte, which indicates the crucial importance of anodic material and composition of the electrolyte on achieving electrocatalytic efficiency (Fig. 6.3 G–I). Cobalt oxalates grown on cobalt foam as pre-catalysts are electrochemically activated to efficient electrocatalysts for both benzylamine oxidation reaction and HER [89]. They achieve

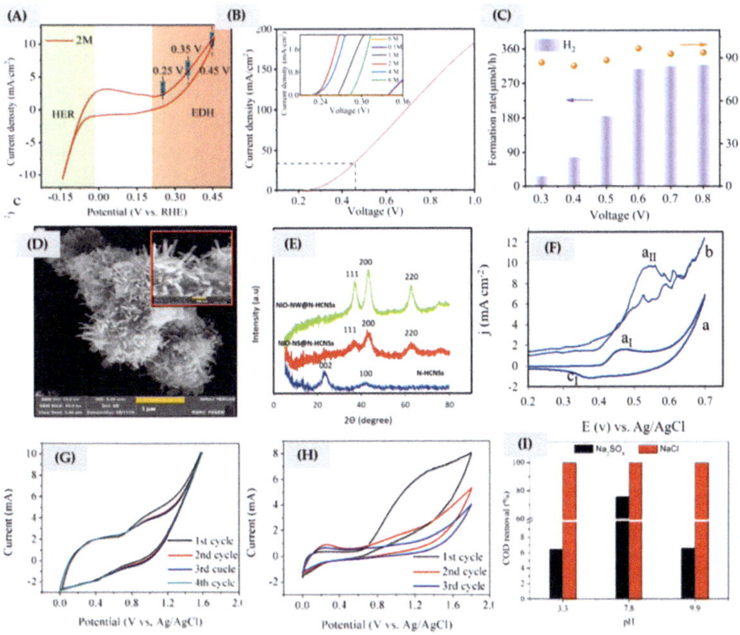

Figure 6.3 (A) Representative CV, normalized to the electrode geometric area, measured on the Pt/C catalyst in Ar-protected 0.1 M KOH + 2M ethylamine aqueous electrolyte at scan rate of 10 mVs^{-1}. (B) Polarization curve of membrane-less ethylamine electrolysis with various ethylamine + 1 M KOH. Inset: Polarization curves obtained with different concentration of ethylamine, zoomed in the low voltage range at scan rate of 10 mVs^{-1}. (C) The measured rate and faradaic efficiency of H_2 generation from the cell cathode at different cell voltage, 1 M KOH + 2 M ethylamine electrolyte is used in experiments. (D) FE-SEM image of NiO-NWs@N-HCNSs with 1 μm magnification. Inset Images magnifications: 500 nm. (E) XRD patterns for N-HCNSs, NiO-NSs@N-HCNSs and NiO-NWs@N-HCNSs. (F) CV curves of NiO-NWs@N-HCNSs/GCE in 0.2 M NaOH (a) in the absence and (b) the presence of 30 mM hydrazine hydrate. (G) cyclicvoltammograms for Ti/RuO_2 and (H) Ti/Ni-Sb-SnO_2 anode in 1000 mg/L NaCl (v = 0.01 V/s, pH = 6.5, aniline concentration was 100 mg/L aniline. (I) Influence of pH on COD removal efficiency in 0.05 mol/L Na_2SO_4 and 0.05 mol/L NaCl, respectively, and the reaction time were 6 h and 1 h (experimental conditions: Ti/RuO_2 was used as the anode, plate distance was 1 cm). Reprinted from Refs. 72 and 87 from Elsevier, and Ref. 150 from Springer Nature BV, Copyright 2023, 2020, and 1994, respectively.

a high current density of 75 $mA\,cm^{-2}$ at the voltage of 1.95 V in a two-electrode configuration, with the high Faraday efficiencies of both BOR (90.0 %) and HER (90.0 %) and the satisfactory yield of benzonitrile (76.8 %). Sergienko et al. reported Mo-doped Mn_xO_y-coated TiO_2 nanotube array NTA anode with high efficiencies and performance in the electro-oxidation of persistent organic contaminants without any formation of chlorinated byproducts when compared to commercial Ti/IrO_x-Pt [90]. The anode allowed for efficient organic pollutant degradation while promoting long-term stability through the generation of oxygen vacancies. Additional work added a sulfur-doped TiO_2 NTA interlayer for a PbO_2 anode and achieved nearly complete degradation of methylene blue (20 mg/L) within 90 min [91]. The Al-doped SnO_2-Sb anode on a porous titanium substrate had a high OEP of 1.57 V, leading to 69.2% phenol mineralization within 2 hours, thus making it a good candidate for removing persistent organic contaminants from industrial wastewater [92].

6.3.3.3 Molecular catalysts

Synthesis and use of molecular electrocatalysts for a range of oxidation reactions have recently gained widespread interest, particularly in the valorization of biomass, its derived alcohols, and upcycling of plastic waste-derived alcohols. Wei et al. reported that a methanol-assisted Cu-Co-based water-splitting process using the MOF-2D provided an electric cell potential value of 1.365 V and 100% faradaic yield efficiency toward hydrogen production [93]. It includes entirely oxidative methanol transformation toward CO_2 by adopting any kind of intermediate steps for conversion into aldehyde along with carbon monoxide end-product. Kuiry et al. studied iron complex mediators, a highly efficient electrochemical method for the selective oxidation of alcohols to aldehydes and ketones using a biomimetic iron complex in an undivided electrochemical cell [94]. This process was based on inexpensive carbon and nickel electrodes, and the high-valent $Fe^{V}(O)$ species was responsible for the oxidation. Solid-state catalysts were utilized in transforming HMF into the building block for sustainable polymers, as FDCA. It involved designing several solid-state catalysts that were mainly aimed at utilizing a ruthenium complex anchored onto mesoporous electrodes for its application in good yield, with an 85% FDCA, and without substrate degradation, as reported by Buhler et al. [95]. A carbon paper-supported nickel–iron layered double hydroxide as reported by Qi et al. converted HMF to target the product 95.35%

with the best-possible stability and good faradaic efficiency at 1 M KOH [96]. Liu et al. reported the electrocatalytic performance of a ternary MOF on nickel foam, so that at 1.35 V, the current density is as high as 100 mA/cm^2, and nearly 100% Faradaic efficiency is reached in FDCA production [97]. Electro-oxidation of lignin has been done using different transition-metal complexes, used as mediators. In work by Xie et al., $Fe(bpy)_3$ is shown to be highly effective for Cα–Cβ cleavage and Cα-carbonylation [98]. Selective oxidative transformation of primary amines to nitriles involved an oxidant-free method with t-Ni/Co MOF as a suitable anodic electrocatalyst, producing such products at low energy loss [99]. The nickel macrocyclic complex-modified carbon electrodes, prepared through anodic oxidation, are excellent electrocatalysts for the oxidation of alcohols and phenolic compounds, with a chance of irreversible passivation during catalytic reactions.

6.3.3.4 Carbon-based support materials as catalysts

The research has witnessed more momentous developments in the electro-oxidation of alcohols and amines over carbon-based anodic materials. Studies thus focus on the efficiency, recyclability, and optimization of catalysts, where specific attributes of carbon materials contribute toward optimal reaction outcomes. NHPI has been used in a very simple setup to lay down the protocol for an electro-oxidation reaction of alcohol to aldehyde and ketone by Zhao et al. [100]. Such a process can be efficient with less stringent conditions applied and works on multiple functional groups without any sort of degradation. The Ni–Ag alloy catalysts on rGO studied by Santos et al. show improved activity for ethanol, ethylene glycol, and glycerol oxidation, and the $Ni_{66}Ag_{34}$/rGO alloy gives the best performance and stability [101]. Sustainable carbon materials from Kraft black liquor are prepared and used as nickel-based electrocatalyst supports by Labat et al., and these show promising results for the oxidation of ethanol, attributed to their micro-mesoporous structure as well as the uniform dispersion of nickel nanoparticles [102]. gCN(H) has been suggested as a highly stable, corrosion-resistant semiconductor and as a potential alternate support for catalysts in polymer electrolyte membrane fuel cells (PEMFCs) by Graczyk et al. [103]. Strong metal–support interactions lead to substantial enhancements in the electrocatalytic activity of the Ni-doped gCN(H) toward alcohol oxidation. The electrocatalytic activity toward alcohol oxidation by the Pd-polypyrrole/nitrogen-doped graphene ternary nanocomposite is high and durable in an alkaline environment, significantly outperforming the state-of-the-art Pd/C catalysts [104]. Dakshinamoorthy reported poly(sulfonyldiphenol/cyclophosphazene/benzidine)–graphene oxide (poly(SDP/CP/BZ)–GO) composite supports for platinum–copper nanoparticles that show excellent catalytic activity toward alcohol electro-oxidation in direct alkaline fuel cells with high power densities [105]. Hybrid water electrolyzers thus come out superlatively in the electro-oxidative conversion of benzylamine to benzonitrile as an excellent illustration. In this regard, the catalyst Ni_2Co-LDH/rGO notably is credited with yielding enhanced current density and faradaic efficiency compared to conventional water electrolysis.

The electrochemical oxidation of organic pollutants has been increasingly recognized as an efficient methodology in the treatment of wastewater from industrial origins, particularly through advanced oxidation processes, such as electro-Fenton. A recent study by Abbas et al. has mainly focused on the development and application of carbon-based anodic materials, which have established significant advantages in pollutant degradation [106]. In the electro-Fenton process, a hybrid PbO_2 anode electrodeposited on a graphite substrate, coupled with a carbon fiber cathode, doped with graphene, showed excellent efficiency for the removal of organic pollutants like phenol. COD was removed by 81.23% after 6 h of electrolysis. Low-cost composite electrodes consisting of graphene nanoplatelets (GNP) and metal oxides (TiO_2, SnO_2) supported on stainless steel substrates have been found to be effective in the degradation of phenol-based dyes, as reported by Ducic et al. [107]. The major degradation products were oxidized aromatic compounds, with mineralization rates at about 30–35%. A porous carbon felt anode (CF/b-PbO_2) was more efficient than the conventional planar graphite anodes (G/b-PbO_2) for electrochemical degradation of diuron, with faster removal rates and lower energy consumption [108]. For the treatment of oil sands process water (OSPW), graphite anodes show effective removal of classical naphthenic acids and aromatic compounds at low current density, though DSA requires higher current density [109]. Nitrogen and boron-doped graphenic sponges removed persistent organic contaminants with high removal rates [110]. Direct electron transfer, ozone generation, and hydroxyl radical formation realized the synergy for these materials to work effectively. It is reported that electrochemical oxidation of toluene to benzaldehyde involved carbon-based electrodes, particularly graphite rods, to ensure high selectivity [111]. It was observed that the high surface area and defect sites at the electrodes prevent over-oxidation of the target product.

6.3.4 Challenges and Future Directions

Anodic materials for organic oxidation reactions, especially the oxidation of alcohols, amines, and organic pollutants, are a competent area with wide scope. Challenges need to be addressed in the optimization of catalyst stability and selectivity under varying operational conditions such as pH and temperature. There are many precious metal-based catalysts that, even though effective, are prone to poisoning by reaction intermediates, primarily in complex substrates such as lignin and various organic pollutants. Some promising progress was reported on non-precious metal catalysts and advanced composite materials; their catalytic performance, however, usually lags behind that of their precious counterparts. Scalability issues are another aspect of novel materials and processes, since most promising laboratory-scale solutions have not easily been implemented at the industrial level. Besides, major energy consumption reduction is still urgently required in most electrochemical processes during wastewater treatment involving COD-containing water.

Future aspects of this research area might involve designing and synthesizing new anodic materials that offer excellent electrocatalytic properties. High-entropy alloys, MOFs, and hybrid nanostructures are involved in it, where the kinetics of electron transfer are improved, along with the active surface area and cost-effectiveness. Further help can be received from advanced characterization techniques and computational modeling to get a deeper understanding of the electro-oxidation mechanism at the most basic level, and thereby guide the catalyst design. Interdisciplinary approaches involving materials science, electrochemistry, and chemical engineering can better facilitate the development of more efficient anodic systems.

In this respect, future work should focus on sustainability in the sense that ways to make use of waste feedstocks as precursors for the catalyst or as a feed to the electrochemical systems need to be developed, and then the oxidation processes would support the more general aims of circular economy and green chemistry. A greater emphasis on the operational stability, reusability, and adaptability of anodic materials will be critical to enabling their exploitation for organic pollutant treatment and alcohol/amine valorization.

6.4 Materials for Ammonia Oxidation Reaction

The role of ammonia oxidation in the development of clean energy and commercially useful products of nitrogen is highly important, especially in quite a number of industrial and environmental contexts [112]. It could, for instance, be exploited in fuel cells. It would then deliver electricity based on a potentially carbon-free resource. It is important in wastewater treatment plants that remove nitrogenous pollutant compounds that might unfavorably affect the environment where they are freely released. The process also promotes sustainability as ammonia converts to harmless nitrogen gas, hence reducing ammonia-related pollution and leading to greener chemical industries. Various catalysts have been developed for oxidizing ammonia, from Pt-based catalysts to non-noble metal catalysts, molecular catalysts, and carbonaceous materials, by tuning the synthetic procedure, morphology, surface area, and various electrochemical parameters in optimizing its yield of the desired product [113]. An attempt has been made to give a concise overview of the ammonia oxidation reaction (AOR), its mechanism, and the development of selected anodic materials in this regard in the following section.

6.4.1 Overview of AOR

The dominant source of energy supply, at present, continues to be fossil fuels: coal, petroleum, and natural gas. Since these resources are non-renewable, their limited availability poses the risk of an energy crisis [114]. This has spurred intense efforts toward developing green, safe, and renewable sources of energy, such as wind, solar, and tidal power. Generally, these energy sources are self-sustaining but pose challenges for direct application. Therefore, there is a need to find appropriate secondary energy carriers and develop them up to optimal levels of production. Of all the options evaluated, hydrogen stands out as one of the most economical and clean, has zero carbon emissions, is readily available in huge quantities in the ground, and has the world's highest energy density at 120 MJ/kg. However, safety issues regarding its storage and transportation have capped the use of hydrogen in practical applications [115, 116].

To overcome these challenges, ammonia has been suggested as a promising hydrogen carrier. Ammonia offers several advantages, such as a high energy density, carbon-free profile, and ease of production, storage, and transportation, supported by an existing global infrastructure. Its theoretical charge efficiency is comparable to that of methanol, making it an attractive alternative fuel [117, 118]. Despite concerns about its toxicity, established research in ammonia handling and use has mitigated these issues, reinforcing ammonia's viability as a clean energy source. Additionally, ammonia is much cheaper as a hydrogen carrier, as it can be liquefied at a reasonably modest pressure close to room temperature, whereas hydrogen gas requires 690 bar of pressure at 15°C. It also does not emit greenhouse gases from consumption and has a higher energy density in liquid form, at 11.5 MJ/kg, compared to liquid hydrogen, which is 8.5 MJ/kg. Besides, ammonia leaks become more detectable, even at 1 ppm concentrations, allowing early response. Another fact is that ammonia is already produced and utilized throughout the world in huge quantities: from the pharmaceutical and fertilizer sectors to the chemical industry's annual output of 176 million tons. This infrastructure makes ammonia a practically feasible source of energy [119, 120].

The electrochemical AOR, therefore, is key to unlocking ammonia for energy uses, be it for the generation of hydrogen through ammonia splitting or electricity through direct ammonia fuel cells (DAFCs), as DAFCs produce water and nitrogen gas only during operation [121]. In addition, theoretically, ammonia fuel cells possess advantages over hydrogen fuel cells in energy density and easier fuel storage. The direct electro-oxidation of ammonia offers significant advantages but is hindered by the current limitations of electrocatalysts, which struggle with slow reaction kinetics and high costs, making large-scale applications challenging. The key issues include the sluggish nature of AORs, which require high overpotentials and lead to efficiency losses. To address these challenges, research has focused on designing electrocatalysts that are cost-effective, highly active, and stable over time. Strategies include alloying platinum (Pt) with cheaper metals like Ir, Ru, and Rh to enhance synergy, increasing the electrochemical activity by optimizing catalyst surface structures, developing Pt-free alternatives, and exploring supportive materials to boost overall performance.

6.4.2 Mechanism of AOR

Understanding the fundamental reaction mechanism of AOR is essential for developing and designing optimum AOR catalysts [122]. In acidic electrolytes, ammonia appears as a positively charged NH_4^+, which is not very good at forming chemical bonds onto electrode surfaces. The ammonia might even be repelled by the positive catalyst surface at the oxidation potential itself, leading to less efficient AOR processes. As a result, AOR occurs in an alkaline medium where ammonia adsorbed onto the catalyst surface reacts with OH^- within the electrolyte during electro-oxidation. Ideally, this is a six-electron pathway with a potential as low as 0.056 V vs. RHE according to Eq. 6.20. However, the reaction is also accompanied by unwanted side products such as hydrazine and nitrogen oxides. Two mechanisms for the AOR reaction pathways involving an alkaline medium have also been suggested: continuous dehydrogenation mechanism and partial dehydrogenation mechanism.

Overall reaction equation:

$$2NH_3 + 6OH^- \rightarrow N_2 + 6H_2O \tag{6.20}$$

Oswin and Salomon (O-S) mechanism:

$$^*NH_3 + OH^- \rightarrow {}^*NH_2 + H_2O + e^- \tag{6.21}$$

$$^*NH_2 + OH^- \rightarrow {}^*NH + H_2O + e^- \tag{6.22}$$

$$^*NH + OH^- \rightarrow {}^*N + H_2O + e^- \tag{6.23}$$

$$2{}^*N \rightarrow {}^*N_2 \rightarrow N_2 \tag{6.24}$$

Gerischer and Maurer (G–M) mechanism:

$$^*NH_3 + (3-x)OH^- \rightarrow {}^*NH_x + (3-x)H_2O + (3-x)e^- (x = 1,2) \tag{6.25}$$

$$^*NH_{x1} + {}^*NH_{x2} \rightarrow {}^*N_2H_y \; (y = x1 + x2) \tag{6.26}$$

$$^{*}N_2H_y + yOH^- \rightarrow {}^{*}N_2 + yH_2O + ye^- \quad (6.27)$$

$$^{*}N_2 \rightarrow N_2 \quad (6.28)$$

In 1960, Oswin and Salomon proposed, based on their experimental studies of ammonia oxidation on platinum black, that ammonia undergoes continuous removal of three hydrogen atoms at the electrode surface to form nitrogen (N) atoms [123]. These nitrogen atoms then couple to form nitrogen molecules (N_2), which is known as the continuous dehydrogenation mechanism or the O–S mechanism (Eqs. 6.21–6.24). This suggests that at high current density, the combination of two *N atoms to form N≡N is regarded as the critical step in AOR, while at low current density, the formation of *NH is the key factor in control of the reaction rate.

Gerischer and Maurer proposed an alternative partial dehydrogenation mechanism in the early 1970s. There, on the catalyst surface, ammonia is first dehydrogenated to form species NH_x with x = 1, 2. These dimerize into hydrazine-like structures, $^*N_2Hy$ with y = 2, 3, or 4 that are then smoothly oxidized to N_2 after converting to $^*N_2$. This mechanism, referred to hereafter as the G–M mechanism (Eqs. 6.25–6.28). It involves very strong chemisorption of nitrogen atoms on the catalyst surface that does not allow further dimerization [124].

Beyond N_2, the AOR products also include NO_2^- and NO_3^-, produced commonly due to overlapping electrochemical windows of AOR and water oxidation. While there is no generally accepted explanation for producing these oxygenated nitrogen species, pathways through which they are formed during AOR should be elucidated.

6.4.3 Catalysts for AOR

AOR anodic materials, significantly, play a crucial role in enhancing both the efficiency and selectivity in electrochemical processes. Traditional favorites of noble metals are platinum-based materials due to their extraordinary catalytic activity and stability; the mentioned metals and their alloys catalyze direct ammonia oxidation to nitrogen with little byproducts. However, due to the high prices of these materials, Ni-based catalysts and Fe-based electrodes have gained more popularity and have an economic advantage. Alternative materials generally have extraordinary properties, including higher corrosion and degradation resistance, which provide for the successful oxidation of ammonia (AO) under different pH levels. Yet another relevant development has been toward the hybrid and molecular catalytic systems, which are either transition metal complexes or carbon-based materials [122]. Such optimization, in AOR, can now be improved further in terms of superior electron transfer and toward products such as nitrates or dinitrogen through high selectivity.

6.4.3.1 Pt group metals

A series of research studies represent a discussion on the progress made in the development of anode materials for the AOR to be carried out with superior performance, lower cost, and increased stability in ammonia-fueled electrochemical systems. Pt is widely used as an anode material for AORs owing to its stability and excellent electronic conductivity in harsh electrochemical conditions. Bimetallic PtZn nanoparticles synthesized by Matin et al. using a glycerol-assisted polyol method exhibited remarkable electrocatalytic performance, which surpassed that of Pt/C by the removal of oxygenated species on Pt surfaces [125]. Further, Rahardjo et al. synthesized Ni and Cu oxides decorated with Pt nanoparticles that enable strong electron transfer in AOR with successful NH_3 oxidation both at +1.1 V and +0.23 V vs. Hg/HgO on a Ni- and Cu-supported sites, respectively, demonstrating efficient ammonia conversion with high selectivity toward nitrogen gas up to 60% [126]. PtIrNi alloy on CNT-COOH/SiO_2 support significantly reduces the onset potential, hence enhances the activity of AOR in alkaline conditions [127]. XRD patterns of the synthesized catalysts Pt/SiO_2-CNT-COOH, commercial PtIr/C, PtIr/SiO_2-CNT-COOH, PtNi/SiO_2-CNT-COOH, $PtIrNi_1$/SiO_2-CNT-COOH, $PtIrNi_2$/SiO_2-CNT-COOH, and $PtIrNi_3$/SiO_2-CNT-COOH has been shown in Fig. 6.4 A. CV curves for different catalysts and AOR activity comparison for the catalysts at 0.5 V vs. RHE can be seen in Fig. 6.4 B and C. Siddhart et al. studied Ir-decorated Pt nanocubes and $PtSnO_2$/C that further enhance AOR by lowering the energy barriers for NH* formation due to Ir and by preventing nitrogen poisoning due to SnO_2, thus increasing peak current densities [128]. Chan et al. investigated nanocubes of Pt alloyed with transition metals, Fe, Co, Ni, or Zn, which exhibit a significantly higher activity for AOR, where PtZn shows particularly enhanced mass and specific activities to as much as a factor of 1.6–1.8 times those for neat Pt nanocuboids through a bifunctional mechanism which enhances activity while suppressing oxophilicity [129]. Pt nanoparticles on 3D CNT films meant for ammonia electrolysis also give a high current density and lower overpotentials by an increase in the electrochemical surface area [130]. Wang et al. studied a plasmon-assisted light-driven Au-Pt system that exhibits significantly enhanced kinetics for ammonia oxidation; but significant progress in methanol-modified non-aqueous electrolytes prevents nitride poisoning at the surface of Pt, leading to stable, high-density ammonia electrolysis [131]. Katayama et al. studied on platinum-based catalysts doped with rare earth oxides (REOs) such as CeO_2, Y_2O_3, La_2O_3, and Sm_2O_3 and have shown that REOs significantly improved the activity of AOR [132]. The CeO_2-modified platinum showed the highest improvement, with the peak current density being 3.5 times higher than that of the unmodified platinum. PtRh alloys prepared by borohydride reduction and studied for a range of Pt ratios 90:10, 70:30, and 50:50 presented the best performance at a ratio of 90:10 [133]. This structure yielded a power density that was 60% higher than the pure Pt value. The effect was ascribed to the electronic synergies of Pt-Rh, diminished adsorption strength of inhibitory intermediates, and the interaction between Pt and Rh. On the other hand, ammonia-containing effluents can also be treated electrochemically through Pt and Ni electrodes; in fact, the electrode of Ni proved highly electroactive for the electro-oxidation of ammonia. Applications of these novel anode materials promise bright futures for ammonia-based energy application and power generation.

6.4.3.2 Non-noble metal catalysts

Ammonia electrolysis depends on rather expensive noble metal anodes such as platinum, ruthenium, and iridium; therefore, it involves high cost, reduced scalability, and less feasible in hydrogen storage applications. In recent times, the search has been extended to inexpensive alternatives, such as iron and nickel, which might be promising as a substitute in spite of complex anodic behavior in liquid ammonia. Iron is susceptible to corrosion under certain conditions; however, when sufficient overpotential is applied, it forms a protective iron nitride film that allows efficient dinitrogen evolution at over 95% efficiency similar to that of platinum [134]. Studies with nickel electrodes show that ammonia oxidation is pH-dependent and occurs mainly at pH values above 7, where ammonia (NH_3) is oxidized rather than ammonium (NH_4^+) [135]. The process works within the redox couple potential range of Ni(II)/Ni(III), with effective oxidation without dehydration of the nickel hydroxide layer. About 11% of the oxidized ammonia forms nitrates, while the rest forms gaseous nitrogen, which is due to direct electron transfer from the anode and oxygen derived from water. Ni-based catalysts like NiOOH-Ni and NiCuOOH which have tremendous potential in the electro-oxidation of ammonia, showing high Faradaic efficiencies of 90.4% for NiOOH-Ni and good ammonia removal rates were studied [136]. Moreover, the foam structures of the Ni support have a large surface area and stability, allowing them to be utilized to a higher extent for the treatment of ammonia and generation of nitrogen. Wang et al. reported a three-dimensional nickel foam electrode coated with petal-like clusters of NiCuOOH synthesized through a facile sonochemical method coupled with surface electrochemical reconstruction [137]. This has a very high surface area and very active sites, which allows for outstanding electrocatalytic activity for AOR. The maximum current density for AOR obtained is 97.8 mA cm^{-2} at 0.60 V vs. a saturated calomel electrode (SCE). Moreover, using the NiCuOOH-coated nickel foam electrode, the removal efficiency of $NH4^+$-N reached up to 96.53%, and total nitrogen (TN) reached 63.12% in an electrolysis system with good stability for more than 24 hours. Generally, this electrode has great potential for the clean treatment of ammonia-nitrogen wastewater (Fig. 6.4 D–F). The ammonia oxidation significantly improves when the Pt-based bimetallic alloys are modified using Rh, Ir, or CeO_2 to further enhance the availability of hydroxyl species at reactive sites. For instance, improvement in the power density up to 60% with Pt-Rh alloys in fuel cells, and this shows how alloying platinum can result in better reactivity besides cost-effectiveness. Transition metal oxides and multi-layered structures like TiO_2-Sb-SnO_2/PbO_2, developed by Zhang et al. have demonstrated remarkable stability and selective ammonia removal, promising sustainability to wastewater treatment [138]. High-entropy oxides such as the generic $(Mn, Fe, Co, Ni, Cu)_3O_4$ likewise show attractive catalytic efficiency, reaching a Faradaic efficiency above 85% by converting ammonia to dinitrogen [139].

Novel nanostructures with vacancy-engineered CeO_x and layered MnO_2 catalysts were revealed as crucial optimizers for surface and electronic structures in ammonia oxidation improvement. Indeed, CeO_x had twice the Faradaic efficiency of Pt/C electrodes, because it has a lower energy barrier in ammonia dissociation [140]. Similarly, intercalated MnO_2 functionalized with Ni and Cu ions showed nearly complete Faradaic efficiency at lower overpotentials and thus reduces NO_x production, which is beneficial for the release of nitrogen [141]. Further development of high-entropy oxides and modified metal oxides will be a step forward toward efficient large-scale ammonia-based fuel systems and successful wastewater treatment solutions.

6.4.3.3 Molecular catalysts

The electrocatalytic oxidation of ammonia (AO) is an alternative route to the conventional methods, which are dependent on noble metals, wherein dinitrogen formation and surface poisoning could be encountered. The "Blue Layer" (BL) is an organometallic–inorganic hybrid anode consisting of $[IrO_2]_x$ nanoclusters, which displayed promise for selective AO in aqueous solutions, achieving as high as 90.5% nitrate selectivity under optimal conditions at pH 8.0 and 1.00 V, with resistance against ammonia-induced degradation [142]. This promising candidate for sustainable energy carriage has been applied in successfully conducted catalytic oxidation in the presence of low-cost, abundant transition metal complexes as studied by Liu et al. [143]. Examples of such species include a macrocyclic ligated complex, namely, the trans-$[LFe^{II}(MeCN)_2][PF_6]_2$ as one that demonstrated a rather high turnover of nitrogen gas, N_2, when coupled with catalyst loading into successful electrocatalysis within its fair efficiency. These functionalized carbon anodes using the oligomer $\{[Ru_2^{2+}$ bpy$]_{10}$ (bda-κ-N_2O_2 (4,4′-bpy)} have demonstrated to be effective in converting NH_4^+/NH_3 to N_2 and H_2 with near-quantitative Faradaic efficiency [144]. Research in ruthenium-based complexes such as [Ru(tda-κ-N_3O)$(py)_2$] was equally impressive in ammonia oxidation [145]. Iron(II)-based complexes, such as $[(TPA)Fe(NH_3)_2]OTf_2$, were developed to provide a low-cost alternative for this kind of catalysis: this compound exhibits excellent electrochemical activity in ammonia oxidation and very promising Faradaic efficiencies [146]. Molecular catalysts, particularly ruthenium polypyridyl complexes, have catalyzed ammonia oxidation into dinitrogen under ambient conditions with significant efficiency and have contributed significantly to a better understanding of nitrogen cycling for a carbon-free economy [147]. The structure of molecular Ammonia oxidation catalysts are shown in Fig. 6.4 G [142, 145, 146]. Overall, studies on alternative anode materials, such as iron and ruthenium complexes, suggest that they hold great potential for advancing ammonia oxidation processes and, therefore, endorse their use in sustainable energy systems.

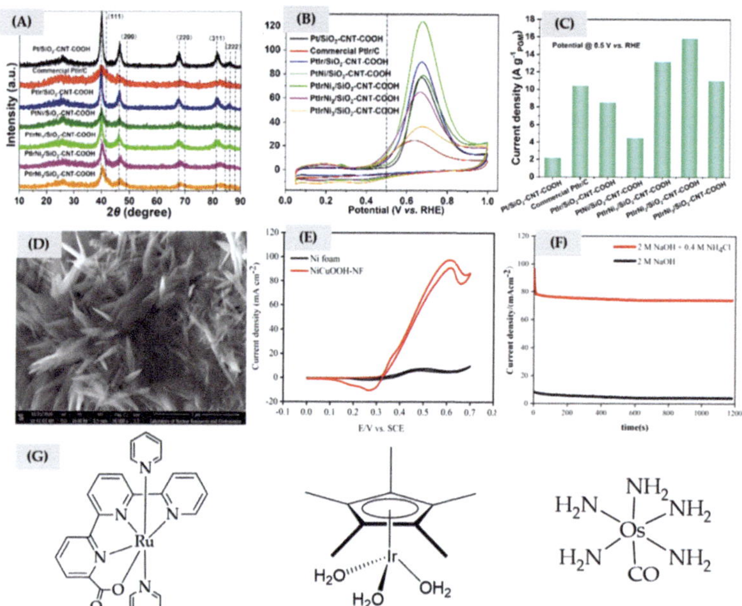

Figure 6.4 (A) XRD patterns of the synthesized catalysts Pt/SiO_2-CNT-COOH, commercial PtIr/C, PtIr/SiO_2-CNT-COOH, PtNi/SiO_2-CNT-COOH, $PtIrNi_1$/SiO_2-CNT-COOH, $PtIrNi_2$/SiO_2-CNT-COOH, and $PtIrNi_3$/SiO_2-CNT-COOH. (B) CV curves for different catalysts in Ar-saturated 1.0 M KOH + 0.1 M NH_3 at 5 mV s^{-1} and 900 rpm under ambient conditions, comparison between Pt, PtIr, PtNi, $PtIrNi_1$, $PtIrNi_2$, $PtIrNi_3$ deposited on SiO_2-CNT-COOH, and commercial PtIr/C. (C) AOR activity comparison for the catalysts in (B) at 0.5 V vs reversible hydrogen electrode (RHE). (D) SEM image of NiCuOOH-NF electrode. (E) The comparison of CV curves between NiCuOOH-NF electrode and Ni foam electrode at 2 M NaOH + 0.4 M NH_4^+ with a scan rate of 20 mV s^{-1}. (F) Comparison of i-t curves obtained on the NiCuOOH-NF electrode at 0.6 V vs. SCE in 2 M NaOH + 0.4 M NH_4Cl or 2 M NaOH. (G) Structures of molecular ammonia oxidation catalysts. Reprinted from Refs. 127 and 137 from the American Chemical Society and MDPI, Copyright 2020 and 2023, respectively.

6.4.3.4 Carbon-based catalysts

Kapalka et al. studied electrochemical AO on BDD electrodes occurs very efficiently at high pH ~8 where free ammonia is available [148]. The oxidation process is optimized for pH levels that are lower where ammonium is dominant by using in situ formed active chlorine as a mediator. This enables fast reaction between active chlorine and ammonia and prevents chlorate and perchlorate as the byproducts. One of the studies by Zollig et al. shows that direct AO may be carried out using graphite electrodes, which removes ammonia well from urine samples at high pH levels (>9) without yielding chlorinated byproducts [149]. The removal rate of ammonia was proven to reach a maximum value of 2.9 ± 0.3 gN/m^2/day in real urine with a pH level of 9.0, coupled with a specific energy consumption of 42 Wh/gN. Graphite electrodes have been recognized as cost-efficient, energetically favorable, and resistant to corrosion and poisoning. In the AOR, a novel porous nitrogen-doped graphene aerogel embedded with Pt nanoclusters (Pt/NGA) is designed by Zhou et al. for electrochemical applications [149]. The material obtained through the self-assembly of graphene oxide has been found to exhibit high conductivity and good AOR activity. The attained mass activity is 1.77 mA/μgPt, and the specific activity is 0.64 mA/cm^2. Pt/NGA produced 8.5 mL of hydrogen at 0.8 V for 3 hours as a working electrode in an alkaline electrochemical cell, and it is promising as an electrocatalyst and as an energy-converting materials. BDD and graphite are good anode materials from which ammonia is oxidized, while innovative structures such as Pt/NGA can improve the efficiency and sustainability of hydrogen production.

6.4.4 Challenges and Future Directions

Development of anodic materials for AOR is associated with challenges that have to be overcome in terms of efficiency, selectivity, and sustainability. The most important challenges of noble metal catalysts, like platinum, are their high cost and scarcity, although having higher catalytic activity, they limit large-scale applications on economic grounds. While nickel and iron offer some promise as substitutes, they often exhibit lowered activity and instability in aggressive electrochemical environments. Non-noble catalysts can deteriorate or corrode quickly under wide pH variation ranges that are typical of real application environments. Unwanted side products like hydrazine and nitrogen oxides complicate the search for optimum catalysts due to their complex reaction mechanism. Bringing materials together with appropriate balances between cost and catalytic performance, durability, and selectivity will be a needed area of research.

Going forward for AOR anodic materials, hybrid and molecular systems should advance as promising ones to enhance the performance without noble metals that could be costly. Engineered high-entropy oxides and modified transition metal oxides represent the most exciting examples of innovation that can be leveraged for the development of more sustainable catalysts. Notably, these new developments are expected to ensure a higher Faradaic efficiency as well as selectivity toward the desired nitrogen products. Integrating carbon-based materials with electrocatalysts also presents the possibility of further improved conductivity and resistance against poisoning, thus enhancing performance as a whole. Future research should look forward to the use of nanostructure and composite materials based on synergetic effects because of the presence of synergism between different components in order to develop very electrochemically active catalytic materials. Further developments at the fundamental level of understanding these mechanisms will be crucial since it will directly go into designing economical advanced materials for large-scale processes in ammonia-based energy conversion systems and wastewater treatments.

6.5 Summary and Outlook

Electrocatalytic hydrogen production is an environmentally friendly approach, aligned with the global renewable energy targets. Recent breakthroughs in the context of anodic oxidation reactions, including OER and hybrid water electrolysis, are focused upon as it overcomes the problems associated with conventional methods like slow kinetics, high overpotential, and risks associated with the mixing of gases. This approach can generate hydrogen efficiently with maximum valorization of organic compounds and treatment of environmental pollutants through the utilization of feedstocks like small alcohols, biomass, and plastic waste. Ammonia oxidation, also obtained from fertilizers and industrial processes, is another essential route toward minimizing environmental impacts by using ammonia as a carrier of hydrogen in fuel cells while facilitating a reduced carbon footprint in chemical manufacturing through improved agriculture. There is a need for the engineering of anodic materials uniquely designed for these specific oxidation processes, so that this trend in improvement in terms of efficiency, cost-effectiveness, and durability of electrodes will continue from energy conversion up to environmental remediation applications. Future research and innovations in materials science will become critical for the resolution of the challenges presented and realized in full exploitation of such technologies for a transition toward a more sustainable and carbon-neutral future.

The development of anodic materials for WOR is critical in the advancement of electrochemical energy conversion technologies, specifically for sustainable hydrogen production by water splitting. Recent innovation shows a transition from the use of precious metal catalysts, such as iridium and RuO_2, toward more inexpensive non-precious metal alternatives, namely transition metal oxides and MOFs. These materials are meant to reach the goal of lower overpotentials, enhanced stability, and more robust performance in both acidic and alkaline environments. Carbon-based catalysts and molecular catalysts exploration are also promising with regard to unique structural properties and functionalities that will enhance catalytic activity. Future directions may well be multifunctional materials that can synergistically enhance performance, stability, and cost-effectiveness but have to overcome the critical challenges of durability and efficiency in real-world applications. With the progress in research, heteroatom doping and single-atom catalysis are likely to revolutionize the landscape of anodic materials for WOR, which would finally benefit a sustainable energy future.

Advances in anodic materials for organic oxidation reactions hold promising efficiency and sustainability in electrochemical processes. The application of precious and non-precious metal catalysts and innovative carbon-based supports opens the possibility for the efficient degradation of persistent organic pollutants and conversion of a large number of organic substrates, such as alcohols and amines, into valuable chemicals. Ru-MMOs and cobalt-based spinel oxides show promising catalytic properties, which ensure excellent selectivity and stability under operational conditions. New techniques such as dynamic potential cycling and the application of advanced hybrid electrodes have been shown to significantly increase the lifetime and efficiency of the catalyst. Research is starting to improve these anodic materials continuously along a clear trajectory toward scalable, environmentally friendly solutions in wastewater treatment and valorization of organic waste, aligned with broader goals of green chemistry and sustainable development. Further enrichments to the toolkit for these challenging oxidation reactions include studies of new doping strategies, composite structures, and applications of molecular and carbon-based catalysts.

In AORs, developing the newest trends in anodic materials that can improve cleaner energies, efficient, and effective solutions for wastewater treatment is of prime interest in the present scenario. This work is primarily devoted to improving catalyst performance along with increasing sustainability, hence it addresses specific challenges: the high cost of very expensive noble metals, among them platinum. Hybrid and molecular catalysts, engineered high-entropy oxides, and modified transition metal oxides are promising means of increasing Faradaic efficiency and selectivity toward nitrogen products. Carbon-based materials can be used to enhance conductivity and mitigate poisoning effects. Non-noble catalysts such as nickel and iron are popular for their cost-effectiveness but often suffer from problems with activity and stability in variable pH conditions. Future directions in the field should be focused on designing nanostructured and composite materials that take advantage of synergistic effects, as well as on a deeper understanding of mechanisms to make possible the manufacture of cost-effective advanced materials applicable to large-scale energy conversion systems based on ammonia and sustainable wastewater treatment purposes.

Development of anodic materials for water oxidation, organic oxidation, and ammonia oxidation has become a rapidly evolving area of research focused on solving some of the global challenges. The future development of electrochemical technologies is stimulated by further innovation in catalyst design and a greater emphasis on sustainability and cost-effectiveness. Discoveries of new materials and optimization of known materials will continue to open up further avenues for scalable applications in wastewater treatment, renewable energy production, and environmental remediation. The future directions should emphasize approaches that are interdisciplinary in nature—by combining insights from materials science, electrochemistry, and environmental engineering to unlock the fuller potential of anodic materials in addressing the complex challenges of modern society.

References

1. L. Tian, Z. Li, M. Song, and J. Li, *Nanoscale* 13, 12088 (2021).
2. R. Li, K. Xiang, Z. Peng, Y. Zou, et al., *Adv. Energy Mater* 11, (2021).
3. E. Brillas and C. A. Martínez-Huitle, *Appl. Catal. B* 166–167, 603 (2015).
4. N. S. Hassan, A. A. Jalil, R. Saravanan, N. M. Izzuddin, et al., *J. Mater. Chem. A Mater.* (2024).
5. L. Li, P. Wang, Q. Shao, and X. Huang, *Adv. Mater.* 33, (2021).
6. Z. Y. Yu, Y. Duan, X. Y. Feng, X. Yu, et al., *Adv. Mater.* 33, (2021).
7. J. Li, C. A. Triana, W. Wan, D. P. Adiyeri Saseendran, et al., *Chem. Soc. Rev.* 50, 2444 (2021).
8. J. Mohammed-ibrahim and H. Moussab, *J. Energy Chem.* 56, 299 (2021).
9. C. Acar and I. Dincer, *Int. J. Hydrogen Energy* 39, 1 (2014).
10. X. Zou and Y. Zhang, *Chem. Soc. Rev.* 44, 5148 (2015).
11. P. C. K. Vesborg, B. Seger, and I. Chorkendorff, *J. Phys. Chem. Lett.* 6, 951 (2015).
12. I. C. Man, H. Su, F. Calle-Vallejo, H. A. Hansen, et al. *Rossmeisl, ChemCatChem* 3, 1159 (2011).
13. F.-Y. Chen, Z.-Y. Wu, Z. Adler, and H. Wang, *Joule* 5, 1704 (2021).
14. T. Maiyalagan, K. A. Jarvis, S. Therese, P. J. Ferreira, et al., *Nat. Commun.* 5, 3949 (2014).
15. J. Fan, Y. Mu, X. Ge, L. Zhang, et al., *ACS Catal.* 13, 4120 (2023).
16. T. Kwon, A. Yu, S. jin Kim, M. H. Kim, et al., *Appl. Surf. Sci.* 563 (2021).
17. Y. Qin, Z. Wang, W. Yu, Y. Sun, et al., *Nano Lett.* 21, 5774 (2021).
18. M. Liu, S. Liu, Q. Mao, S. Yin, et al., *J. Mater. Chem. A Mater* 10, 2021 (2022).
19. S. M. Galani, A. Mondal, D. N. Srivastava, and A. B. Panda, *Int. J. Hydrogen Energy* 45, 18635 (2020).

20. J. Z. Y. Seow and T. D. Nguyen, *Electrochim Acta* 341 (2020).
21. Y. Li, B. Wei, Z. Yu, O. Bondarchuk, et al., *ACS Sustain. Chem. Eng.* 8, 10193 (2020).
22. N. K. Shrestha, S. A. Patil, J. Han, S. Cho, et al., *J. Mater. Chem. A Mater.* 10, 8989 (2022).
23. W. Adamson, X. Bo, Y. Li, B. H. R. Suryanto, et al., *Catal. Today* 351, 44 (2020).
24. Q. Qian, Y. Li, Y. Liu, and G. Zhang, *Appl. Catal. B* 266 (2020).
25. L. Wang, X. Duan, X. Liu, J. Gu, et al., *Adv. Energy Mater.* 10 (2020).
26. S. Pan, H. Li, D. Liu, R. Huang, et al., *Nat. Commun.* 13 (2022).
27. M. Retuerto, L. Pascual, F. Calle-Vallejo, P. Ferrer, et al., *Nat. Commun.* 10 (2019).
28. J. Xu, R. Yang, T. Hou, J. Chen, et al. *Chen, ChemCatChem* (2024).
29. A. Meena, P. Thangavel, D. S. Jeong, A. N. Singh, et al., *Appl. Catal. B* 306 (2022).
30. G. Qian, J. Chen, T. Yu, J. Liu, et al., *Nanomicro Lett.* 14 (2022).
31. Q. Liu, J. Huang, L. Cao, K. Kajiyoshi, et al., *ACS Sustain. Chem. Eng.* 8, 6222 (2020).
32. F. Te Tsai, Y. T. Deng, C. W. Pao, J. L. Chen, et al., *J. Mater. Chem. A Mater.* 8, 9939 (2020).
33. J. Yang, L. Wang, S. Zhan, H. Zou, et al., , *Nat. Commun.* 12 (2021).
34. X. Li, X. P. Zhang, M. Guo, B. Lv, et al., *J. Am. Chem. Soc.* 143, 14613 (2021).
35. D. Yang, Y. Chen, Z. Su, X. Zhang, et al., *Coord. Chem. Rev.* 428 (2021).
36. M. A. Hoque, M. Gil-Sepulcre, A. de Aguirre, J. A. A. W. Elemans, et al., *Nat. Chem.* 12, 1060 (2020).
37. S. Gutiérrez-Tarriño, J. L. Olloqui-Sariego, J. J. Calvente, G. M. Espallargas, et al., *J. Am. Chem. Soc.* 142, 19198 (2020).
38. M. Kondo, H. Tatewaki, and S. Masaoka, *Chem. Soc. Rev.* 50, 6790 (2021).
39. D. Pangotra, L. I. Csepei, A. Roth, C. Ponce de León, et al., *Appl. Catal. B* 303 (2022).
40. X. Gu, Z. Chen, Y. Li, J. Wu, et al., *ACS App. Mater. Interfaces* 13, 24814 (2021).
41. X. Zhao, H. Su, W. Cheng, H. Zhang, et al., *ACS Appl. Mater. Interfaces* 11, 34854 (2019).

42. R. Gao, Q. Dai, F. Du, D. Yan, et al., *J. Am. Chem. Soc.* 141, 11658 (2019).
43. K. Mamtani, D. Jain, D. Dogu, V. Gustin, et al., *Appl. Catal. B* 220, 88 (2018).
44. Y. Zhu, T. Zhang, and J. Y. Lee, *ChemElectroChem* 5, 583 (2018).
45. M. Panizza and G. Cerisola, *Chem. Rev.* 109, 6541 (2009).
46. C. A. Martínez-Huitle and M. Panizza, *Curr. Opin. Electrochem.* 11, 62 (2018).
47. C. Tang, Y. Zheng, M. Jaroniec, and S. Qiao, *Angew. Chem. Int. Ed.* 60, 19572 (2021).
48. J. Hao, J. Liu, D. Wu, M. Chen, et al., *Appl. Catal. B* 281, 119510 (2021).
49. H. Huang, C. Yu, X. Han, H. Huang, et al., *Energy Environ. Sci.* 13, 4990 (2020).
50. C. H. Lam, A. J. Bloomfield, and P. T. Anastas, *Green Chem.* 19, 1958 (2017).
51. B. Zhao, J.-W. Liu, Y.-R. Yin, D. Wu, et al., *J. Mater. Chem. A Mater.* 7, 25878 (2019).
52. L. Huang, X. Zhang, Q. Wang, Y. Han, et al., *J. Am. Chem. Soc.* 140, 1142 (2018).
53. J. Qiu, Y. Han, K. Zhang, Y. Deng, et al., *ACS Appl. Energy Mater.* 5, 4439 (2022).
54. S. Behera, S. Ganguly, C. Loha, B. Mondal, et al., *Energy Fuels* 37, 7603 (2023).
55. K. M. Lee, H. J. Lee, J. Seo, T. Lee, et al., *ACS ES&T Eng.* 2, 1797 (2022).
56. B. Mondal, N. Karjule, C. Singh, R. Shimoni, et al., *Adv. Energy Mater.* 11, (2021).
57. S. Ghosh, B. Mondal, S. Roy, M. Shalom, et al., *J. Mater. Chem. A Mater.* 10, 8238 (2022).
58. Y. He, H. Lin, Z. Guo, W. Zhang, et al., *Sep. Purif. Technol.* 212, 802 (2019).
59. J. Sun, H. Lu, H. Lin, L. Du, et al., *Sep. Purif. Technol.* 88, 116 (2012).
60. J. Iniesta, *Electrochim. Acta* 46, 3573 (2001).
61. E. Mostafa, P. Reinsberg, S. Garcia-Segura, and H. Baltruschat, *Electrochim. Acta* 281, 831 (2018).
62. J. Cai, M. Zhou, Y. Liu, A. Savall, et al., *Chemosphere* 204, 163 (2018).
63. C. A. Martínez-Huitle and S. Ferro, *Chem. Soc. Rev.* 35, 1324 (2006).
64. N. Zaman, N. Iqbal, and T. Noor, *Int. J. Hydrogen Energy* 63, 737 (2024).

65. P. Zhu, Y. Shen, L. Dai, Q. Yu, et al., *ACS Appl. Mater. Interfaces* 14, 1452 (2022).
66. V. K. Puthiyapura, D. J. L. Brett, A. E. Russell, W. F. Lin, et al., *ACS Appl. Mater. Interfaces* 8, 12859 (2016).
67. F. Munoz, C. Hua, T. Kwong, L. Tran, et al., *Appl. Catal. B* 174–175, 323 (2015).
68. M. Zhiani, S. Majidi, H. Rostami, and M. M. Taghiabadi, *Int. J. Hydrogen Energy* 40, 568 (2015).
69. S. R. Kubota and K. S. Choi, *ChemSusChem* 11, 2138 (2018).
70. K. Beliaeva, M. Elsheref, D. Walden, F. Dappozze, et al., *J. Electrochem. Soc.* 167, 134511 (2020).
71. G. Zhao, J. Lin, M. Lu, L. Li, et al., *Nat. Commun.* 15, 8463 (2024).
72. J. Li, J. Tang, D. Wu, L. Yao, et al., *Int. J. Hydrogen Energy* 48, 36286 (2023).
73. A. Mirshafiee, M. Nourollahi, and A. Shahriary, *Sci. Rep.* 14, (2024).
74. Z. Yin, Y. Zheng, H. Wang, J. Li, et al., *ACS Nano.* 11, 12365 (2017).
75. P. Cox and D. Pletcher, Electrosynthesis at Oxide Coated Electrodes Part 2. The Oxidation of Alcohols and Amines at Spinel Anodes in Aqueous Base (1991).
76. M. Panizza and G. Cerisola, *Electrochim Acta* 49, 3221 (2004).
77. R. M. Rodrigues and A. Varghese, *Electrochim Acta* 498, (2024).
78. S. E. Michaud, M. M. Barber, K. E. Rivera Cruz, and C. C. L. McCrory, *ACS Catal* 13, 515 (2023).
79. W. Jud, C. A. Salazar, J. Imbrogno, J. Verghese, et al., *Org. Process Res. Dev.* 26, 1486 (2022).
80. M. K. Adak, A. Rajput, L. Mallick, and B. Chakraborty, *ACS Appl. Energy Mater.* 5, 5652 (2022).
81. Y. Lu, C. L. Dong, Y. C. Huang, Y. Zou, et al., *Angew. Chem.– Int. Ed.* 59, 19215 (2020).
82. S. Barwe, J. Weidner, S. Cychy, D. M. Morales, et al., *Angew. Chem.– Int. Ed.* 57, 11460 (2018).
83. A. M. B. Honorato, M. Khalid, A. A. da S. Curvelo, H. Varela, *Polymers* (Basel) 14, (2022).
84. F. Bateni, R. Ghahremani, and J. A. Staser, *J. Appl. Electrochem.* 51, 65 (2021).
85. H. Zhou, Z. Li, S. M. Xu, L. Lu, et al., *Angew. Chem.– Int. Ed.* 60, 8976 (2021).

86. S. Behera, S. Dinda, R. Saha, and B. Mondal, *ACS Catal.* 13, 469 (2023).
87. S. Taghaddosi, S. Rezaee, and S. Shahrokhian, *J. Electroanal. Chem.* 873 (2020).
88. X. Zhu, W. Hu, C. Feng, N. Chen, et al., *Chemosphere* 269 (2021).
89. G. Zhang, J. Hu, W. Zhang, K. Yu, et al., *ChemSusChem* (2024).
90. N. Sergienko, E. C. Lumbaque, N. Duinslaeger, and J. Radjenovic, *Appl. Catal.* B 334 (2023).
91. C. Yang, S. Shang, and X. yan Li, *Sep. Purif. Technol.* 258, (2021).
92. J. Zhang, X. Wei, J. Miao, R. Zhang, et al., *Mater. Today Commun.* 24, (2020).
93. X. Wei, S. Wang, Z. Hua, L. Chen, et al., *ACS Appl. Mater. Interfaces* 10, 25422 (2018).
94. H. Kuiry, D. Das, S. Das, S. Chakraborty, et al., *Faraday Discuss.* 234, 42 (2022).
95. J. Bühler, A. Muntwyler, H. Roithmeyer, P. Adams, et al., *Chem.– Eur. J.* 30 (2024).
96. Y. F. Qi, K. Y. Wang, Y. Sun, J. Wang, et al., *ACS Sustain. Chem. Eng.* 10, 645 (2022).
97. B. Liu, S. Xu, M. Zhang, X. Li, et al., *Green Chem.* 23, 4034 (2021).
98. B. Xie, Y. Tobimatsu, K. Narita, S. Yokohata, et al., *ACS Sustain. Chem. Eng.* 10, 16701 (2022).
99. M. Xiang, Z. Xu, Q. Wu, Y. Wang, et al., *J. Power Sources* 535 (2022).
100. J. Zhao, C. Deng, L. Zhang, J. Zhang, et al., *J. Org. Chem.* (2024).
101. J. R. N. dos Santos, I. C. B. Alves, A. L. B. Marques, and E. P. Marques, *Electrocatalysis* 13, 713 (2022).
102. G. Amaral-Labat, E. L. da Silva, A. Cuña, C. F. Malfatti, et al., *Waste Biomass Valor.* 12, 2507 (2021).
103. I. S. Pieta, A. Lewalska-Graczyk, P. Pieta, G. Garbarino, et al., *ACS Sustain. Chem. Eng.* 8, 7244 (2020).
104. S. Xie, L. Deng, H. Huang, J. Yuan, et al., *J. Colloid. Interface Sci.* 608, 3130 (2022).
105. P. Dakshinamoorthy and S. Vaithilingam, *RSC Adv.* 7, 34922 (2017).
106. R. N. Abbas and A. S. Abbas, *Environ. Res. Technol.* 7, 175 (2024).
107. M. Ječmenica Dučić, A. Krstić, N. Zdolšek, D. Aćimović, et al., *Crystals* (Basel) 13 (2023).
108. A. Rahmani, A. Seid-mohammadi, M. Leili, A. Shabanloo, et al., *Chemosphere* 276 (2021).

109. A. S. Abdalrhman and M. Gamal El-Din, *Chem. Eng. J.* 389 (2020).
110. L. Baptista-Pires, G. F. Norra, and J. Radjenovic, *Water Res.* 203 (2021).
111. B. Seo, W. H. Lee, Y. J. Sa, U. Lee, et al., *Appl. Surf. Sci.* 534 (2020).
112. N. S. Hassan, A. A. Jalil, R. Saravanan, N. M. Izzuddin, et al., *J. Mater. Chem. A Mater.* (2024).
113. C. Zhong, W. B. Hu, and Y. F. Cheng, *J. Mater. Chem. A Mater.* 1, 3216 (2013).
114. F. Chang, W. Gao, J. Guo, and P. Chen, *Adv. Mater.* 33 (2021).
115. S. S. Rathore, S. Biswas, D. Fini, A. P. Kulkarni, et al., *Int. J. Hydrogen Energy* 46, 35365 (2021).
116. N. M. Adli, H. Zhang, S. Mukherjee, and G. Wu, *J. Electrochem. Soc.* 165, J3130 (2018).
117. J. A. Herron, P. Ferrin, and M. Mavrikakis, *J. Phys. Chem. C* 119, 14692 (2015).
118. T. V. Choudhary, C. Sivadinarayana, and D. W. Goodman, *Chem. Eng. J.* 93, 69 (2003).
119. K. Endo, K. Nakamura, Y. Katayama, and T. Miura, *Electrochim. Acta* 49, 2503 (2004).
120. J. Liu, B. Chen, Y. Kou, Z. Liu, et al., *J. Mater. Chem. A Mater.* 4, 11060 (2016).
121. R. Lan and S. Tao, *Electrochem. Solid-State Lett.* 13, B83 (2010).
122. Y. Tian, Z. Mao, L. Wang, and J. Liang, *Small Struct.* 4, (2023).
123. H. G. Oswin and M. Salomon, *Can. J. Chem.* 41, 1686 (1963).
124. H. Gerischer and A. Maurer, *J. Electroanal. Chem. Interfacial Electrochem.* 25, 421–433 (1970).
125. M. A. Matin, S. Kim, Y. Kim, C., *J. Power Sources* 591 (2024).
126. S. S. Prabowo Rahardjo and Y. J. Shih, *ACS Sustain. Chem. Eng.* 10, 5043 (2022).
127. Y. Li, X. Li, H. S. Pillai, J. Lattimer, et al., *ACS Catal.* 10, 3945 (2020).
128. K. Siddharth, Y. Hong, X. Qin, H. J. Lee, et al., *Appl. Catal. B* 269 (2020).
129. Y. T. Chan, K. Siddharth, and M. Shao, *Nano Res.* 13, 1920 (2020).
130. N. Hanada, Y. Kohase, K. Hori, H. Sugime, et al., *Electrochim. Acta* 341 (2020).
131. J. Wang, J. Heo, C. Chen, A. J. Wilson, et al., *Angew. Chem.– Int. Ed.* 59, 18430 (2020).
132. Y. Katayama, T. Okanishi, H. Muroyama, T. Matsui, et al., *J. Phys. Chem. C*

119, 9134 (2015).

133. Mô. H. M. T. Assumpção, R. M. Piasentin, P. Hammer, R. F. B. De Souza, et al., *Appl. Catal. B* 174–175, 136 (2015).

134. D. J. Little, D. O. Edwards, M. R. Smith, and T. W. Hamann, *ACS Appl. Mater. Interfaces* 9, 16488 (2017).

135. A. Kapałka, A. Cally, S. Neodo, C. Comninellis, et al., *Electrochem. Commun.* 12, 18 (2010).

136. H. Liu, C. J. Yang, C. L. Dong, J. Wang, et al., *Adv. Energy Mater.* (2024).

137. H. Wang, L. Zhou, W. Chen, X. Tong, et al., *Processes* 11, (2023).

138. Z. Zhang, J. Liu, H. Ai, A. Chen, et al. Yuan, *J. Environ. Chem. Eng.* 11, (2023).

139. S. He, V. Somayaji, M. Wang, S. H. Lee, et al., *Nano Res.* 15, 4785 (2022).

140. X. Yang, L. Sun, X. Liu, Z. Yang, et al. Chen, *ACS Catal.* 14, 6236 (2024).

141. K. Nagita, Y. Yuhara, K. Fujii, Y. Katayama, et al., *ACS Appl. Mater. Interfaces* 13, 28098 (2021).

142. H. Y. Liu, J. A. Jayworth, R. H. Crabtree, and G. W. Brudvig, *ACS Catal.* 14, 2842 (2024).

143. L. Liu, S. I. Johnson, A. M. Appel, and R. M. Bullock, *Angew. Chem. Int. Ed.* (2024).

144. A. M. Beiler, A. Denisiuk, J. Holub, F. J. Sánchez-Baygual, et al., *ACS Energy Lett.* 8, 172 (2023).

145. J. Holub, N. Vereshchuk, F. J. Sánchez-Baygual, M. Gil-Sepulcre, et al., *Inorg. Chem.* 60, 13929 (2021).

146. M. D. Zott, P. Garrido-Barros, and J. C. Peters, *ACS Catal.* 9, 10101 (2019).

147. F. Habibzadeh, S. L. Miller, T. W. Hamann, and M. R. Smith, *Proc Natl. Acad. Sci. U S A* 116, 2849 (2019).

148. A. Kapałka, L. Joss, Á. Anglada, C. Comninellis, et al., *Electrochem. Commun.* 12, 1714 (2010).

149. H. Zöllig, C. Fritzsche, E. Morgenroth, and K. M. Udert, *Water Res.* 69, 284 (2015).

150. X. Zhu, Y. Deng, W. Hu, H. Chen, C. Feng, and N. Chen, *Environmental Science and Pollution Research* 30, 109691 (2023).

Chapter 7

Materials for Cathodic Reduction Reactions: Proton/Water Reduction, $CO_2/N_2/NO_x$ Reduction

Souradip Ganguly[a] **and Sirshendu Ghosh**[b]
[a]*CSIR–Central Mechanical Engineering Research Institute (CMERI), Durgapur, West Bengal, India*
[b]*School of Mathematics and Natural Sciences, Chanakya University, Bengaluru, Karnataka, India*
sirshendu.g@chanakyauniversity.edu.in

7.1 Introduction

Electrification using renewables is a potential approach to decarbonize the chemical industry by the end of the 21st century. Besides the development of human society, such as, electrifying invention and technologies along with convenient transportation all over the world, in the transition from human and horse power to horsepower since the first industrial revolution, excessive carbon emission from modern industries contributed to a sharp increase in CO_2 and hazardous wastes. The resultant environmental issues

Materials Design for Electrocatalysis
Edited by Sagarika Bhattacharya and Biswarup Chakraborty

ISBN 978-981-5352-32-0 (Hardcover), 978-1-042-02558-9 (eBook)
www.jennystanford.com

DOI: 10.1201/9781042025589-7

possess a serious menace to the human race. CO_2 emissions are expected to increase up to ~ 43 Gt yr^{-1} under the current policies,[1] which will intensify the global energy crisis along with a rise in global temperature and acidification of the ocean.[2] To mitigate the global CO_2 emission, a defossilization of our economy is inevitable. The use of hydrogen (H_2) energy from renewable sources is now a significant global development trend. Because of its lack of pollution (during combustion), efficiency, plentiful sources, and versatility, and high gravimetric energy, H_2 energy is considered as the energy source of the future. The reduction of fossil resources, both as fuel (in the energy sector) and feedstock (in the chemical sector), facing environmental challenges, demands an emphasis on the development of new and sustainable technologies toward fuel and chemical production with greener and carbon-neutral emission balance.

Environmental concern over unprecedented global energy consumption provokes the urgent need of clean, sustainable energy conversion and storage systems, such as fuel cells, dye–sensitized solar cells, metal air batteries, Li–CO_2 battery, etc. It is of particular interest to use electrochemistry to maximize the utilization of renewable energy via directly converting renewable energy into chemical energy.

Recently, ammonia (NH_3) has drawn a lot of interest for the storage of hydrogen because of its high weight H content (17.6% by weight), high volumetric H_2 energy density (10.7 kg H_2/100 L), and ease of liquefaction, handling, storage, and transportation.[3] The commercial industrial–scale synthesis of NH_3 still relies on the over a century–old Haber-Bosch technique, which releases a large amount of greenhouse gases and uses a lot of energy and natural gas.[4] Considering the current worldwide NH_3 production rate and its demand, it is a significant challenge to develop techniques and means to convert atmospheric N_2 into NH_3 at ambient temperature and pressure.

Moreover, the introduction of anthropogenic nitrogen species for agricultural and industrial purposes has significantly disrupted the nitrogen cycle, negatively impacting both local and global ecosystems. Typically found in concentrated streams from industrial processes and nuclear waste, or more diluted streams from agricultural runoff, nitrogen oxide (NO_x) species—especially NO_3^-, are highly harmful to marine environments. These detrimental effects of NO_x species underscore the need for efficient removal or remediation technologies. Among the existing technologies (including separation through concentrated brine solution[5] and bioremediation[6]), the promising alternative is electrochemical reduction to value-added products. NO_x could be converted to green ammonia, construct new C–N bonds, and synthesis of various nitrogenous compounds.[7] Thus, these N- and O-containing NO_x species are no longer a pollutant in air and water but also a potential renewable source for value-added nitrogenous species through electrocatalysis.

7.2 Importance of Cathodic Reactions

Electrocatalytic cathodic reactions occur at the electrode–electrolyte interface, where the electrode acts both as catalyst and an electron donor or acceptor. To limit the use of nonrenewable fossil fuels, the development of rechargeable metal-ion/air batteries, supercapacitors, and flow batteries is one of the possible ways for the conversion and storage of electricity generated from renewables. [8] But they all suffer from low-energy density and high cost, which limits their commercialization. Another way of storing electrical energy from the renewables is in the form of chemical bonds. Water (H_2O), carbon dioxide (CO_2), oxygen (O_2), and nitrogen (N_2) are readily reduced at the cathode, and the energy is stored in the form of hydrogen (H_2), carbon monoxide (CO), formic acid (HCOOH), hydrocarbons, alcohols, and ammonium (NH_3).[9] These value-added products could be directly applied for electricity generation via fuel cell technology or serve as feedstock in the chemical industry.

Unlike the CO_2 reduction reaction (CO_2RR) where it is converted to detrimental value-added products, the nitrogen reduction reaction (NRR) helps the environment in an indirect way via low-energy production of ammonia, replacing the Haber–Bosch process.

NO_x is one of the most significant contributors to air pollution and causes respiratory, and cardiovascular diseases, lung cancer, and even stroke. Again, NH_3 is a crucial feedstock for fertilizers. This NO_x could be reduced to NH_3 in a much easier and sustainable way, owing to the lower bond energy and polar N=O bond (compared to thermodynamically stable N_2). The integration and development of such technologies (reducing overall carbon footprint) can also contribute to overall grid stability and thereby enable the use of renewables.

7.3 Thermodynamic Potential of Cathodic Half-Cell Reactions

In electrochemistry, each half-cell reaction (either oxidation or reduction reaction) consists of thermodynamic electrode potentials. All the electrode potential are measured against the standard hydrogen electrode (SHE) potential, which is taken as zero (in a 1 M aqueous solution at standard temperature, 298 K, and a 1 atm pressure of hydrogen gas).

Hydrogen Evolution Reaction (HER)

Acidic medium	Potential vs. SHE	Alkaline medium	Potential vs. SHE
$2H^+ + 2e^- \rightarrow H_2$	0.00	$2H_2O + 2e^- \rightarrow H_2 + 2OH^-$	-0.828

Oxygen Reduction Reaction (ORR)[10]

Acidic medium	Potential vs. SHE	Alkaline medium	Potential vs. SHE
$O_2 + 4H^+ + 4e^- \rightarrow 2H_2O$	1.229	$O_2 + 2H_2O + 4e^- \rightarrow 4OH^-$	0.401
$O_2 + 2H^+ + 2e^- \rightarrow H_2O_2$	0.695	$O_2 + H_2O + 2e^- \rightarrow HO_2^- + OH^-$	-0.076
$H_2O_2 + 2H^+ + 2e^- \rightarrow 2H_2O$	1.776	$HO_2^- + H_2O + 2e^- \rightarrow 3OH^-$	0.878

Carbon Dioxide Reduction Reaction (CO_2RR)[11, 12]

	Potential vs. SHE
$CO_2 + e^- \rightarrow CO_2^{\bullet-}$	- 1.900
$CO_2 + 2H^+ + 2e^- \rightarrow HCOOH$	- 0.610

$CO_2 + 2H^+ + 2e^- \rightarrow CO + H_2O$	- 0.520
$3CO_2 + H_2O + 2e^- \rightarrow CO + 2HCO_3^-$	- 0.560
$2CO_2 + 2H^+ + 2e^- \rightarrow H_2C_2O_4$	- 0.500
$CO_2 + 4H^+ + 4e^- \rightarrow H_2C(OH)_2 + H_2O$	- 0.490
$CO_2 + 6H^+ + 6e^- \rightarrow CH_3OH + H_2O$	- 0.380
$CO_2 + 8H^+ + 8e^- \rightarrow CH_4 + 2H_2O$	- 0.240
$2CO_2 + 12H^+ + 12e^- \rightarrow CH_2CH_2 + 4H_2O$	0.064
$2CO_2 + 12H^+ + 12e^- \rightarrow CH_3CH_2OH + 3H_2O$	0.084
$2CO_2 + 14H^+ + 14e^- \rightarrow C_2H_6 + 4H_2O$	- 0.140

Nitrogen Reduction Reaction (NRR)[13]

Acidic medium	Potential vs. RHE	Alkaline medium	Potential vs. RHE
$N_2 + H^+ + e^- \rightarrow N_2H$	- 3.20	$N_2 + e^- \rightarrow N_2^-$	- 3.37
$N_2 + 2H^+ + 2e^- \rightarrow N_2H_2$	- 1.10	$N_2 + 4H_2O + 6e^- \rightarrow N_2H_4 + 4OH^-$	- 1.16
$N_2 + 4H^+ + 4e^- \rightarrow N_2H_4$	- 0.36	$N_2 + 6H_2O + 6e^- \rightarrow 2NH_3 + 6OH^-$	- 0.736
$N_2 + 6H^+ + 6e^- \rightarrow 2NH_3$	0.55 NHE		

Nitrogen Oxides Reduction Reaction (NO_xRR)[14]	
	Potential vs. RHE
$NO_3^- + 2H^+ + 2e^- \rightarrow NO_2^- + H_2O$	0.94
$NO_3^- + 4H^+ + 3e^- \rightarrow NO + 2H_2O$	0.96
$NO_3^- + 7H^+ + 6e^- \rightarrow NH_2OH + 2H_2O$	0.73
$NO_3^- + 9H^+ + 8e^- \rightarrow NH_3 + 3H_2O$	0.88
$2NO_3^- + 10H^+ + 8e^- \rightarrow N_2O + 5H_2O$	1.12
$2NO_3^- + 12H^+ + 10e^- \rightarrow N_2 + 6H_2O$	1.25
$NO_2^- + 2H^+ + e^- \rightarrow NO + H_2O$	1.00
$NO_2^- + 5H^+ + 4e^- \rightarrow NH_2OH + H_2O$	0.62
$2NO_2^- + 6H^+ + 4e^- \rightarrow N_2O + 3H_2O$	1.29
$NO_2^- + 7H^+ + 6e^- \rightarrow NH_3 + 2H_2O$	0.86
$2NO_2^- + 8H^+ + 6e^- \rightarrow N_2 + 4H_2O$	1.45
$2NO + 2H^+ + 2e^- \rightarrow N_2O + H_2O$	1.59
$NO + 3H^+ + 3e^- \rightarrow NH_2OH$	0.38
$2NO + 4H^+ + 4e^- \rightarrow N_2 + 2H_2O$	1.68
$NO + 5H^+ + 5e^- \rightarrow NH_3 + H_2O$	0.71

7.4 Hydrogen Evolution Reaction

7.4.1 Mechanism of HER

The electrochemical HER process for the reduction of protons in acidic media or water in alkaline media to produce H_2 on the surface of the catalyst proceeds via three possible principal steps. In the first step, Volmer reaction occurs via formation of an adsorbed hydrogen atom (H*), reacting proton with an electron at the surface of the electrode. Subsequently, H_2 formation occurs via Heyrovsky reaction or the Tafel reaction, or it may be both. Briefly, in the Heyrovsky reaction step, another proton diffuses to the H* and reacts with an electron to produce H_2. In the Tafel step, two H* atoms come into close proximity at the electrode surface to produce H_2. The overall HER pathway in both acidic and basic media follows:

Acidic medium	Alkaline medium
$H_3O^+ + M + e^- \rightarrow M\text{–}H^* + H_2O$	$H_2O + M + e^- \rightarrow M - H^* + OH$- (Volmer reaction)
$H^+ + e^- + M\text{–}H^* \rightarrow H_2 + M$	$H_2O + e^- + M\text{–}H^* \rightarrow H_2 + OH^- +$ M (Heyrovsky rection)
$2M\text{–}H^* \rightarrow H_2 + 2M$	$2M\text{–}H^* \rightarrow H_2 + 2M$ (Tafel reaction)

7.4.2 Materials for HER

7.4.2.1 Metal-based catalyst

According to the Sabatier principle,[15] the interaction between the catalyst and reactive intermediates should be optimum. In the case of too weak an interaction, very few intermediates bind to the catalytic surface and, as a result, the reaction slows down; while a very strong interaction blocks the active sites and fails to produce the desired product. The H* adsorption and H_2 desorption in HER process are evaluated by measuring ΔG_{H^*}. Ideally, ΔG_{H^*} must be zero with the highest j_0 value for HER. The volcano type trend correlates the experimental and DFT calculated ΔG_{H^*}. The volcano peak lies at $\Delta G_{H^*} = 0$; a strong H* adsorption happens at $\Delta G_{H^*} < 0$ and vice versa.

A volcano plot is also studied to measure the strength of M–H bonds of various metals.[16] According to the volcano curve, platinum group metals (viz., Pt, Pd, Ru, Ir, and Rh) located close to the peak and shows excellent HER activity.[17] With an increase in surface-area-to-volume ratio, catalytic activity increases, but an opposite trend was found with Pt nanoparticles ranging from 1 to 3 nm. Owing to the lower activation energy of the face site compared to the edge site, the edge-to-face ratio increases with decreasing size, and hence, lower in catalytic activity was found.[18] In order to increase surface area, especially with catalytically active planes, exposing a particular plane or facet leads to an improvement in catalytic activity. For instance, the (100) facet of Pt was proved to be the most active for HER. Likewise, the dimension and morphologies of Pd nanoparticles were seen to affect catalytic activity. The octahedral Pd nanoparticle was seen to absorb more H atoms compared to that of a cubic Pd nanoparticle or bulk Pd.[19] As a cheaper alternative, Ru exhibits moderate Ru–H binding strength and a lower energy barrier for H_2 desorption.[20] But because of aggregation, Ru nanoparticles suffers long-term stability. In order to address the stability, suitable supports or cocatalysts stabilizing the Ru nanoparticles are developed, which includes MoO_2, CeO_2, graphite, NiCoP, etc. With tungsten (W) support, adsorption of H* on Ru shows behavior similar to Pt, leading to improved HER activity.

7.4.2.2 Metal oxides

Transition metal oxides are widely studied, owing to their cost-effectivity and stability, but lack in intrinsic low-electric conductivity, sluggish kinetics, and limited H adsorption sites. But fortunately, MoO_2 and WO_2, having a distorted rutile structure, show metallic character and good electrical conductivity and eventually improve HER. Nevertheless, MoO_2 suffers from aggregation during the HER process.[21] Various strategies have been employed to improve HER performance as well as stability, which includes decreasing the size of the structure, employing porosity in the structure, integrating with conductive carbon supports, introducing O-vacancies or surface defects, and doping with other atoms. For instance, porous MoO_2 on nickel foam exhibited better HER activity compared to MoO_2, owing to the higher surface area and exposed catalytic sites. Moreover, MoO_2 encapsulated in phosphorus-doped porous carbon embedded on rGO (MoO_2@PC–rGO) exhibited an overpotential as low as 64 mV (η_{10}) and Tafel slope of 41 mV dec^{-1} along with remarkable long-term stability in an acidic medium.[21]

HER activity was improved by introducing oxygen vacancies. For instance, O–vacancies were introduced in MnO_2 nanosheets where a low overpotential (η_{10}) of 197 mV and a Tafel slope of 62 mV dec^{-1} were exhibited for HER owing to a large number of Mn^{3+} active sites.[22] Similarly, the HER activity of mesoporous MoO_{3-x} was improved due to the presence of a large number of Mo^{5+} species (suitable adsorption/desorption sites for H^+). Metallic WO_2–carbon mesoporous nanowires with high concentration of O–vacancies possess a prominent HER performance with a low overpotential (η_{10}) of 58 mV. The O–vacancies not only open up a large number of active sites but also enhance electron mobility.[23] In addition, NiO, CoO_x, $NiFeO_x$, $CuCoO_x$, $ZnFe_2O_4$, etc., have demonstrated promising HER activity.

7.4.2.3 Metal nitrides

The expansion of the metal lattice and broadening of the metal d-band caused by the intercalation of negatively charged nitrogen atoms into TMs raises the density of states (DOS) close to the Fermi level and the d-band contraction.[24] Nitride of Mo, W, Co, Ni, Fe are employed as HER electrocatalyst. The pioneer studies with δ-MoN nanosheets and γ-Mo_2N nanoparticles were not seen to be efficient for HER. To this regard, MoN encapsulated with an N-doped carbon support was reported to show remarkable HER performance with low overpotential (η_{10}) of 62 mV.[25] Similarly, the initially low-active WN was improved for HER activity by fabricating porous WN nanowire on carbon cloth, which showed an improved overpotential (η_{10}) of 134 mV and 130 mV in acidic and basic media, respectively.[26] On a similar note, Ni_3N has been studied enormously for HER and was seen to exhibit remarkable HER performance. It is noteworthy that a Pt/C-like HER activity was observed when Ni_3N nanosheets were downsized to atomic thickness owing to the strong metallic nature of the latter (Fig. 7.1A).[27, 28] To improve the catalytic activity, several binary and ternary nitrides have been reported. For instance, the divergent intrinsic properties of two metals in $NiMoN_x$ nanosheet revealed a high HER electrocatalytic activity with a low overpotential of 78 mV with a long-term durability, specifically that

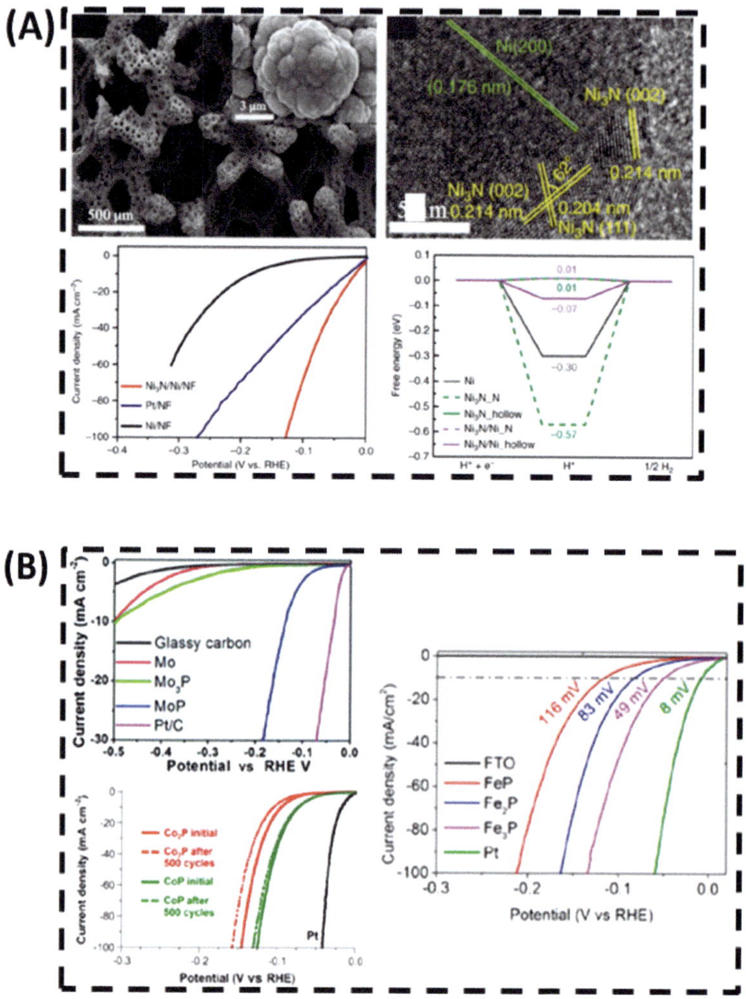

Figure 7.1 (A) SEM image, HRTEM image, LSV plots for HER performance and calculated H–adsorption free energy (ΔG_{H*}) of Ni_3N/Ni/NF. Reproduced with permission from [28], copyright 2018, Springer. (B) Comparative HER LSV plots for metal phosphides (metal = Mo, Co, and Fe). Reproduced with permission from [29–31], copyright 2014, RSC; copyright 2018, ACS, respectively.

nickel binds hydrogen weakly while molybdenum binds hydrogen aggressively, were utilized to achieve elevated HER electrocatalytic activity. The layered structure of cobalt-molybdenum nitride

($Co_{0.6}Mo_{1.4}N_2$) is believed to facilitate the adjustment of electronic states of Mo at the surface without compromising catalytic activity, whereas different replacements at the octahedral location enhance HER action. Similarly, other binary or ternary nitrides that show improved HER activity include Ni_3FeN/r-GO, Ni foam@Ni-$Ni_{0.2}Mo_{0.8}N$, NiCoN@C, MoVN, FeNiMoN, etc.

7.4.2.4 Metal phosphides

Initially, Ni_2P was predicted through DFT studies to show excellent HER activity, outperforming that of noble metal-based electrocatalysts. Later on, it was validated experimentally that Ni_2P is highly active toward HER in an acidic medium.[32] Other transition metal-based phosphides were grown on carbon cloth, such as FeP, WP, and CoP, and were reported to show excellent HER activity with long-term stability.[33, 34, 35] Along with the improvement in HER activity, higher P content in transition metal phosphides also resists corrosion during HER in an acidic medium and thereby improves stability. In fact, MoP was seen to have better HER activity and stability over Mo_2P.[29] Similarly, Ni_5P_4 among the nickel phosphides (viz., Ni_2P, $Ni_{12}P_5$, Ni_5P_4)[36] and CoP among cobalt phosphides (viz., CoP and Co_2P)[30, 37] were seen to have better HER activity along with stability. But interestingly, the reverse trend was followed by iron phosphides—Fe_3P > Fe_2P > FeP (Fig. 7.1B).[31] The superior activity of Fe_3P was attributed to the higher H* coverage, lower ΔG_{H^*}, and higher electric conductivity of Fe_3P compared to Fe_2P and FeP. Metallic doping was enabled to boost HER activity. Transition metal (viz., Fe, Co, Mn, and Mo)-doped nickel phosphides were synthesized and were found to exhibit different HER activities owing to the different H* coverage of nickel phosphides. Depending on the electron filling of their metallic bands and H* coverage, it was found that Mo-doped nickel phosphide exhibited the best HER performance. Similarly, W doping on $Ni_{12}P_5$ displayed excellent catalytic activity with an overpotential of 172 mV (η_{10}) in 1 M KOH.[38] Doping significantly increases the intrinsic HER activity by lowering the energy barrier for electron-coupled water dissociation (*Volmer step*). Bimetallic phosphides that exhibit excellent HER activity includes NiCoP, MoWP, $NiMoP_2$, NiFeP, CoMoP, $Co_{0.68}Fe_{0.32}P$, FeMnP, and Cu_xCo_yP.

7.4.2.5 Metal sulfides

Among the sulfides, molybdenum sulfide (MoS_2) is the most studied catalyst. In acidic media, metallic MoS_2 exhibited an overpotential of 180 mV (η_{10}) with a Tafel slope of 42 mV dec^{-1}.[39] Incorporating first row transition metals (viz., Fe, Co, and Ni) to MoS_2 significantly improves the HER activity of MoS_2 in alkaline medium with robust stability. Nickel sulfides (NiS, NiS_2, and Ni_3S_2) were studied as HER electrocatalysts, but no such significant activity was found. Improvements were made with defect-rich heterointerface of MoNi sulfides which showed an overpotential of 68 mV at 10 mA cm^{-2}.[40] The hierarchical porous structure of h–NiS_x showed even better activity with a low overpotential of 60 mV at 10 mA cm^{-2}.[41] No morphological or compositional changes were seen even after 10 h of electrolysis. At neutral pH, $Ni_{0.33}Co_{0.67}S_2$ was reported to exhibit a low overpotential of 72 mV at 10 mA cm^{-2}, outperforming that of CoS_2 and NiS_2.[42]

7.4.2.6 Metal carbides

Owing to the outstanding electrical conductivity, chemical and thermal stability, and electrochemical activity comparable to that of noble metal catalysts, transition metal carbides have been studied for the past few years as HER electrocatalysts. Initially, Mo_2C was seen to possess good HER activity. Even several crystal phases of MoC (α–MoC_{1-x}, α–Mo_2C, β–Mo_2C, η–MoC, γ–MoC, and δ–MoC) were reported to exhibit different HER performances owing to the difference in atomic stacking. Among them, β-Mo_2C was found to show the highest HER activity. Similarly, tungsten carbide (WC) was proven to show Pt–like catalytic activity.[43] Porous WC_x nanowire on carbon cloth were reported to exhibit excellent HER activity along with stability in both acidic and alkaline media. Modifications have been made to enhance the catalytic activity of transition metal-based carbides. For instance, heteroatom-doped carbon substrates were seen to be better HER electrocatalyst along with long-term stability in both alkaline and acidic media, like Mo_2C/N–CNT, MoO_x@g–C_3N_4, Mo_2C@NPC/NPRGO, β–Mo_2C@BCN, W_2C/MWCNT, etc. On this path, transition metal-based (M = Fe, Co, Ni, W, and Mo) graphene nanoribbon-supported carbides showed robust HER activity. Later on, carbon-supported high-index (222) faceted tantalum carbide (TaC) nanocrystals were reported to be more active than bulk TaC.

7.4.2.7 MXenes

MXenes with O*/OH* terminals exhibited metallic properties and thus are efficient in charge transfer process. An effective HER electrocatalyst was created by using ultrathin O–functionalized Ti_3C_2 MXenes. The basal plane's equivalent F–functionalized MXenes hampered HER by slowing down the rate of hydrogen adsorption. For Ti_3C_2, Mo_2C, and $Mo_2Ti_2C_3$ MXenes, higher F content on the basal plane indicated lower HER activity, but O–terminated Ti_2C_3 MXenes are ideal for HER. Additionally, compared to $Ti_3C_2T_x$ flakes (385 mV), $Ti_3C_2T_x$ NFs demonstrated superior HER catalytic activity, as evidenced by an overpotential of 169 mV at 10 mA cm^{-2}. For the $Ti_3C_2T_x$ flakes and NFs, the Tafel slope was 188 and 97 mV dec^{-1}, respectively.[44] $MoS_2/Ti_3C_2T_x$ composite was even better in showing HER activity with an initial overpotential of 30 mV at 10 mA cm^{-2}.[45]

7.4.2.8 Surface modification of non-oxide/hydroxide materials

Surface hydroxylation of non-oxide or hydroxide materials is a common practice in OER catalysis. Exposing selected non-precious metal chalcogenides and phosphides to an alkaline and oxidative environment may result in anion exchange (formation of corresponding hydroxides) and surface amorphization, thereby increasing the overall catalytic activity of the modified materials. However, because the activity of hydroxide-based catalysts for the HER is generally unsatisfactory, this method has been overlooked in HER catalysis. Recently, Ni–S–OH/Ni foam was reported to exhibit an overpotential of 305 mV at 100 mA cm^{-2}, which is 110 V advantageous over pristine Ni–S/Ni foam.[46] Similarly, modified $Co(OH)_x$@CoP exhibit a decent overpotential of 100 mV at 10 mA-cm^{-2}. The improvement was attributed to the formation of low-valent Co species on the CoP surface and the synergistic effect between $Co(OH)_x$ and CoP.[47]

7.4.2.9 Interfacial engineering for HER

The construction of (hetero) interfaces serves as a method to concurrently tackle several issues while leveraging each component to create sophisticated interfaces that can synergistically improve HER activity in acidic, alkaline, and neutral conditions. A 2D Pt layer was synthesized on NiFe-LDH nanosheet, which synergistically promoted the HER activity with an exceptional operational stability of 50 h and a 6.1-fold increase in mass activity compared to 20% Pt/C.[48] MoS_2–NiS_2 on 3D N-doped graphene acted as an excellent HER catalyst. An electron-rich high-quality interface was created with cNiP_2–mNiP_2 to exhibit significant better HER activity.[49] Se–MoS_2/$CoSe_2$ interface helps in efficient electron transfer via strong Co–S bonds, which eventually favor *OH chemisorption for the water adsorption/desorption process. WS_2, a relatively underexplored HER material, established interfaces with several metal sulfides, including CoS_2, and exhibited exceptional HER performance as a core-shell structure in KOH.[50] Additionally, Mo_2C/MoP heterostructure enhanced the active sites and surface area, improving HER catalytic activity with an overpotential of 88 mV at 10 mA cm^{-2}.[51] Interestingly, Ni_5P_4/NiP_2/Ni_2P on carbon cloth was seen to achieve a stability of 168 h in an acidic medium with an overpotential of 63 mV (η_{10}). Similarly, many other heterostructures were studied as efficient HER catalysis, including Cu_3P-Co_2P, Co_xP-Fe_2P/NF, Ni_3N/Co_3N, etc. Apart from that, heterostructures with alloy/intermetallic phases were also studied to show excellent HER catalysis. Examples include AL–Pt/Pd_3Pb,[52] ($Pt_{1-x}Co_x$)Al/Pt-$Co(OH)_2$,[53] Ir_3V,[54] etc.

7.5 Oxygen Reduction Reaction

Because of its enormous importance in electrochemical energy conversion and storage devices, such as fuel cells, proton exchange membrane fuel cell, and metal air batteries, ORR is designed as the dominating cathodic process. The mechanism involves various intermediates, such as superoxide ion (O_2^-), dianion (O_2^{2-}), and peroxide ion (HOO^-).

7.5.1 Mechanism of ORR

Generally, ORR proceeds via either a "direct" 4e$^-$ pathway or an "alternative" 2e$^-$ pathway to produce OH^- or HO_2^-, respectively. The O_2 adsorption ability of the active site triggers the different ORR pathway. When the active site has a strong affinity to bind O_2, a bridged oxygen configuration is formed, which subsequently gets protonated to an OH* intermediate and further reduced to H_2O (Fig. 7.2).[55]

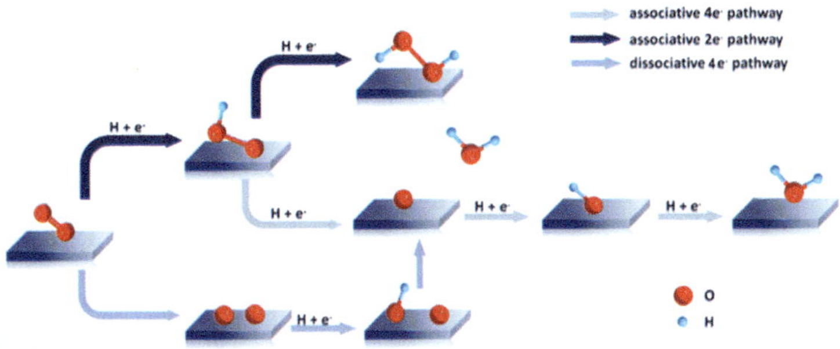

Figure 7.2 Mechanistic pathway of ORR: $2e^-$ and $4e^-$ pathway. Reproduced with permission from [55], copyright 2024, RSC.

7.5.1.1 $2e^-$ ORR pathway

The $2e^-$ ORR pathway differs in alkaline or acidic conditions to produce H_2O_2. The acquisition of an electron by the intermediate to generate HO_2^- and H_2O_2 plays a major role in both the conditions. In an alkaline medium, O_2 first binds to the active site, then gets protonated to form OOH*, which again accepts an electron to form HO_2^-, whereas in acidic condition, the protonated OOH* intermediate readily gets converted to H_2O_2 in the presence of excess H^+.

Alkaline medium: $O_2 + H_2O + 2e^- \rightarrow HO_2^- + OH^-$

$E° = -0.076$ V (vs. SHE)

Acidic medium: $O_2 + 2H^+ + 2e^- \rightarrow H_2O_2$

$E° = 0.695$ V (vs. SHE)

The peroxide ion produced in alkaline medium further gets reduction or disproportionated as follows:[10]

$$HO_2^- + H_2O + 2e^- \rightarrow 3OH^-$$

$$E° = 0.878 \text{ V (vs. SHE)}$$

$$2HO_2^- \rightarrow 2OH^- + O_2$$

$$E° = -0.076 \text{ V (vs. SHE)}$$

7.5.1.2 4e⁻ ORR pathway

In an alkaline medium, the adsorbed O_2 molecule gets protonated to form OOH* intermediate, which accepts an electron to form O*. The O* again gets protonated to form OH*, which further gets reduced to OH^- accepting an electron. In an acidic medium, the pathway occurs via the same intermediate and produces H_2O as the final product:

Alkaline medium: $O_2 + 2H_2O + 4e^- \rightarrow 4OH^-$
$E° = 0.401$ V (vs. SHE)

Acidic medium: $O_2 + 4H^+ + 4e^- \rightarrow 2H_2O$
$E° = 1.229$ V (vs. SHE)

7.5.2 Materials for ORR

7.5.2.1 Metal-based catalyst

Various noble metal (viz., Pt, Pd, Ag, Au)–based electrocatalysts, which lie on the top of volcano plot (Fig. 7.3A), have been studied for ORR in order to acquire desire output owing to reduced sluggish kinetics. The noble metals tend to follow the trend: Pt > Pd > Ir > Rh for ORR activity (Fig. 7.3A).[56] Various works with Pt/C in acidic medium showed excellent ORR activity;[57] although work on alkaline media gained more attention due to increasing interest in alkaline fuel cells and metal air-batteries. Initial studies with Pt electrodes revealed that the ORR activity in alkaline medium alters with the particle size of Pt. A three-fold decrease in activity was observed with a reduction of particle size from 24 nm to 2 nm in 1 M NaOH solution due to the desorption of excess HO_2^- into the bulk solution. Facet-dependent ORR activity with Pt-based electrocatalysts, especially with low-index facets, was studied.[58] In strongly adsorbed electrolyte (viz., H_2SO_4), the ORR activity of Pt(111) < Pt(100), while in weakly adsorbed electrolyte, the ORR activity order follows the order of Pt(100) Pt(110) ≈ Pt(111).[57] Similarly, Au(100) was found to be the most active facet for effective ORR activity; also, Au electrodeposited on a Ni substrate showed desirable activity for HO_2^- reduction, indicating ORR on the Au surface could follow either a $2e^- + 2e^-$ pathway or a $4e^-$ pathway.[10] Approaches have been made to reduce the cost, involving core–shell structure (with Pt as shell), non-

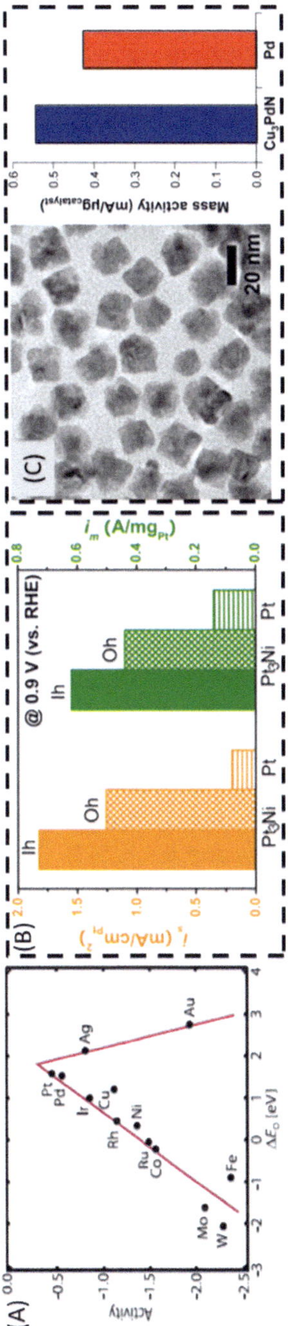

Figure 7.3 (A) Trend in ORR activity of various metals. Reproduced with permission from [56], copyright 2004, ACS. (B) Mass- and area-specific activities of the Pt_3Ni icosahedral and octahedral nanocrystal. Reproduced with permission from [59], copyright 2012, ACS. (C) TEM image of Cu_3PdN and comparative ORR mass activity plot for Cu_3PdN and Pd. Reproduced with permission from [65], copyright 2014, ACS.

precious metal oxide supports for Pt, coating (of Pt) on tin-doped indium oxide (ITO), and alloying with transition metals have been studied effectively as ORR electrocatalysts. Several PtM (M = La, Y, Zr, Ti, Ni, and Co) films were studied for ORR and were concluded to follow the order of $Pt_3Ti < Pt < Pt_3Zr < Pt_3Co < Pt_3Ni < Pt_3Y < Pt_3La$ as per the activity considered. Later on, Pt_5La, Pt_5Ce, Pt_5Gd, and Pt_3Y were reported to show much higher ORR activity than Pt_3La (activity only compared to Pt). Additionally, the diffuse elastic strain of the icosahedral Pt_3Ni nanocrystal boosted the ORR activity compared to its octahedral counterpart.[59] Theoretical study also supports Pt_3Ni (111) to show higher specific activity compared to Pt(111) for ORR.[60]

Au nanoparticle along with suitable carbon support was also studied for ORR and was found out to be more effective compared to Au nanoparticle. Cost-effective AgAu Janus particle was reported to be 6 to 7 times more active for ORR compared to Ag nanoparticle. Similar studies with Pd and Ag showed effective ORR activity owing to good HO_2^- eliminating ability of Ag.

7.5.2.2 Metal oxides

Metal oxides of Groups IV and V are chemically stable in harsh acidic conditions and hence are usually proposed for ORR studies; but the lack of adsorption sites and low electrical conductivity result in low ORR activity. Various metal oxides were studied in H_2SO_4 medium for good ORR activity due to the generated O-vacancies on the surface, which include ZrO_{2-x}, Co_3O_{4-x}, TiO_{2-x}, SnO_{2-x}, and Nb_2O_{5-x}.[61] Further, binary metal oxide systems were studied for ORR in 0.5 M H_2SO_4 to improve ORR activity compared to TiO_x, such as $Ti_{0.7}Zr_{0.3}O_x$, and $Ti_{0.5}Ta_{0.5}O_x$. Thus, in extension, study with RuM (M = La, Mo, V) and IrM (M = La, Ru, Mo, W, V) revealed that Ru–LaO_x and Ir–VO_x show much better catalytic ORR activity compared to RuO_x and IrO_x in acidic media.

Inspired by biological catalysts in photosystem, MnO_2 was reported with $E_{1/2}$ of 0.73 V, which is 130 mV more negative compared to Pt/C in an alkaline medium.[62] In addition to that, improved ORR activity was achieved with the introduction of oxygen deficiencies on MnO_2 surface. Further, tabular monolayer superlattices of hollow Mn_3O_4 nanocrystals were reported to deliver outstanding ORR activity with an onset potential of 0.91 V and $E_{1/2}$ of 0.84 V (vs.

RHE).[63] Initially, Ag–Mn_3O_4–C composite was studied for ORR and zinc–air battery applications.[64] Later on, the enhanced ORR activity with similar Ag–Mn_3O_4 on Vulcan carbon was noticed because of the covalent coupling between Ag and Mn_3O_4 and O_2 spillover from Ag_2O/Ag to Mn_3O_4. A number of Fe_3O_4 based spinels such as 3D Fe_3O_4/NGR, Fe_3O_4/N–doped carbon, Fe_3O_4-Co_3O_4 yolk-shell nanostructure, and Co/Fe_3O_4 were reported as ORR electrocatalysts.

Similarly, rGO-supported Co_3O_4 was studied in an alkaline medium with $E_{1/2}$ of 0.83 V (vs. RHE) similar to that of Pt/C.[66] The synergistic effect between the metal oxide and graphene of the hybrid structure encouraged researchers to study it with carbon nanotube (CNT) support, where an onset potential of 0.93 V was achieved in 1 M KOH. In perovskites, the antibonding orbital (σ^*) occupation and the extent of covalency between the B-site metal and oxygen enhance the ORR activity. In this regard, $Mn_xCo_{3-x}O_4$/N–rmGO was synthesized varying the Mn/Co ratio for Mn–Co spinel oxides, which exhibited improved ORR activity. Among them, $MnCo_2O_4$/NrmGO showed better ORR performance where Mn^{3+} was present majorly along with a small amount of Mn^{4+}. A dual–phase $MnCo_2O_4$ (i.e., 77 wt% of cubic $(Mn_{1/3}Co_{2/3})[Mn_{1/3}Co_{2/3}]_2O_4$ and 23 wt% of tetragonal $(Co,Mn)[Mn,Co]_2O_4$) was reported where the Mn^{3+}/Mn^{4+} located at the octahedral B-site was more active than Co species and served as the active site for ORR.[67] Likewise, $NiCo_2O_4$ and $CuCo_2O_4$ were also reported to show good ORR performance.

7.5.2.3 Metal nitrides

For acidic ORR, metal nitrides have been reported to show lower ORR activity compared to metal oxides. For illustration, MoN was reported to show an onset potential of 0.58 V in 0.5 M H_2SO_4. A synergistic effect of Fe_xN and N-doped graphene aerogel was reported for ORR.[68] Later on, the electrocatalytic activity of a binary metal nitride ($Co_{0.6}Mo_{1.4}N_2$) was studied and reported to be higher than that of monometallic Mo_xN and highly stable in an acidic medium. In alkaline solution, the ORR activity of Cu_3N[69] was further optimized to show much better ORR activity (comparable to that of Pd) with Cu_3PdN binary nitrides (Fig. 7.3C).[65] The highest activity was achieved with a TiN/CNT composite, with an onset potential of 0.85 V.[70] Similar to surface modification, N-doping plays an important role for promotional ORR activity of metal oxides by lowering band gap and increasing electrical conductivity. Several oxynitrides, namely TiO_xN_y, $Co_xMo_{1-x}O_yN_z$, and HfO_xN_y were reported to exhibit excellent ORR activity in the potential window of 0.7 to 0.8 V. Carbon-supported NbO_xN_y was studied where the ORR onset potential shifted positively by 80 mV from 0.78 V compared to NbO_x.[71] Further, the promotional ORR activity with binary (Ba–Nb–O–N)[72] and ternary (Ba–Nb–Zr–O–N)[73] oxynitrides was achieved. The increased content of Nb^{5+} on the surface suppressing Nb^{4+} with the addition of Ba and Zr promotes ORR and proceeds via a $4e^-$ transfer pathway.

7.5.2.4 Metal carbides

Iron carbides were studied enormously for ORR in various pH values. N–Fe/Fe_3C@C nanorod comprised of Fe and Fe_3C to show superior ORR activity (compared to traditional Pt/C) in neutral medium.[74] Additionally, bamboo-like CNT/Fe_3C was reported to show excellent ORR activity in acidic and alkaline media along with outstanding methanol tolerance and superior robustness. The promotional activity is associated with the synergistic effect of Fe_3C and Fe/N/C active sites.[75] Tungsten carbide (W_2C) was seen to exhibit Pt-like behavior for HER. Also, W_2N had been developed for OER and ORR. Thus, the heterostructure of W_2N/WC was conveyed to exhibit significant ORR activity compared to their individual counterparts.[76]

7.5.2.5 Metal carbonitrides

Inspired by the high ORR activity of transition metal–N modified carbon and transition metal carbonitrides, Fe–C–N, Co–C–N, Cr–C–N, and Ta–C–N were studied to show increased ORR onset potential dramatically from 0.6 to 0.82 V when the N content was increased from 0 to 3%. To further increase the ORR activity of these carbonitrides, partial oxidation may create a metal carbonitride core and oxide shell structure. With an increase in the degree of oxidation of TaC_xN_y from 0 to 0.2, the ORR onset potential sharply increases.[77] The existence of oxygen vacancies may act as active sites for O_2 adsorption and electron transfer. A similar study was reported with ZrC_xN_y with an onset potential of 0.97 V.

Generally, the ORR onset potential trend follows: metal nitrides < metal carbonitrides < metal oxynitrides < metal oxides < partially oxidized metal carbonitrides.[78]

7.5.2.6 MXene-based electrocatalyst

MXenes are two-dimensional (2D) carbide and nitride materials having the general formula $M_{n+1}X_nT_x$ (where, M, X, and T stand for early d-bock transition elements, C and/or N atoms, and surface-terminating groups, such as –F, –OH, and/or –O–). The Ti_3C_2 MXene family was solely reported for ORR, where a single-layer Ti_3C_2 exhibited an onset potential of 0.85 V vs. RHE in 0.1 M KOH. Owing to the better conductivity and charge-transfer efficiency of MXenes, they could potentially behave as scaffolds or support M–N–C catalysts. Various works on MXene-supported M–N–C electrocatalysts were reported to exhibit excellent ORR activity, including FeCo–N–d–Ti_3C_2 MXene, MXene/NW–$Ag_{0.9}Ti_{0.1}$, $Ti_3C_2T_x$ MXene-based ORR electrocatalysts, where incorporation of antioxidants to reduce surface oxidation needs to be considered.[79]

7.5.2.7 Non-noble metal (M/N/C)-based electrocatalyst

To replace noble metal-based ORR electrocatalysts, non-precious metal–N–C composites were explored for ORR. For the first time, ORR activity with cobalt phthalocyanine (CoPc) was investigated. As a solution to the low stability and for identifying the active site to enhance their activity, bimetallic PANI–Fe–CoC was reported to show excellent ORR activity with an onset potential of 0.93 V vs. RHE (close to that of Pt-based electrocatalysts) in acidic medium.[80] The monometallic electrocatalysts were less stable, and the necessity of Fe and Co for ORR was reported. Also, enhanced ORR catalytic activity was achieved with nitrogen-doped CNT/carbon nanoparticle–Fe composite. Additionally, it was noted that in acidic medium, the presence of halide or low-valent sulfur species, such as SCN^-, SO_2, and H_2S on Fe/N/C electrocatalysts (derived from polyphenylenediamine) readily suppress the catalytic activity of ORR. Even the ORR catalysis of Fe/N/C electrocatalysts follows a different pathway in the presence and absence of Fe atom. Where Fe–N–C is promising to follow a $4e^-$ pathway, the N–C electrocatalysts prefer to follow a $2e^- + 2e^-$ reduction pathway. After the incorporation of Fe and Co on N-doped porous graphene, it was observed that the ORR onset increases from 0.88 V to 0.93 V vs. RHE in 0.1 M KOH. Incorporating Ni along with Fe in nitrogen-doped reduced graphene oxide (Ni–Fe–NrGO) exhibited 0.14 times better $E_{1/2}$ than commercially available Pt/C.[81] Further, Fe, Co, and Cu were incorporated in a π-extended ligand framework (2–NAPD@CV), and the resultant ORR onset gradually increases from 0.82 V to 0.87 V vs. RHE.

Likewise, the incorporation of multi-hetero metallic and nonmetallic atom (viz., S, P, B) enhances the intrinsic electronic configuration and eventually promotes ORR activity. For instance, S–Fe/N/C, Co@Co_9S_8/S–N–C[82] and CoO–NSC were reported to show excellent ORR activity.

7.6 Carbon Dioxide Reduction Reaction

Conceptually, CO_2 can be converted to various value-added products at potentials ranging from –0.25 V to +0.169 V vs. RHE. CO_2 being a thermodynamically stable molecule (viz., the cohesion energy of C=O bond is 750 kJ mol^{-1}), dissociation of CO_2 molecule becomes an energy-efficient process that has a high kinetic potential barrier.

7.6.1 Strategies to Improve CO_2RR: Suppression of HER

The major competitive/side reaction of CO_2RR is HER, particularly at higher potentials. Owing to the lower solubility (of 34 mM) in aqueous electrolytes and slow interfacial transport of CO_2, CO_2RR gets forbidden. Strategies to supress the HER and increase the selectivity of CO_2RR include increasing device pressure, lowering reaction temperature, replacing aqueous with solid electrolyte, and the use of a flow-cell device.

7.6.2 Mechanism of CO_2RR

In aqueous medium, the electrochemical CO_2 reduction (CO_2RR) follows a complicated multistep pathway. Depending on the properties of the electrocatalyst, the pH of the electrolyte, and the applied electrode potential, CO_2RR follows various electron transfer pathways to produce either C_1 products (viz., HCOOH, CO,

CH_4, and CH_3OH) or carbon-rich ($C_{\geq2}$) products (viz., C_2H_4, C_2H_5OH, 1-propanol) with high added values. When there are up to 16 different product formations possible from CO_2RR, the modes of adsorption and activation of the CO_2 molecule on the catalytic site (surface of the catalyst) are the deciding factors toward product selectivity. Pioneering work by Hori et al. found C_1 products at lower overpotential and $C_{\geq2}$ products at higher overpotentials.[83, 84] With applied potential bias, an anion radical – $CO_2^{\bullet-}$ species is formed[85] (from one electron transfer to the CO_2 molecule at E = – 1.9 V vs. SHE), the adsorption of which on the surface of the catalyst determines the reaction pathways. The thus-generated anion radical – $CO_2^{\bullet-}$ species is so unstable that it either gets readily protonated or stabilized through adsorption on the catalytic site of the electrocatalyst (Fig. 7.4).

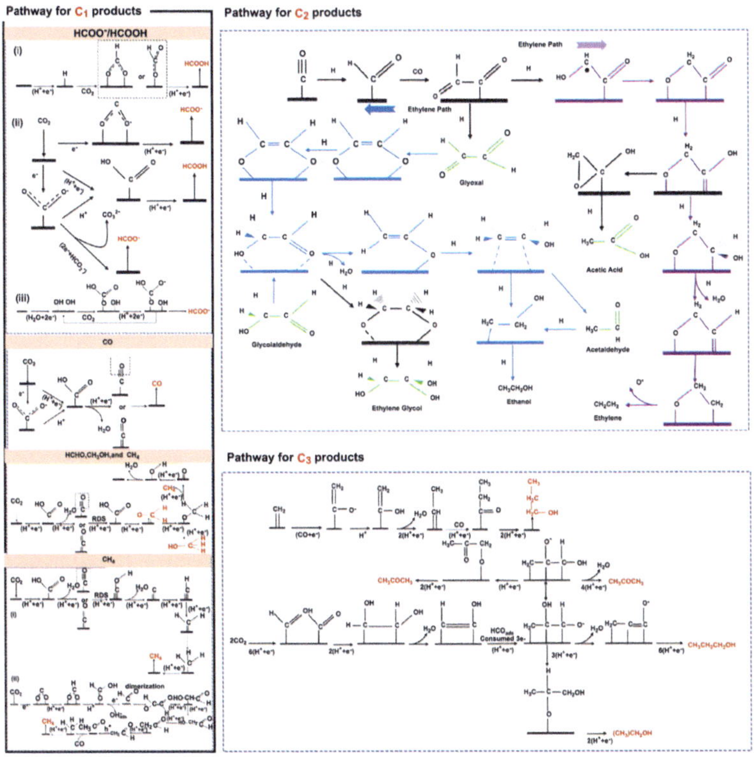

Figure 7.4 Mechanistic pathways of CO_2RR toward various C_1, C_2, and C_3 products. Reproduced with permission from [86], copyright 2021, Springer.

7.6.2.1 C_1 pathway

The final product from CO_2RR at lower (reduction) potential bias includes CO, HCOOH, CH_3OH, and CH_4, following simpler C_1 pathway. CO and HCOOH are easily formed at much lower overpotential and with high selectivity (> 90%). The CO and HCOOH formation are only two electron ($2e^-$) transfer process and has shown highest economic potential among other CO_2RR products. In most of the current studies, the intermediate involved in formation of CO and HCOOH are suggested to be – *COOH and *OCHO (or HCOO*). Following a proton-coupled electron transfer (PCET) step, *COOH is generated first, followed by the generation of *CO (through dehydrogenation of *COOH), which finally get desorbed from the catalytic site to release CO. In C_1 pathway, the adsorption strength of so generated anion radical – $CO_2^{\bullet-}$ species decides the selectivity between CO and HCOOH formation. When the intermediate is strongly bound to the catalytic site through C–atom (*COOH), CO is majorly formed as reduced product; while HCOOH is found to be formed with the weakly bounded (bidentate) O-atom (*OCHO or HCOO*) intermediate, owing to the medium adsorption energy. Other intermediates such as *CO, *CHO or *COH (by protonation of *CO) and *CH_3O (formed via protonation of *CHO) are also formed in the pathway, which is known to be PDS (potential determining step). Further hydrogenation of these intermediate to *CH_2O or *CH_3O yields methane (CH_4) and methanol (CH_3OH) through protonation at the methyl end and at the oxygen end via co-adsorbed *H, at the same time following five and three proton/electron transfer, respectively. But the catalytic site may be poisoned with strong binding of *CO and eventually HER predominates. From kinetic viewpoint, CH_4 formation through CHO* intermediate requires 1.21 eV (@ – 1.15 V vs. RHE) compared to CH_3OH formation where only 0.15 eV is required at same potential. But experimentally, CH_4 production is much feasible owing to the fact that the protonation at the C-site is favored over O-site, that is, reduced intermediate formation of *CHO is favored (viz., lower energy barrier) over CHO* (from *CO).

7.6.2.2 C_2 pathway

Through several protonation and electron transfer of *CO, several C_2 products are formed, which include C_2H_4, C_2H_6, C_2H_5OH, CH_3COOH,

$C_2H_6O_2$, etc. The C_2 pathway proceeds via three major steps: the stabilization of the so-generated *CO intermediate through protonation, the C–C coupling, and the post C–C coupling reactions. The intermediate *CO can either be protonated (at C-site or O-site) or get coupled via a C–C bond for stabilization. Different pathways are followed yielding different C_2 products via C–site or O–site protonation. When a proton gets bound to the C-site, $CH_2OH–CH_2OH$ is formed through CO insertion along with CH_4 (via a parallel five H^+/e^- pathway). Although CH_4 yields at higher amount owing to the lower energy barrier than that of $CH_2OH–CH_2OH$, on the contrary, protonation at the O-site yields a *CH_2 intermediate, which produces CH_3COOH and C_2H_6 through CO insertion and H^+/e^- step followed by C–C coupling of the *CH_3 intermediate, respectively.[87]

7.6.2.3 C_3 pathway

Long chain products formed through CO_2RR still remains a challenge and are limited to propyl alcohol ($CH_3CH_2CH_2OH$), iso-propyl alcohol ($(CH_3)_2CHOH$), and acetone ($(CH_3)_2CO$). The most proposed pathway toward C_3 or higher products involves intermolecular C–C coupling with an adjacent C_1 intermediate and subsequent H^+/e^- transfer. Also, three H^+/e^- transfers to CO_2 direct HCO_{ads} intermediate formation, which gets coupled with a C_2 intermediate species to form a C_3 adsorbed intermediate, from which 1-propanol/2-propanol and acetone are formed later on.[88]

7.6.3 Factors Affecting Catalytic Activity/Selectivity for CO_2RR

Despite having major pathways for CO_2RR, several physical factors (viz., pH of electrolyte, cation/anion of electrolyte, and overpotential) and chemical factors (surface adsorbates, structure, morphology, facets, defects, etc.) affect the adsorption energy and kinetic barrier of the reaction, which lead to different product selectivity through alteration of reaction pathways.

7.6.3.1 Effect of pH

Unlike water splitting, CO_2RR involves several protons/electrons (H^+/e^-) transfer and multiple elemental reduction steps. Depending on the number of proton availability and proton transfer ability (i.e., local pH near electrode/electrolyte interface), stability of the intermediates, and the relative kinetics of each pathway lead to product distribution in CO_2RR. Recently, a study on planar polycrystalline Cu in basic medium revealed the pH dependency for C_1 product formation, whereas C_{2+} products formation is pH independent. Another pH-dependent study on Cu (111) summarizes that at lower pH (=1), C_1 pathway (toward CH_4 formation via *COH intermediate) predominates over C_2/C_3 pathway (kinetically suppressed); at near neutral pH, C_1 and C_2/C_3 pathways share common intermediate (*CHO), and at higher pH (=12), C_2/C_3 pathway predominates by kinetically blocking C_1 pathway, thereby accelerating C_{2+} products.

7.6.3.2 Overpotential

Besides local pH, DFT calculation shows that the preferable pathway toward product selectivity depends on the applied overpotential. With an applied more negative potential (−0.6 to −0.8 V (vs. RHE)), the active *H and *CO intermediates compete to get adsorbed on the catalytic sites, which decrease the CO coverage and inhibits CO dimerization and *CO–COH is dominant, which eventually prevents C_2 product formation. When the potential is raised toward lower overpotential, *CO–CO dimerization pathway leads to the formation of C_2H_4 over CH_4 owing to the lower energy barrier for CO dimerization (ΔG = 0.69 eV). Moreover, regulating the potential to be more positive than −0.7 V (vs. RHE) leads to get CO and $HCOO^-$ significantly, whereas C_2H_4 and C_2H_5OH are produced between −0.8 V and −1.1 V (vs. RHE). Beyond −1.1 V (vs. RHE), CH_4 is selectively formed.

7.6.3.3 Nature of Electrolytes

7.6.3.3.1 *Aqueous electrolyte*

Mostly CO_2RR was studied either in weakly acidic or alkaline electrolytes containing HCO_3^-, SO_4^{2-}, and Cl^- as the anionic part along with alkali metal cations like Na^+, K^+, etc. The catalysts' reactivity and selectivity are regulated by the cation present in the electrolyte with alteration of hydrogen coverage on the electrode surface. When the cations are in the close proximity to the electrode, the potential at the outer Helmholtz layer differs, and as a result the hydrogen coverage at the electrode surface gets affected. On a Cu electrode, with an increase in Stokes radii of the hydrated cations ($Li^+ > Na^+ > K^+ > Cs^+$), selectivity of C_2H_4 production over CH_4 was studied. The rate of CO_2RR increases with increasing size of the anions of the electrolyte, such as $Cl^- > Br^- > I^-$. The electron from the halide ion (X_{ad}^-) is transferred to the vacant orbital of CO_2, facilitating charge transfer to CO_2 and increasing the rate of CO_2RR. It is also found that while the addition of Cl^- and Br^- in the electrolyte favors CO selectivity, the addition of I^- suppresses CO formation, and CH_4 was formed exclusively in a halide-free electrolyte. Additionally, the rate of the CO_2RR increased at lower overpotential toward CH_3OH selectivity with increasing surface charge of (multivalent) cation of the supporting electrolyte in the order: $Na^+ < Mg^{2+} < Ca^{2+} < Ba^{2+} < Zr^{4+} < Nd^{3+} < La^{3+}$. Large cation suppresses the competing HER and favors the C_1 product (viz., HCOOH on Hg electrode, CO on Ag electrode), except for a Cu electrode, where C_2H_4 predominates.

7.6.3.3.2 *Non-aqueous/organic electrolytes*

Non-aqueous solvents (DMF, DMSO, acetonitrile (AN), and CH_3OH) have more advantages for CO_2RR, be it the higher solubility or less availability of protons which helps in suppressing HER and eventually promotes CO_2RR. It is found out that DMF, DMSO, and AN facilitates $CO_2^{\bullet -}$ dimerization with free CO_2 to produce $C_2O_4^{2-}$ on Sn, Pb, In, and Hg electrodes, whereas CH_4 is selectively formed in CH_3OH using Cu electrode. When Cr-Mo-Ni steel was used as cathode in AN and propylene carbonate (PC) electrolyte, oxalic acid was major product. Interestingly, addition of slight amount of water gives formate exclusively.

7.6.3.4 Facet dependency

Recent studies on Cu electrode clearly demonstrate the facet dependency in CO_2RR reactivity and product selectivity. Cu catalyst with higher exposure of Cu(100) showed C_2 product selectivity compared to bulk polycrystalline Cu surface. Cu(100) favors C_2H_4, while Cu(111) favors CH_4 selectivity.[89] Later on, facet-dependent similar selectivity (toward C_2 products) was found with differently shaped Cu NPs (viz., Cu_{sph}, Cu_{cub}, and Cu_{oh}) (Fig. 7.5A).[90] The result was further supported by DFT study where the reason stated being the intermediate $*CH_2$ formation is favored on Cu(111) for which CH_4 is formed, whereas Cu(100) is favorable for C–C coupling and yields C_2H_4 majorly. Similarly, facet dependency was seen with Cu_2O electrocatalyst for CH_3OH formation. They observed $Cu_2O(110)$ facet is selective for CH_3OH formation, whereas $Cu_2O(100)$ is inert.

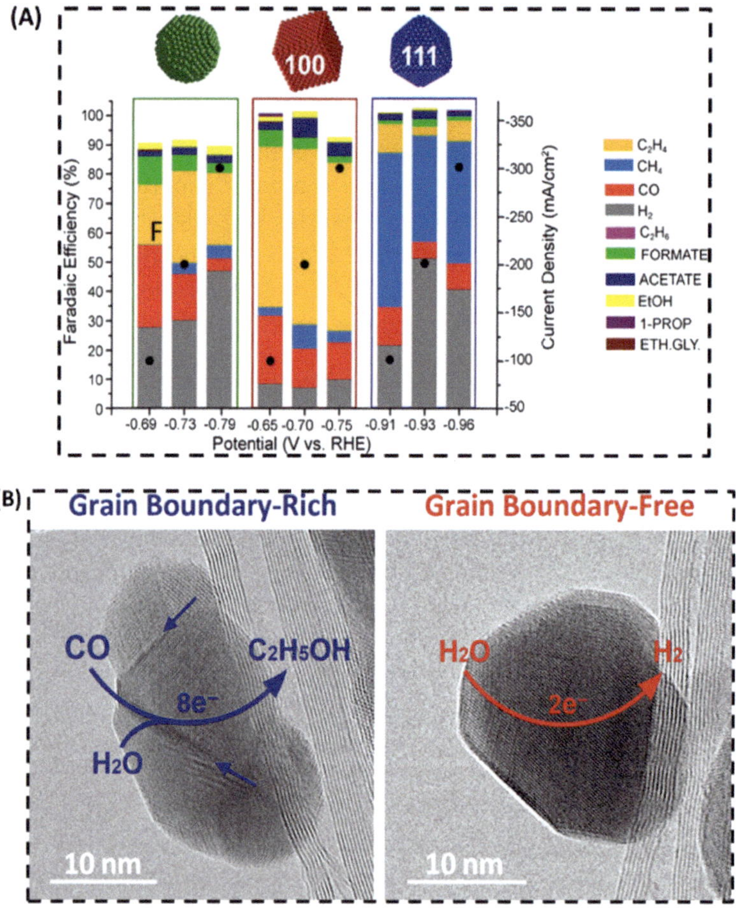

Figure 7.5 (A) Faradaic efficiency of various CO_2RR products at different potentials with sphere, cube and octahedral Cu nanocrystals. Reproduced with permission from [90], copyright 2020, ACS. (B) Grain boundary rich Cu NP/CNT electrochemically converting CO to C_2H_5OH. Reproduced with permission from [91], copyright 2016, ACS.

7.6.3.5 Defect site

Inducing defects from vacancies or grain boundaries leads to improve CO_2RR activity and product selectivity, generating active sites on the surface. With an increase in the density of grain boundaries in Cu NP CO electroreduction follows a pathway toward C_{2+} products (C_2H_5OH, CH_3COO^-) (Fig. 7.5B).[91] Similarly, the presence of grain boundaries increased formic acid production catalyzed by SnO_2 quantum wires and nanotubes via CO_2RR.

7.6.4 Materials for CO_2RR

7.6.4.1 Metal-based materials

The $CO_2^{\bullet-}$ anion radical intermediate generated initially in C_1 pathway is very unstable. Its stabilization is crucial to gain high-rate, energy-efficient CO_2RR activity. HCOOH was found selectively in early stages of CO_2RR with metal electrodes. For instance, indium (In) electrode was studied in 0.2 M $LiHCO_3$ to get the highest yield of HCOOH at −1.4 V vs. SHE.[92] Later on, measurable amounts of CO and CH_4 along with HCOOH were revealed, and it was stated that the product selectivity is highly dependent on the purity of the metal electrode as well as the electrolyte solution. Based on the early-studied metal electrodes, the availability of vacant orbitals and active *d* electrons in these metals tends to bind with intermediates and they are classified in four groups depending on their binding efficiency:[93]

Group 1: Do not bind $CO_2^{\bullet-}$ intermediate (either *OCHO or *COOH) and cannot reduce CO (includes Pb, Hg, In, Sn, Cd, Bi, and Tl).

Group 2: Bind $CO_2^{\bullet-}$ intermediate (*COOH) but cannot reduce CO (includes Au, Ag, Zn, and Ga).

Group 3: Bind $CO_2^{\bullet-}$ intermediate and effectively reduce CO (includes Cu).

Group 4: Bind *H instead $CO_2^{\bullet-}$ intermediate and active for HER (includes Ni, Fe, Pt, and Ti).

7.6.4.1.1 *Group 1 metals: Selectivity toward HCOOH*

HCOOH is the major reduced product formed from the Group 1 metal (selectivity toward 2e⁻ product) as the pathway beyond CO are suppressed owing to weak binding with H and *OCHO intermediate get stabilized. In situ CO_2RR study with Sn electrode revealed $Sn_6O_4(OH)_4$ and SnO_2 nanoparticles to be catalytically active for selective HCOOH production. A surface-bound carbonate intermediate (formed via reaction of Sn^{II} and CO_2) is responsible to form HCOOH via $2e^-/H^+$ transfer. Additionally, HCO_3^- ions as electrolyte were found to boost HCOOH production with these metals. The HCOOH selectivity with Sn electrode were further supported with Sn dendrite, electrodeposited Sn, Sn-Pd alloy with improvement in FE. Further, to increase stability, 3D-porous structure of Cu on Sn was prepared which not only showed HCOOH selectivity (FE = 100%) but was also stable for 15 h in aqueous $KHCO_3$. Most commonly used metal electrodes for selective HCOOH production are Sn, In, and Bi; but, FE forbids with HER at higher potential. To address that, Sn-Bi-In alloy family was prepared for HCOOH selectivity. Among them, $In_{0.6}Bi_{0.2}Sn_{0.2}$, and $In_{0.4}Bi_{0.6}$ alloys showed 95 to 98% FE for HCOOH in the potential range of −1.2 V to −1.3 V vs. RHE.

7.6.4.1.2 *Group 2 metals: Selectivity toward CO*

On the contrary, because of the higher O affinity and weak *CO-binding ability of Group 2 metals, *COOH gets strongly bound for further reduction and *CO desorbs easily to form CO as a major product. An increase in pH or pressure boosts *COOH formation over *CO. Au and Ag, being on top of volcano plot for CO_2RR, have optimal binding energy for *COOH intermediate, which leads to the formation of CO with high FE with a lower overpotential. A limitation arises with low abundance and high cost of these metals. One of the effective ways to address this is by using them as NPs. For instance, the catalytic activity was studied by reducing the size of Au NPs from 7.7 nm to 1.1 nm. Although the performance improved, a lowering in CO formation (viz., FE) with increasing HER was observed as the edge-to-corner ratio decreases. Similar results were found for Ag NPs. Although, this finding was further contradicted when edge promoting HER while corner shows CO evolution in case of Au, whereas edges are responsible for CO evolution for Ag. Further, size effect of Au trisoctahedron (Au TOH) NP exhibited 1.5 times increment in FE of CO compared to colloidal Au. They hypothesized that edge site of Au TOH is the active catalytic site which balances *COOH stability and *CO desorption free energy, and as a result, with an increase in particle size, edge-to-face ratio decreases and gives rise to deterioration in performance. Similarly, improvement of Ag NP was studied with high surface area Ag nanocoral, which reported a 32-fold increment in catalytic activity. Additionally, Ag nanocubes were also studied, which showed 99% FE for CO in ultralow overpotential and with higher energy efficiency and durability.

Apart from Au and Ag, due to its ability to selectively pick either CO or HCOOH, Pd comes out to be an emerging alternative. Previously, based on in situ and DFT studies on Pd, two phases of hydrogen-adsorbed Pd surface, namely α- and β-palladium hydride (PdH) were found, which are responsible for product selectivity. While α–PdH core promotes HCOO* intermediate and yields HCOOH, β–PdH selectively forms CO via COOH* instead. However, a better catalytic activity of Pd core with Pd-N_4 active surface site was reported disagreeing with this agreement. Although, facet-dependent (experimental and DFT) study found out that high index (310) facet of concave cubes stabilizes *COOH intermediate and facilitates *CO desorption from the facet, which displayed the best activity and FE selectivity (91%). Several other Ag electrodes, including Ag nanowire, Ag/Ni foam, etc., are reported to selectively form CO.

Zn comes out to be an alternative to noble metals like Au and Ag for selective production of CO from CO_2RR. Zn electrocatalyst is seen to show better performance in NaCl solution compared to $NaHCO_3$ solution. The FE for CO increase from 60% in $NaHCO_3$ to 93% in NaCl, the reason being the presence of Cl^- suppresses the competing HER and helps boosting CO formation stabilizing $CO_2^{\bullet-}$ anion radical intermediate. The trend was also validated with Ag NPs. Recently, owing to the local pH effect and surface area increment, porous Zn electrocatalyst are being used for promotional performance.

7.6.4.1.3 *Group 3 metals*

In Group 3, Cu metal having optimal *CO binding energy, which allows hydrogenation of *CO to follow a pathway for hydrocarbons and enables C–C coupling between the intermediates and C_{2+} products are formed.

So far, Cu metal has been in the forefront for CO_2RR toward alcohol products and especially C_2 or C_{2+} products. For the very first time, increased catalytic activity for CO_2RR with decreasing size of Cu NPs was reported, where the product selectivity was favored toward H_2 and CO with decreasing size of Cu NPs. Interestingly, especially pretreated Cu foil for CO_2RR yields CO and HCOOH at lower overpotentials but with rise in overpotential, ethylene and ethane were found replacing CO and HCOOH. Similarly, Cu^+ (present in plasma-activated Cu electrode) was claimed to lower the onset potential and selectively produce ethylene. Simultaneous presence of Cu(0) and Cu^+ at the catalyst surface enhanced the multi-carbon products (FE for C_2 > 60%) formation when Cu_3N nanoparticles was employed as an electrocatalyst.

7.6.4.1.4 *Group 4 metals*

In an interesting study with 32 metal electrodes for CO_2RR, it was shown that Ni and Pt (which were believed to participate in HER at ambient temperature and pressure) can reduce CO_2 to CO or HCOOH under elevated pressure (60 atm).

7.6.4.1.5 *Anomalies with other metal electrodes*

Other than selective CO or HCOOH formation, formation of higher carboxylic acids in aqueous acid was found. For instance, an initial report on formation of malate and glycolate from Hg and Pb electrodes, respectively, was contradicted by the study of Kaiser and Heitz. They demonstrated that oxalic acid is produced initially (in aqueous medium in presence of quaternary ammonium salts), which gets reduced to glycolic, glyoxylic, and maleic acids at Hg, Pb, Ni, and Al electrodes.[94] In a similar electrolyte (aqueous tetraethylammonium), when Pb, Sn, and In were used as electrodes, oxalic as well as propionic acid were found to be formed.

Bismuth (Bi), being cheap and nontoxic, is a potential electrocatalyst for CO_2RR to HCOOH in an aqueous medium. However, the selectivity changes to CO when used in an ionic liquid electrolyte. The promotional activity of Bi catalyst in imidazolium-based ionic liquid is mainly because of possible interaction between Bi^0 and Bi^{3+}, which helps in stabilizing $CO_2^{\bullet-}$ anion radical intermediate.

7.6.4.2 Metal oxide

Similar to Sn, SnO_2 selectively produces HCOOH. The activity along with FE was improved via imposing defects (grain boundary), mesoporosity, etc. Similarly, defect-rich Bi nanotubes (derived from β-Bi_2O_3 double-walled nanotube) showed excellent HCOOH selectivity. To improve the activity of oxides, in situ electroreduction of metal oxides to oxide-derived (OD) metal NPs were strategized. OD–Au, Ag selectively yield CO; whereas OD–Sn, Pb, and Bi are selective toward HCOOH. Meanwhile, OD–Cu was found effective for CO reduction.

Interestingly, for Cu/SnO_x heterostructure the selectivity alters toward CO initially (for 6.2 % SnO_x) but shifts toward HCOOH as the amount of SnO_x increases. Similarly, In_2O_3 is known to be selective for HCOOH formation, but a heterostructure with CuO (CuO/In_2O_3) outperformed several metal oxide electrocatalyst for CO_2 to CO conversion.

7.6.4.3 Metal nitride

Nickel nitride (Ni_3N) was found to show more selectivity toward CO in 0.5 M NaCl compared to $NaHCO_3$ solution. Similarly, Cu_3N-derived Cu nanowire with high-density grain boundaries exhibited excellent selectivity toward C_2 products with high FE.

7.6.4.4 Metal phosphide

Han's group was first to report 96% FE for HCOOH formation with MoP-supported Indium-doped porous carbon using ionic liquid (1-butyl-3-methylimidazolium hexafluorophosphate) as electrolyte. C_3 and C_4 selectivity was reported with various phases of nickel phosphides (Ni_3P, Ni_2P, $Ni_{12}P_5$, Ni_5P_4, and NiP_2) in aqueous media. The phosphorus-rich phase NiP_2 yields 2,3-furandiol (FE = 71%) and methyl glyoxal (FE = 84%) at 0 V and –0.1 V (vs. RHE), respectively. Unlike radical coupling of *CO intermediate for C–C coupling, a hydride transfer mechanism to O-bounded intermediate was proposed followed by a self-coupling condensation. This mechanism was also supported when 1-butanol was traced along with HCOOH and CH_3COOH with CuP_2. After that, CO_2RR was studied (in 0.1 M $KHCO_3$) with boron phosphide (BP) to produce CH_3OH with 92% FE. The B and P synergy promotes the activation of CO_2 and binding of intermediates. Similar result was found for earth-abundant FeP. Further, for Cu_3P HER was the major reaction along with trace HCOOH; Ni_2P favors 2,3-furandiol exclusively, while ethylene glycol (majorly) along with HCOOH, methylglyoxal and 2,3-furandiol was formed with isostructural Fe_2P. The loosely bound surface hydrides on Fe_2P are more reactive compared to Ni_2P (owing to the lesser number of *d* electron for Fe), which shifts product selectivity toward ethylene glycol for Fe_2P. Interestingly, Ni_2P/Ho_2O_3 core-shell nanoparticles yield C_3 product (acetone with FE = 25.4%) where the heterostructure modulates the binding of intermediates and leads to C_3 pathway. While Ni_2P and Ho_2O_3 individually were mainly prone to HER and produced very less C_1 products (viz., HCOOH, CH_3OH), the Ni_2P/Ho_2O_3 core-shell structure enhanced C_3 selectivity (Fig. 7.6A).[95]

7.6.4.5 Metal sulfide

Metal sulfide cluster on Ir and Co was first studied to produce oxalate selectively. Incorporation of sulfur not only improves selectivity but also alters the selectivity too. Initial report with CuS on nickel foam revealed CH_4 selectivity along with CO, while formate selectivity was shown by carbon-supported CuS. Recently, CuS nanosheet on brass exhibited excellent HCOOH selectivity. It is concluded that Cu^0 and residual S collectively lowered the binding energy of *COOH and HCOO* intermediate and thereby directed toward HCOOH selectively. The above reports illustrate the C_1 product formation with copper sulfides, whereas Cu is known to form C_2 products. Similarly, Cd selectively produces HCOOH, while CdS produces CO exclusively. Other studied metal sulfides include MoS_2, NiS, ZnS, and Ag_2S, which are selective toward CO formation; contrarily, Co_3S_4, Cu_2S, In_2S_3, SnS_2, and SnS are selective toward HCOOH. In a recent study, SnS_2 nanosheets decorated with in situ generated single Sn atoms coordinated with oxygen atoms showed maximum selectivity toward ethanol formation with FE of more than 80%. Presence of dual active sites comprising Sn and O atoms promotes C–C bond formation.

7.6.4.6 High entropy alloy

Owing to the high overpotential and limited selectivity of (single) metal-based catalyst, HEAs are better alternatives to maximize catalytic activity and product selectivity. Intrinsic properties (viz., lattice distortion, high-entropy effect, sluggish diffusion, cocktail effect) in HEA could be tuned and thereby optimizing reaction kinetics of rate-determining step improves selectivity especially for multicarbon products. CO_2RR with AuAgPtPdCu HEA resulted 100% FE for gas products (H_2, CO, CH_4, and C_2H_4). Compared to

Cu, the Cu^{2+}/Cu^0 redox and the synergy between other metals in HEA mainly optimize the binding of $*OCH_3$ and *O intermediates in a much better way and promote better performance. Recently, a layered high-entropy transition metal sulfide (with group V and VI transition metals—Mo, W, V, Nb, and Ta) was reported to be highly selective toward CO formation and also showed high catalytic activity. Similarly, a highly stable oxide socketed (CNF@TiO_2) HEA showed selectivity toward CO. Besides, PdCuAuAgBiIn HEA aerogel achieved 100% FE for HCOOH.

7.6.4.7 Interfacial engineering for CO_2RR

The limitations of CO_2RR such as high energy input, low solubility, low selectivity, etc., are encountered through interface engineering more effectively. Apart from bulk phase atom, active sites located at the interface are more effective for adsorption or desorption of the intermediates, which regulate the binding energy along with number of electron transfer. Owing to the unique electronic versality, Cu is often integrated with other metals for improved activity and selectivity. As an illustration, Ag/Cu or Ag/Cu nanodimers were studied to have even better activity and C_2H_4 selectivity compared to individual metals or Ag–Cu alloy. Additionally, Cu/Sn and Zn/Bi interface was seen to enhance the FE toward HCOOH. Further, the activity toward HCOOH production was promoted through SnO_2/Sn_3O_4, SnO_2/Bi_2O_3,[97] etc. The electron transfer at the heterointerface enhances the adsorption and activation of CO_2 stabilizing $CO_2^{\bullet-}$ intermediate. Bi_2O_3 was found to favor the adsorption of HCOO* intermediate and improve the conversion of CO_2 into HCOOH on SnO_2/Bi_2O_3.

Furthermore, Au/Cu and Au/Ti were addressed to improve far better activity toward CO formation by stabilizing *COOH intermediate through oxygen–metal interaction. Likewise, atomically dispersed Fe atoms on Au to tune the electronic structure (Fe_1/Au), improve CO selectivity. Interestingly, 4H Au@Cu and 4H/fcc Au@Cu were exhibited to have much better current density along with better C_2H_4 selectivity rather than CO (Fig. 7.6B). The DFT understanding supports the low-energy barrier for *CHO formation, leading to a kinetically feasible formation of C_2H_4.[96] The synergetic effect at Cu–O–Mo between Mo and Cu was investigated to yield CH_3COOH with FE of 48.68 %.

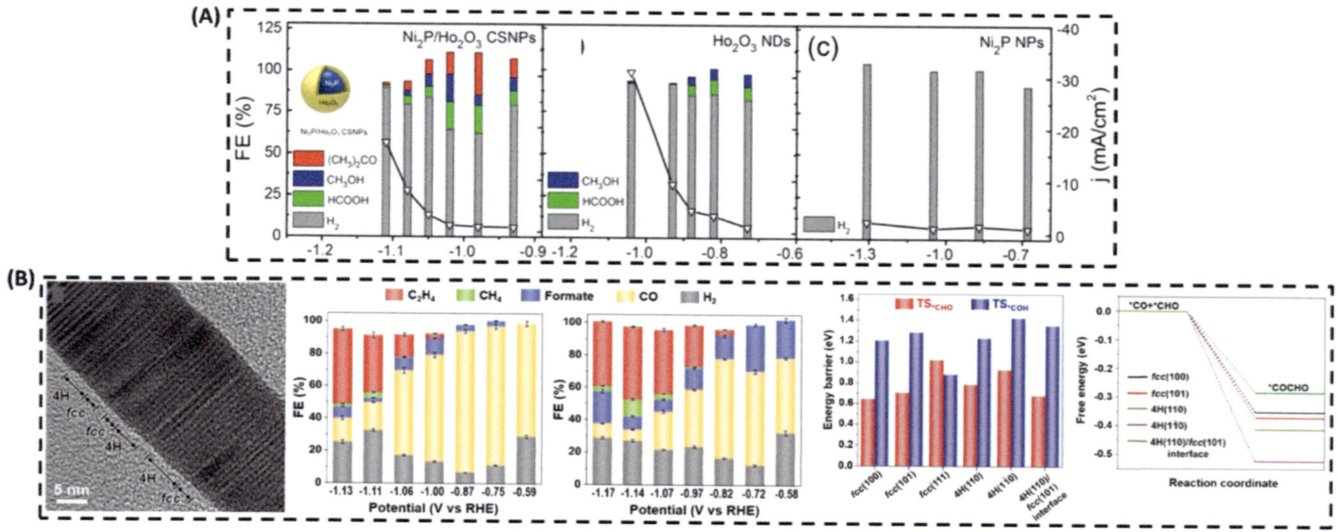

Figure 7.6 (A) Faradaic efficiency of various CO_2RR products at different potentials with Ni_2P/Ho_2O_3, Ho_2O_3 and Ni_2P as electrocatalysts. Reproduced with permission from [95], copyright 2020, ACS. (B) HRTEM image of 4H/fcc Au NR, Faradaic efficiency of different CO_2RR products varying potentials with 4H Au@Cu and 4H/fcc Au@Cu, energy barrier for *CO hydrogenation on various Cu surface and free energy diagram for C–C bond formation. Reproduced with permission from [96], copyright 2020, ACS.

Besides, Cu_1^0 – Cu_1^{x+} atom pair was reported to show FE of 92% for CO production with complete inhibition of HER. Interestingly, Cu^0 and Cu^{x+} interface of g-C_3N_4 supported tetrahedral Cu_4 cluster (Td-Cu_4@g-C_3N_4) prepared showed selective conversion of CO_2 to C_2H_5OH. Another Cu^+/Cu^{2+} interface on Cu sandwich electrode was also found to produce C_2H_5OH with 31% FE.

Moreover, the in situ-generated $Sn_{2.7}Cu$ from hierarchically heterogeneous Sn–Cu alloy/Sn metallic core catalyst exhibited highest CO_2RR activity (FE ≥ 99%) among all the Sn–Cu-based electrocatalysts.[98] Recently, in an investigation through a DFT study, thin hexagonal boron nitride (*h*–BN) on Ni and Co interface was found to lower H binding. Thus, suppressing HER, the predicted catalyst would selectively produce HCOOH.

7.7 Nitrogen Reduction Reaction

7.7.1 Thermodynamics and Kinetics of NRR

Nitrogen (N_2) is a thermodynamically stable molecule. According to Molecular Orbital Theory (MOT), the electronic configuration of N_2 is $(1\sigma_g)^2(1\sigma_u^*)^2(2\sigma_g)^2(2\sigma_u^*)^2\,(1\pi_u)^4(3\sigma_g)^2$. The energy gap between the $3\sigma_g$ (highest occupied molecular orbital, HOMO) and $2\pi_g^*$ (lowest unoccupied molecular orbital, LUMO) is as high as 10.82 eV, for which transition of electron is not energetically feasible. From kinetic point of view, N_2 is having high bond dissociation energy of 941 kJ mol^{-1} and first H addition to N_2 is endothermic ($\Delta H° = 37.6$ kJ mol^{-1}) along with the negative electron affinity (−1.9 eV) as well as high ionization potential (15.85 eV) limits effective NRR.[13]

Naturally, nitrogen fixation occurs from N_2 to NH_3 through nitrogenase (containing FeFe, FeMo, FeV cofactors) metalloenzyme at ambient condition. But owing to tough extraction techniques, poor recycle ability, stability, and inertness (in presence of oxygen) of these nitrogenases, an alternative toward nitrogen fixation catalyst need to be developed. Among them, FeMo nitrogenase consists of two fused Fe–S cluster with a C atom at spiro position and a Mo in one of the apical positions. A minimum of 16 equivalent of adenosine triphosphate (ATP) or 244 kJ mol^{-1} is required to reduce N_2 to NH_3:

$$N_2 + 8H^+ + 8e^- + 16ATP \rightarrow 2NH_3 + H_2 + 16ADP + 16P_i$$

where ADP is adenosine diphosphate and P_i is a mixture of HPO_4^{2-} and $H_2PO_4^-$ ions, also known as inorganic phosphate.[99] The rate of NH_3 production follows first-order kinetics with respect to N_2, while the rate of H_2 production is first order with respect to both proton and electron concentration.

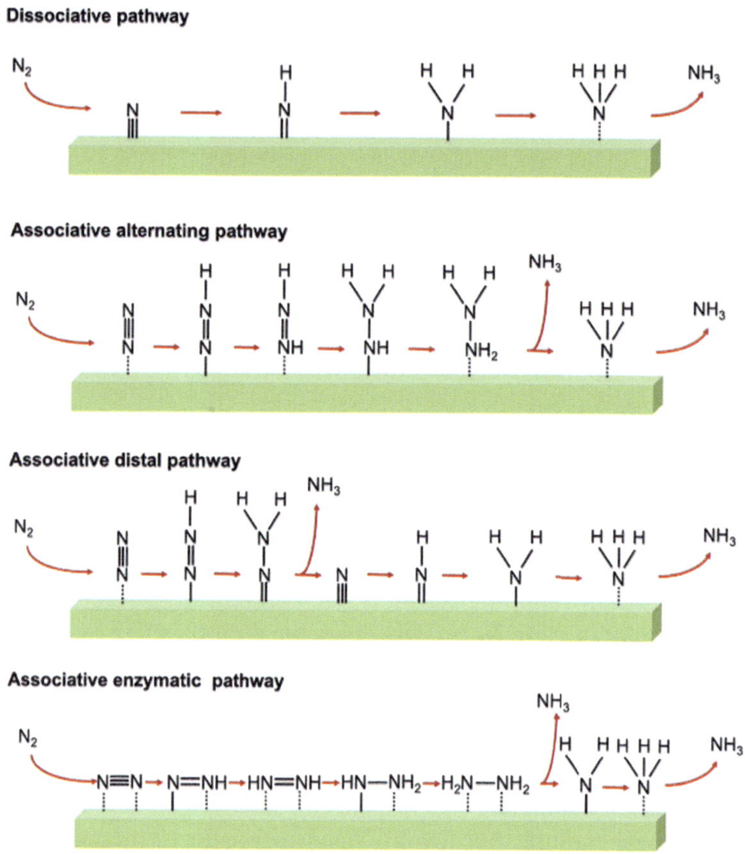

Figure 7.7 Mechanistic pathways of NRR: (A) Dissociative pathway, (B) Distal and alternating associative pathway, and (C) Enzymatic pathway. Reproduced with permission from [100], copyright 2021, Sichuan University and John Wiley & Sons Australia, Ltd.

7.7.2 Mechanism of NRR

For heterogeneous catalysts, N_2 reduction follows two major pathways, such as *Associative* and *Dissociative* pathway (Fig. 7.7).[100]

In *associative pathway*, N_2 molecule gets adsorbed to the surface with an end-on mode. The triple-bonded N≡N gets hydrogenated with simultaneous bond breaking to release NH_3. Now, depending on the hydrogenation of N centers, this associative pathway is further classified into: (a) distal pathway and (b) alternating pathway. In *associative distal* pathway, hydrogenation preferentially occurs at distal N atom (N atom not attached to the catalyst surface) until the first NH_3 molecule gets released. After that, the remote N (attached to the catalyst surface) similarly gets hydrogenated and releases NH_3. On the other hand, in *associative alternating* pathway, both the N atoms are hydrogenated simultaneously which results simultaneous desorption of two NH_3 molecules. In addition, there is another associative pathway, that is, the *enzymatic pathway* (pathway iii), during which both N atoms are anchored on the catalyst surface with a side-on mode.

On the contrary, in *dissociative pathway*, N≡N bond breaking occurs at first. The individual N atom gets hydrogenated on the surface of the catalyst followed by desorption NH_3 molecule.

7.7.3 Materials for NRR

7.7.3.1 Metal-based catalyst

At the earliest, molecular Ti–Al system was reported for electrochemical NRR. In a glyme solution of Ti and Al, electrochemically generated Ti(II) species serve as active sites, where they produce 10.2 mmol of NH_3 with FE of 19.7% by applying a cell potential of 40 V for 11 days. Later in 1983, a heterogeneous metallic Fe-based catalyst for NRR was reported in 6M KOH.

Although being noble metals, Ru and Au have been studied and showed promising NRR activity. For instance, Ru (001) is reported to have a strong binding ability for N≡N and showed excellent electrochemical activity for NRR. Tetrahexahedral Au (THH Au) nanorods with exposed high-indexed (730) crystal facet were studied for efficient NRR electrocatalyst. The (730) facet is a combination of (210) and (310) sub-facets which undergo an alternative hydrogenation pathway to form NNH* at the rate-determining step. Again, hollow Au nanocages (Au HNC) were studied and found to exhibit better NRR activity compared to nanorods (NR), nanosphere (NS), and nanocubes (NC).[101] Defect-rich Au/TiO_2 exhibited surprising activity with FE of 8.11% at –0.2 V vs. RHE when studied in 0.1 M HCl for NRR. The high production rate of 21.4 mg h^{-1} cm^{-2} was attributed to the Au–O–Ti bonding, where the partial positive charge on Au facilitates more N_2 adsorption almost surpassing positively charged H^+. Interestingly, when Au/C_3N_4 was studied for NRR in H_2SO_4 electrolyte NH_4^+ was found to be formed at a rate of 305 mg h^{-1} mg_{Au}^{-1} (FE = 11.1 % @ –0.1 V vs. RHE), which is 22.5 times higher compared to Au NP/C_3N_4. A good balance of crystallinity was introduced by CeO_x in a-Au/CeOx–rGO, where low crystalline CeO_x was found to play crucial role in NRR. An NH_3 yield rate of 8.3 mg h^{-1} mg_{cat}^{-1} and FE of 10.10% at –0.2 V vs. RHE in 0.1 M HCl were obtained, which were far better than its crystalline counterpart, c–Au/rGO. The NRR activity of Pd/C was far ahead compared to Au/C and Pt/C at –0.05 V. The in situ-formed α-Pd hydride phase facilitates N_2 hydrogenation to *N_2H (via a lower energy pathway) through a Grotthuss-like proton-hopping mechanism (Fig. 7.8A).[102] Moreover, Pt/C was studied as NRR electrocatalyst in 0.5 M H_2SO_4 with a mixed NH_4^+/H^+ pre-exchanged Nafion membrane. A maximum NH_3 yield rate of 1.14 × 10^{-5} mol m^{-2} s^{-1} and was obtained at a cell voltage of 1.6 V, which was further modified using mixed Li^+/NH_4^+/ H^+ Nafion membrane and 0.1 M Li_2SO_4 electrolyte at 80°C. Similarly, Pt/C and Ir/C-loaded gas diffusion electrodes were tested for NRR in a PEM type electrochemical cell. Imposing oxygen vacancies on Ru stabilises *NNH (destabilizing *H) intermediate, and tripodic PdRu was reported to further improve NH_3 production.[103]

Theoretical studies revealed that Mo, Fe, Rh, and Ru are effective metals toward NRR owing to their greater π back bonding from d orbitals of N_2, placing on top of the volcano plot for catalytic activity. Apart from that, Sc, Y, Ti, and Zr shows greater affinity toward N atom compared to H adatoms but lack in terms of higher overpotential.[104] Mo(110) nanofilm also showed high activity for NRR (NH_3 formation rate of 3.09 × 10^{-11} mol $s^{-1}cm^{-2}$ @ –0.49 V vs. RHE) compared to commercially available Mo foil in 0.01 M H_2SO_4. For over a century, in the industrial Haber–Bosch process, Fe catalyst has been used for NH_3 production. But low NRR kinetics in alkaline medium, high corrosion, and dissolution in acidic medium limits Fe-based electrocatalysts for NRR. Although, bismuth (Bi) is widely studied for photochemical nitrogen fixation, owing to the strong interaction between Bi 6p and N 2p, nitrogen adsorption on Bi boosts NRR selectivity with higher intrinsic activity. But the semiconductor nature limits the fast electron supply. Taking advantage of poor binding ability for H adatom, remarkable NH_3 production (5.453 μg mg_{Bi}^{-1} h^{-1} at −0.6 V vs. RHE) with FE of 11.68 % was achieved in 0.2 M Na_2SO_4 with defect-rich semimetal Bi (110) nanoplates.[105] Further, metallic Bi was explored for NRR by in situ-generated mosaic Bi nanosheets (from electrochemical reduction of bismuth-oxyiodide) (Fig. 7.8B).[106] Unexpectedly, so-generated Bi nanosheets exhibited NH_3 yield rate of ~13.23 μg mg_{cat}^{-1} h^{-1} and FE of 10.46 % in 0.1 M Na_2SO_4 at −0.8 V vs. RHE. Following a distal pathway, the surface-delocalized p electron on the surface eases end-on adsorption of N_2 for selective reduction to NH_3. Additionally, Bi nanocrystal and K^+ cation came up with the greatest breakthrough in NRR.[107] The K^+ cation stabilizes the key intermediates which regulate proton transfer (from bulk solution to catalyst surface) to increase selectivity. An astonishingly high FE of 66% along with the NH_3 yield rate of 200 mmol g^{-1} h^{-1} was reported.

7.7.3.2 Metal oxides

Electrochemical NRR with transition metal oxides was introduced by Licht et al. who improved a nano Fe_2O_3 system by replacing nano-Fe_2O_3 with active carbon-supported Fe_2O_3 to get maximum NH_3 yield of 8.27×10^{-9} mol cm^{-2} s^{-1}.[108] Various transition metal oxides were reported to reduce N_2 effectively. A series of metal oxides with rutile structure was screened for NRR where they discovered two possible reacting sites, such as cus-sites (top sites of coordination) and br-sites (bridging sites) (Fig. 7.8C).[109, 110] Having higher binding energy, occupancy of br-sites tunes the surface activity compared to cus-sites. NbO_2, ReO_2, and TaO_2 were the most potential NRR catalysts having lower onset potential (−0.57 V, −1.07 V, and −1.21 V vs. RHE, respectively). With this concept, in a comparative study of NbO_2 and Nb_2O_5 for NRR, Nb^{4+} is seen to be more effective in promotional backdonation toward π_{2p}^* antibonding of N≡N bond and results in efficient NRR, whereas the absence of 4d electrons in Nb^{5+} lacks such backdonation. In another study with $CeO_2/Bi_4V_2O_{11}$

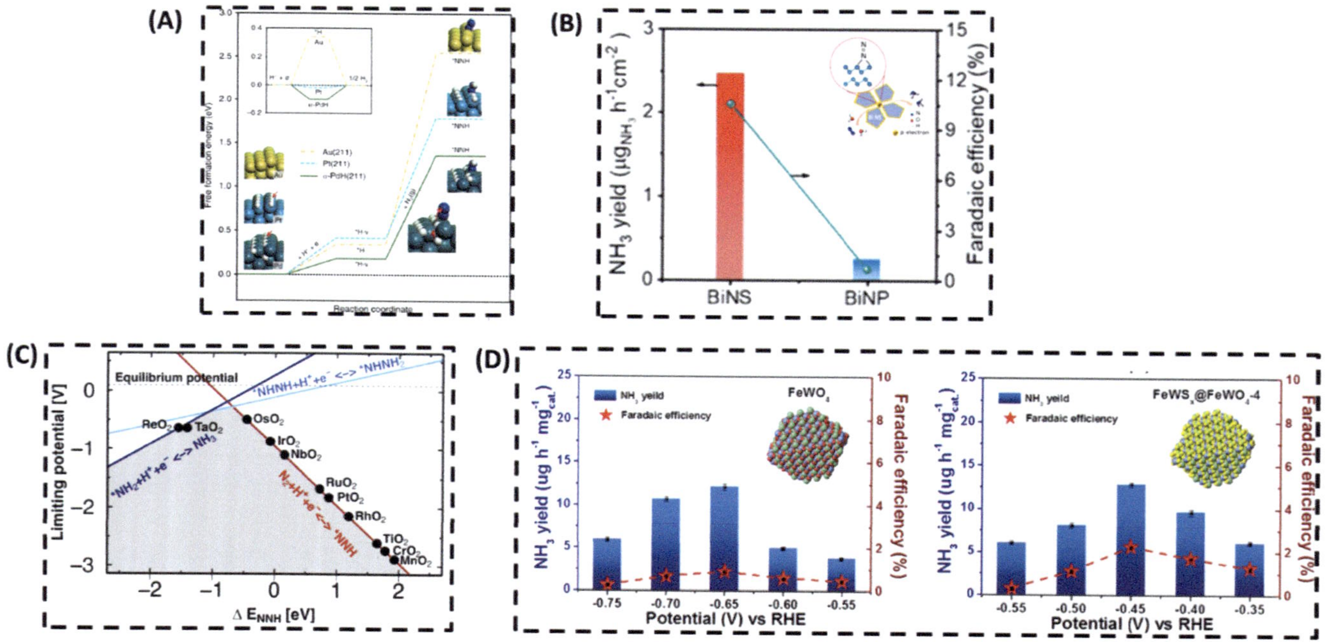

Figure 7.8 (A) DFT-calculated free energy pathways of HER (inset) and the relevant steps of NRR on the (211) surfaces of Au, Pt, and α-PdH at surface potential of 0 V vs. RHE under 298.15 K (atomic structures shown in the insets). Reproduced with permission from [102], copyright 2018, Springer. (B) NH_3 yield and Faradaic efficiency of Bi nanosheet and Bi nanoparticles. Reproduced with permission from [106], copyright 2019, ACS. (C) Volcano plot for NH_3 formation of metal oxides, Reproduced with permission from [109, 110], copyright 2017, ACS. (D) NH_3 yield and Faradaic yield of $FeWO_4$ and $FeWS_x@FeWO_4$. Reproduced with permission from [114], copyright 2021, Wiley – VCH GmbH.

hybrid, disordered amorphous structure of $Bi_4V_2O_{11}$ reduced the energy barrier of the intermediates for accelerating NRR. Au–Fe_3O_4 was demonstrated to be promising electrocatalyst compared to core Au@Fe_3O_4 shell for NRR owing to the presence of d–band near Fermi level.

Many other oxides are reported to be effective for NRR including $LiMn_2O_4$, Fe_3O_4, γ–Fe_2O_3, MoO_3, Mn_3O_4, NiO, C@NiO@Ni, VO_2, MoO_{3-x}, Cr_2O_3, ZrO_2, La–doped TiO_2, Cu–doped TiO_2, Fe–doped TiO_2, B–doped TiO_2, MnO, Y_2O_3, WO_{3-x}. Experimentally, the activity of metal oxides toward the NRR in alkaline solution follows the order PbO– TiO_2 > Sb–SnO_2 > ZnO > Sn–In_2O_3 > SnO_2 > Fe_2O_3 > TiO.[111]

More interestingly, oxygen vacancies (of the metal oxides) could facilitate π back bonding from metal d-orbital to π* of N_2, which eventually influences the promotional NRR activity. For example, a high rate (32.13 mg h^{-1} mg_{cat}^{-1}) of NH_3 production was observed with oxygen vacancy-rich Fe_2O_3 nanocubes at only –0.3 V vs. RHE.[112] Along with the experimental studies, DFT studies for oxygen vacancy-rich WO_3 and MoO_2 reported a high yield of NH_3 at –0.12 V vs. RHE. The DOS calculation suggested that the lowering in activation energy is mainly due to the oxygen vacancy which leads to distal/alternative pathway selectively stabilizing N_2H^* intermediate and destabilizing $N_2H_2^*$ intermediate during N_2 reduction process. Recently, $W_{18}O_{49}$ nanowire was reported to exhibit excellent NRR performance with a high NH_3 yield rate of 24.7 μg h^{-1} mg_{cat}^{-1} and FE of 20% at a very low potential of –0.15 V vs. RHE. Fe intercalation remarkably improved the binding efficiency of W (in $W_{18}O_{49}$) with N_2 molecules.[113]

7.7.3.3 Metal nitrides

Various transition metal mononitrides with rocksalt or zinc-blend structure, such as VN, ZrN, NbN, and CrN had been studied for NRR. Selective NH_3 formation was revealed with a rate of 8.40 × 10^{-11} mol s^{-1} cm^{-2} (suppressing N_2H_4 formation) in 0.1 M HCl with VN nanosheet on Ti mesh at –0.50V vs. RHE (FE = 2.25%). Theoretical studies support that VN catalyst reduced NRR via *Mars–van Krevelen* mechanism. Briefly, *Mars–van Krevelen* mechanism follows a lower energy pathway for nitrides where the N atoms within the nitrides are responsible for NH_3 formation. The N atom of nitrides reacts with the H of electrolyte to form N–H bonds from which diffusion of NH_3 occurs to the electrolyte. The so-generated N-vacancy is replenished by the external N_2 supply to continue the catalytic process.[115] Interestingly, tetragonal Mo_2N was examined to undergo fast chemical decomposition in aqueous electrolytes and resulted in no catalytic activity for NRR. Selectively NH_3 was detected via UV–Vis spectroscopy using Indophenol-blue assay with a highest FE of 42.3% at 0.05 V vs. RHE.[116] Similarly, MoN nanosheet array on carbon cloth was reported to exhibit a high-performance NRR electrocatalyst in 0.1 M HCl. Here also, the DFT studies revealed it follow *Mars–van Krevelen* pathway where the so-formed N vacancy aids in lowering the energy barrier for NH_3 production.[117] Additionally, highly stable CrN nanocubes exhibited NH_3 yield of 31.11 μg h^{-1} $mg_{cat.}^{-1}$ with an FE of 16.6% in 0.1 M HCl electrolyte. Various physical and structural characterization revealed that partial oxidation of pristine CrN nanocubes resulted in an O-rich core and N-rich core during the course of the catalytic process.[118] Additionally, introducing N vacancies on polymeric carbon nitride significantly improves NRR activity with 11 times higher FE compared to defect-free polymeric carbon nitride.[119] B_4C nanosheet also exhibited to yield NH_3 at a rate of 26.57 mg h^{-1} mg_{cat}^{-1} in 0.1 M HCl following an alternative associative pathway.[120]

7.7.3.4 Metal phosphides

Superior NRR activity was studied with a noble metal-free cobalt phosphide hollow nanocages (CoP HNC), which possesses a three-leveled nanoparticle-nanosheet-nanocage structure. A good selectivity toward NH_3 with an FE of 7.36 % at a low negative potential (0 V vs. RHE) was examined; but, with an increase in negative potential, the FE decreases due to the competing HER.[121] Several other pentagonal transition metal phosphides (M = Ti, Zr, and Hf) are predicted to be dynamically, mechanically, and thermally stable through DFT and ab initio studies. The metallic property and quasi-planar structures present in these metal phosphides prefer N_2 chemisorption and ease the activation of N≡N bond, which is beneficial for NRR. Through Gibbs free energy change calculation of each elementary step, penta-TiP was further predicted to prefer the distal associative pathway to exhibit a low overpotential of 0.56 V.[122] Recently, black phosphorus nanosheet was reported to exhibit excellent NRR activity in 0.01 M HCl.

7.7.3.5 Metal sulfides

Based on the theoretical studies, MoS_2 was first utilized as a high selective NRR catalyst to achieve an NH_3 yield of 8.08×10^{-11} mol s^{-1} cm^{-1} at −0.5 V with an FE of 1.17% in 0.1 M Na_2SO_4.[123] Even in an acidic medium, it showed better NRR activity surpassing HER. The introduction of defects to MoS_2 further improved NRR activity. As per DFT studies, NRR on a defect–MoS_2 surface was followed to proceed via the distal associative pathway (lowering an energy barrier of 0.60 eV for *NH → *NH_2) with the defect push the *d*-band center toward Fermi level, enhancing the interaction between MoS_2 surface and N_2 molecule. Additionally, incorporating Fe on MoS_2 exhibited an NH_3 yield rate of 8.63 μg mg_{cat}^{-1} h^{-1} with an FE of 18.8%.

7.7.3.6 Metal carbides

Early theoretical studies on MXene demonstrated that the M atom of the terminal surface is prone to N_2 absorption and thus promotes NRR. The terminal oxygen on MXenes is active toward HER and CO_2RR and was thought to exhibit low NRR activity. MXene ($Ti_2C_2T_x$) nanosheet attached to a vertically aligned metal host was reported to achieve an FE of 5.78% for NH_3 production at an ultralow potential.[124] Based on the theoretical and experimental results, the basal plane of the MXene is relatively inert and a greater number of active edge sites and a metal host with poor HER activity are crucial for efficient NRR to produce NH_3, thereby retarding HER. Mo_2C/C was studied for electrochemical NRR in acidic 0.5 M Li_2SO_4 to achieve 11.3 mg h^{-1} mg Mo_2C^{-1} with an FE of 7.8 % at −0.3 V.[125]

7.7.3.7 Interfacial engineering for NRR

To improve selectivity toward NH_3 over HER or N_2H_4, interfacial regulation is an effective strategy to improve interaction at the solid–gas interface to further promote electrochemical activity. Interestingly Fe/Fe_3O_4 (prepared by pre-oxidized Fe foil followed by reduction) acted as an effective catalyst in 0.1 M phosphate buffer solution (PBS) electrolyte to selectively produce NH_3, supressing HER effectively. The Fe/Fe_3O_4 composite was reported to outperform the NRR activity of Fe (foil), Fe_2O_3, and Fe_3O_4. It is proposed that H-terminated molybdenum carbides are potent toward electrocatalytic nitrogen reduction. Very recently, hydrophobic octadecanethiol was grown on Pd metallene in the form of a self-assembling monolayer by Pd–S bond (Pdene@C18/C), which was reported to show superior NRR activity, with an ammonia production rate of 27.97 μg h^{-1} mg_{cat}^{-1} and an FE of 14.29%, than Pdene/C (14.69 μg h^{-1} mg_{cat}^{-1} and 1.93%).[126] The hydrophobic nature inhibits proton formation and transfer, while also aiding gas transport to Pdene's surface, thus enhancing the NRR and restricting the HER. Recently, inspired by the bimetallic sulfides of the N_2–fixing cofactors, $FeWS_x$@$FeWO_4$ was designed for active NRR electrocatalyst. The electrocatalyst was reported to yield NH_3 at a yield rate of 30.2 μg h^{-1} mg_{cat}^{-1} and an FE of 16.4% at -0.45 V vs. RHE (Fig. 7.8D).[114]

7.8 Nitrogen Oxides Reduction Reaction

Owing to the non-polarity, low proton affinity, and thermodynamic stability of the N_2 molecule, the prospect of simultaneous reduction of other nitrogenous oxyanions (NO_x^{n-}) emerges.

7.8.1 Thermodynamics and Kinetics of NO_xRR

The N=O bond of nitrogenous oxoanions has lower bond dissociation energy (activation energy of 204 kJ mol^{-1}) compared to that of the N≡N (activation energy 941 kJ mol^{-1}) bond of N_2 and thus can readily be reduced electrochemically to produce ammonia.

7.8.2 Mechanism of NO_xRR

Electrochemical NO_x reduction produces various products, including hydroxylamine (NH_2OH), ammonia (NH_3), nitrous oxide (N_2O), and nitrogen (N_2) depending on the number of electrons transferred (Fig. 7.9).[127, 128]

Initially, NO adsorption on the catalyst occurs either via N–end, O–end or side on and thereby forming various intermediates. The production of NH_3 from NO occurs either by an associative or dissociative pathway.[129] In case associative pathway, both "N" and "O" atoms in the NO molecule will be protonated consequently to produce a "H_xNOH_y" intermediate, which will eventually reduce to NH_3. On the contrary, the first step is splitting the N–O bond on the active site. Then, each adsorbed atom (N* or O*) will be

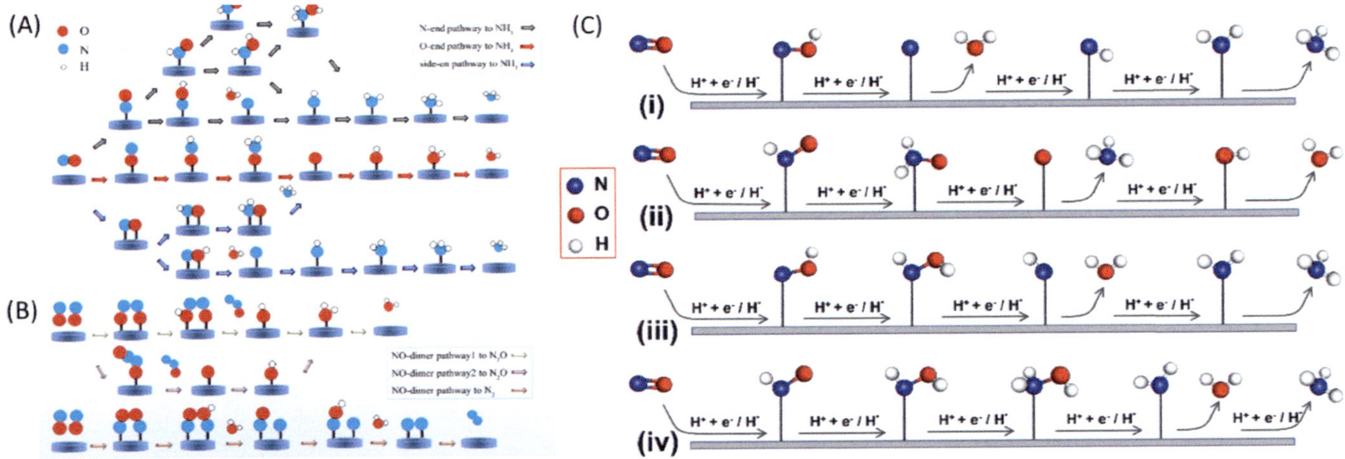

Figure 7.9 Mechanistic pathways of the NORR process toward NH_3 (A) and N_2O/N_2 (B) formation. Reproduced with permission from [128], copyright 2022, ACS. (C) Pathways of the associative mechanism—distal–O (i), distal–N (ii), alternating–O (iii), and alternating–N pathways (iv). Reproduced with permission from [127], copyright 2023, RSC.

hydrogenated separately to yield NH_3 and H_2O. Both pathways involve the hydrogenation process occurring through either the Tafel or Heyrovsky–type mechanism. Before the "NO" hydrogenation process, solvated protons will adhere to the catalyst's surface to create H* in the first scenario, while the direct catalyst surface hydrogenation of "NO" molecules or intermediates takes place in the latter. The dissociative and associative pathways are classified into four categories: dissociative-Tafel (D–T), associative-Tafel (A–T), dissociative-Heyrovsky (D–H), and associative-Heyrovsky (H–A) mechanisms based on the type of hydrogenation. Additionally, the associative processes are grouped into distal–O, distal–N, alternating–O, and alternating–N routes. In the distant route, either N or O in NO will undergo full hydrogenation to form NH_3 or H_2O. Afterward, O* or N* will undergo hydrogenation. On the other hand, both nitrogen and oxygen atoms experience different hydrogenation steps in the alternate route.

7.8.3 Materials for NO_xRR

7.8.3.1 Metal-based catalyst

Similar to other cathodic reduction reactions, theoretical and experimental results revealed promising electrochemical NO_3RR activity (toward NH_3 formation) by noble metals, such as Pt, Pd, Ru, Rh, etc. The catalytic surface possesses different Gibbs free energy of NO* and/or H* adsorption and as a result of which selectivity of product diverges. Recently, Ru nanosheets with low coordination number (Ru–LCN) were studied in both H-cell and flow electrolyzer and concluded that NH_3 formation over Ru surface follows alternating N-mechanistic pathway, where Ru facilitates hydrogenation of NO* (to form HNO*) along with adsorption of NO* for nitric oxide reduction reaction (NORR) activity. At a potential of –0.2 V vs. RHE, Ru–LCN showed an NH_4^+ yield rate of 45.03 μmol h^{-1} mg^{-1} with a FE of ~66%. Additionally, NORR activity with a RuGa intermetallic compound yielded NH_4^+ at a yield rate of 320.6 μmol h^{-1} mg_{Ru}^{-1} and a FE of 72.3% (Fig. 7.13 A).[130] The Ga incorporation into Ru altered the crystal structure from fcc to bcc for which an electron-rich Ru center was created and a superior NORR activity with an NH_4^+ yield of 320.6 μmol h^{-1} mg_{Ru}^{-1} (FE = 72.3%) was obtained. Again, an alternating N mechanistic pathway was reported to be followed for NH_3 production with the $Ru_{0.05}Cu_{0.95}$ alloy.[131] Likewise, increasing the size of Ru unit cell in Ru/oxygen-doped Ru core/shell nanoclusters, induces tensile strain for effective NO_3RR toward selective NH_3 production (NH_3 formation rate of 5.56 mol g^{-1} h^{-1}).[132] Additionally, when Ru was dispersed on Cu nanowire, the catalyst displayed an industrial-level current density of ~1 A cm^{-2} for NO_3RR with a maximum FE of 96%.[133] Similarly, a-Ru–CNTs was reported to deliver an NH_3 yield of 145 μg h^{-1} mg^{-1} with a FE of 80.62% at -0.2 V vs. RHE.[134] Besides, Ag was reported to show 100% FE in NO saturated EDTA–Fe^{2+} electrolyte.[135] In the course of the reaction, Ag stabilized the reaction intermediates through donor-acceptor interaction by binding with the oxygen atom of the intermediates. Au was also studied to show superior NORR activity among the noble metals. For instance, at a lower potential of -0.3 V, excellent NH_4^+ production was obtained via Au-supported Vulcan carbon. Similarly, improved NORR activity was exhibited by Au/rGO in ferrous citrate electrolyte where the NH_3 yield increased linearly with the electrolyte concentration. The stabilization of *N and *NH intermediates over the Au surface may produce NH_3 selectively. Ir was also found to be active for preferential binding of NO over H and yielded NH_3 at a higher rate by lowering the energy barrier for the rate determining *NHO → *NHOH step.

On the contrary, transition metal electrodes are efficient for electrochemical NO_xRR performance but lack in product selectivity. The d orbital of Cu and π* orbital of NO have similar energy levels as a result of which Cu has been explored as an electrocatalyst for NO_xRR. For instance, Fe and Pd are known to produce N_2O as the major product in alkaline medium, whereas Cu was reported to be selective for NH_3 from NORR. Cu was predicted to be ideal for selective NH_3 formation through NORR, which was also validated from porous Cu foam with a maximum NH_3 formation rate of 517 μmol cm^{-2} h^{-1}.[136] Later on, a high NH_3 yield of 3520.80 mg h^{-1} mg^{-1} was achieved with Cu/TiO_2 in NO saturated 0.1 M K_2SO_4.[137] On the contrary, metallic Cu mostly produces N_2 from NO_3RR instead of NH_3.

Owing to the facile protonation of the intermediates, NH_3 formation rates are significantly higher in acidic medium, but transition metal electrodes get degraded and tend to dissolve in an acidic environment. In this regard, Ni@NC exhibited an NH_3 yield rate of 34.6 μmol cm^{-2} h^{-1} at a lower overpotential of 550 mV in 0.1 M HCl.[138] Further improving was achieved with N-doped carbon-wrapped nickel nanoparticles supported onto carbon fiber (NiNC@CF), where an NH_3 yield rate of 94 mmol cm^{-2} h^{-1} at 0.5 V vs. RHE in PBS electrolyte was quantified.[139] Meanwhile, NH_3 yield rate of 27 mmol cm^{-2} h^{-1} with an FE of 80% with hcp–Co nanosheets was also calculated.[140] The donation of e^- through the d–π* orbital activates the adsorbed NO molecule for the rate determining step (NO* → NOH*), as unveiled from DFT calculations.

7.8.3.2 Metal oxides

Fe-based electrocatalyst are involved in biological reduction of NO to N_2O (viz., NO reductase). Similar to that, Fe_2O_3 nanorods were employed for NORR, the activity of which was improved by introducing Fe(II)EDTA in 0.1 M Na_2SO_4 electrolyte. It is also concluded that NO_3^- contamination in electrolyte reduces NORR activity. Further, the NORR activity was boosted with oxygen vacancy-rich MnO_2 nanowire in 0.2 M Na_2SO_4, where the NH_3 yield rate was 27.51 × 10^{-10} mol s^{-1} cm^{-2} with an FE of 82.8%.[141] The oxygen vacancy in MnO_{2-x} (211) surface opens up for low-energy NO adsorption. Similarly, oxygen vacancy-rich TiO_2 on Ti plate was reported to adsorb NO in a more facile manner. Later, CoNi(5:5)O_x@Cu was reported for exceptional NH_3 yield of 20 mg h^{-1} cm^2 at -0.68 V vs. RHE.[142] The NO gets adsorbed on Co site and follows an O-end distal pathway for efficient NORR to produce NH_3. Hollow Cu_2O@-$CoMn_2O_4$, where the electron transfer efficiency was improved via a void-confinement effect, yielded NH_3 with a yield rate of 94.18 mmol g^{-1} h^{-1} and an FE of 75.05 %.[143] Very recently, Fe_2O_3/ZnO heterojunction is reported to exhibit a high FE of 97.4 % with NH_3 yield of 6327.2 μg h^{-1} at −1.0 V vs. RHE in 0.1 M KNO_3 and 0.1 M PBS. The heterojunction effectively decreases the energy barrier of *NO → *NHO and accelerates the reaction kinetics.[144]

7.8.3.3 Metal phosphides

Transition metal phosphides, such as Ni_2P, CoP, and FeP are reported to enhance NORR activity at low overpotentials. The N–O bond gets elongated due to the charge transfer between Ni_2P and NO and gets weakened on the Ni_2P surface, which eventually facilitates hydrogenation of the intermediates and directs towards the distal pathway for NH_3 pathway.[145] Additionally, BP was reported to deliver a high NH_3 production rate of 96.6 μmol h^{-1} cm^2 with an FE of 83.3% at -0.7 V vs. RHE. The NO gets activated from both B and P.[146]

7.8.3.4 Metal sulfides

2H-MoS_2 flakes were studied for NORR to NH_3 in acidic electrolyte. NO adsorption on the MoS_2 (101) surface occurs through an acceptor-donor mechanism, where the electron transfer occurs from MoS_2 to the NO antibonding orbital. Although, low NH_3 FE was obtained at higher potential owing to the competitive HER. The problem was addressed by using an ionic liquid (1-butyl-1-methypyrrolidinium tris(pentafluoroethyl)-trifluoro phosphate) and P doping on MoS_2.[147] The FE was improved up to 70% with an NH_3 yield rate of 237.7 μg h^{-1} mg_{cat}^{-1} at -0.6 V vs. RHE. Similar to oxygen vacancy, sulfur vacancy was reported to enhance the NORR activity toward NH_3 production, as in the case of CoS_{1-x} nanosheets where the localization of electrons around the sulfur vacancies lowered the activation energy for NO adsorption; unsaturated Sn sites were induced in SnS_{2-x} for NO adsorption and were activated via N=O bond elongation. A strong NO activation (through back donation) as well as selective adsorption (over H_2O) was observed with Sb/Sb_2S_3, where NO is selectively reduced to NH_3 in neutral medium.

7.8.3.5 Metal carbides

Amorphous $B_{2.6}C$ was studied for efficient NORR where the electron transfer arose from B-C bond to π_{2p}^* of NO. The electrocatalyst was further modified with TiO_2 support to yield NH_3 at a rate of 3678.6 μg h^{-1} cm^2 with FE of 87.6%.[148]

7.9 Future Aspects of Electrochemical Cathodic Reactions

The in situ-generated ORR intermediates offer an opportunity to be utilized for organic synthesis in an organic medium. The so-generated superoxide ion (O_2^-) can act as base, nucleophile, oxidant, reductant, as well as a free radical. In aprotic medium, many hydroxylation, decyanohydroxylation, epoxidation of α, β-unsaturated ketones, denitrooxylation, and various fragmentation reactions were evidenced via a one-electron process affording $O_2^{-\bullet}$ during electrochemical ORR in aprotic solvents.

Other than that, ORR also have influence in emerging from electronic applications, such as Polymer Light Emitting Electrochemical Cells (LECs), microbial fuel cells to wearable devices that can easily be integrated in human body.

The industrial synthesis of urea needs harsh conditions (200°C, 150–250 bar) from industrially produced NH_3 and CO_2 with large energy consumption. One-ton production of urea consumes 0.8 ton of the global NH_3.[149] Electrochemical formation of C–N bonds was first time reported in the recent past by co-electrolysis of CO and NH_3 toward acetamide with a selectivity of 40% using Cu nanoparticles.[150] PdCu alloyed nanoparticles embedded on TiO_2 NSs were reported to generate urea by co-electrolysis of CO_2 and N_2. Simultaneous adsorption of N_2 and CO_2 and activation are essential for C–N bond formation. New catalysts like Cu_3PdN, FeS_2/Pd, VS_2/Cu, FeS_2/CuAg, and NbS_2/CuAg heterostructure could be potent candidates for urea synthesis where the interface could provide simultaneous binding sites for CO_2 and N_2. Side-on adsorption of N_2 accelerates the back π bonding by electron transfer from *d*-orbitals of Fe, Nb, or Pd to π* orbitals of N_2, reducing the N–N bond order, whereas Cu/Ag/Zn could selectively bind with CO_2 and can convert it to CO. These two reactions are expected to happen at the interface simultaneously and could lead C–N bond formation by the new heterostructure (Fig. 7.10). Additionally, various C–N coupling scope could be achieved involving in situ-generated NO_x species from electrocatalytic reduction reactions, such as cyclohexanone oxime (an important precursor for Nylon 6) from electrocatalytic reduction of nitrate beyond NH_3, the co-reduction of CO_2 and NO_3^- yielding methylamine and ethylamine, and glycine was electro-synthesized from oxalic acid and NO_3^-.[151]

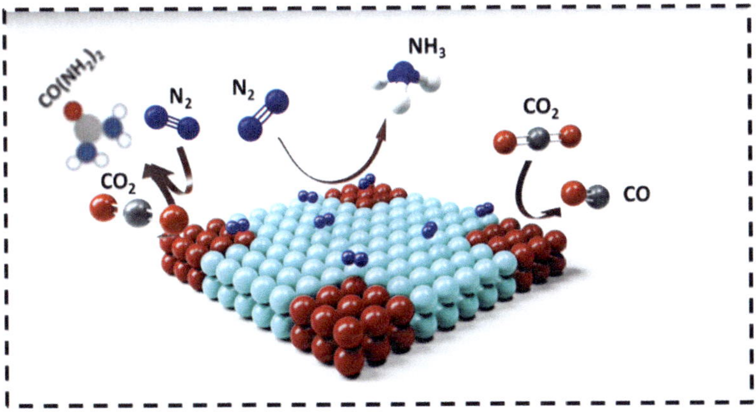

Figure 7.10 Schematic diagram of simultaneous CO_2RR, NRR, and production of urea.

7.10 Conclusion

In summary, recent advancements in cathodic reduction reactions have demonstrated the possible conversion of renewable energy to chemical energy stored in the chemical bonds of value-added products that are generated. A sustainable and powerful approach to access electrons as a clean reductant for cathodic reduction enables organic transformation, exhibiting a greener approach along with several advantages over traditional chemical methods. An electrocatalyst with superior stability, selectivity, and low cost would finally be useful for C- and/or N-containing compounds/precursors useful in chemical and pharmaceutical industries.

References

1. S. Chu, A. Majumdar, *Nature* 2012, 488, 294.
2. D. Y. Chung, J. M. Yoo, and Y.-E. Sung, *Adv. Mater.* 2018, 30, 1704123.
3. Y. Jeong, G. Janani, D. Kim, T.-Y. An, et al., *ACS Appl. Mater. Interfaces* 2023, 15, 52342.
4. C. Guo, J. Ran, A. Vasileff, and S.-Z. Qiao, *Energy Environ. Sci.* 2018, 11, 45.
5. A. Kapoor and T. Viraraghavan, *J. Environ. Eng.* 1997, 123, 371.

6. S. Ghafari, M. Hasan, and M. K. Aroua, *Bioresour. Technol.* 2008, 99, 3965.
7. X. Peng, L. Zeng, D. Wang, Z. Liu, et al., *Chem. Soc. Rev.* 2023, 52, 2193.
8. W. Duan, R. S. Vemuri, J. D. Milshtein, S. Laramie, et al., *J. Mater. Chem. A* 2016, 4, 5448.
9. J. W. Ager and A. A. Lapkin, *Science* 2018, 360, 707.
10. X. Ge, A. Sumboja, D. Wuu, T. An, et al., *ACS Catal.* 2015, 5, 4643.
11. Y. Matsubara, D. C. Grills, and Y. Kuwahara, *ACS Catal.* 2015, 5, 6440.
12. J. Qiao, Y. Liu, F. Hong, and J. Zhang, *Chem. Soc. Rev.* 2014, 43, 631.
13. L. Shi, Y. Yin, S. Wang, and H. Sun, *ACS Catal.* 2020, 10, 6870.
14. H. Wan, A. Bagger, and J. Rossmeisl, *Angew. Chem. Int. Ed.* 2021, 60, 21966.
15. P. Yu, F. Wang, T. A. Shifa, X. Zhan, et al., *Nano Energy* 2019, 58, 244.
16. S. Trasatti, *J. Electroanal. Chem. Interfacial Electrochem.* 1972, 39, 163.
17. E. Skúlason, V. Tripkovic, M. E. Björketun, S. Gudmundsdóttir, et al., *J. Phys. Chem. C* 2010, 114, 18182.
18. T. L. Tan, L.-L. Wang, J. Zhang, D. D. Johnson, et al., *ACS Catal.* 2015, 5, 2376.
19. A. Zalineeva, S. Baranton, C. Coutanceau, and G. Jerkiewicz, *Sci. Adv.* 3, e1600542.
20. U. Joshi, S. Malkhandi, Y. Ren, T. L. Tan, et al., *ACS Appl. Mater. Interfaces* 2018, 10, 6354.
21. Y.-J. Tang, M.-R. Gao, C.-H. Liu, S.-L. Li, et al., *Angew. Chem. Int. Ed.* 2015, 54, 12928.
22. Y. Zhao, C. Chang, F. Teng, Y. Zhao, et al., *Adv. Energy Mater.* 2017, 7, 1700005.
23. R. Wu, J. Zhang, Y. Shi, D. Liu, et al., *J. Am. Chem. Soc.* 2015, 137, 6983.
24. X. Peng, C. Pi, X. Zhang, S. Li, et al., *Sustain. Energy Fuels* 2019, 3, 366.
25. Y. Zhu, G. Chen, X. Xu, G. Yang, et al., *ACS Catal.* 2017, 7, 3540.
26. B. Ren, D. Li, Q. Jin, H. Cui, et al., *J. Mater. Chem. A* 2017, 5, 19072.
27. D. Gao, J. Zhang, T. Wang, W. Xiao, et al.. Ding, *J. Mater. Chem. A* 2016, 4, 17363.
28. F. Song, W. Li, J. Yang, G. Han, et al. Sun, *Nat. Commun.* 2018, 9, 4531.
29. P. Xiao, M. A. Sk, L. Thia, X. Ge, et al., *Energy Environ. Sci.* 2014, 7, 2624.
30. F. H. Saadi, A. I. Carim, E. Verlage, J. C. Hemminger, et al., *J. Phys. Chem. C* 2014, 118, 29294.

31. D. E. Schipper, Z. Zhao, H. Thirumalai, A. P. Leitner, et al., *Chem. Mater.* 2018, 30, 3588.
32. E. J. Popczun, J. R. McKone, C. G. Read, A. J. Biacchi, et al., *J. Am. Chem. Soc.* 2013, 135, 9267.
33. J. Tian, Q. Liu, A. M. Asiri, X. Sun, *J. Am. Chem. Soc.* 2014, 136, 7587.
34. L. Tian, X. Yan, X. Chen, *ACS Catal.* 2016, 6, 5441.
35. Z. Pu, Q. Liu, A. M. Asiri, X. Sun, *ACS Appl. Mater. Interfaces* 2014, 6, 21874.
36. Y. Pan, Y. Liu, J. Zhao, K. Yang, et al., *J. Mater. Chem. A* 2015, 3, 1656.
37. J. F. Callejas, C. G. Read, E. J. Popczun, J. M. McEnaney, et al., *Chem. Mater.* 2015, 27, 3769.
38. S. Ghosh, S. R. Kadam, S. Kolatkar, A. Neyman, et al., *ACS Appl. Mater. Interfaces* 2022, 14, 581.
39. M. A. Lukowski, A. S. Daniel, F. Meng, A. Forticaux, et al., *J. Am. Chem. Soc.* 2013, 135, 10274.
40. Y. Xie, Y. Liu, and Z. Yang, *Int. J. Hydrogen Energy* 2020, 45, 6500.
41. B. You and Y. Sun, *Adv. Energy Mater.* 2016, 6, 1502333.
42. Z. Peng, D. Jia, A. M. Al-Enizi, A. A. Elzatahry, et al., *Adv. Energy Mater.* 2015, 5, 1402031.
43. R. B. Levy and M. Boudart, *Science* 1973, 181, 547.
44. W. Yuan, L. Cheng, Y. An, H. Wu, et al., *ACS Sustain. Chem. Eng.* 2018, 6, 8976.
45. J. Liu, Y. Liu, D. Xu, Y. Zhu, et al. Fan, *Appl. Catal. B* 2019, 241, 89.
46. S. Anantharaj, H. Sugime, and S. Noda, *Chem. Eng. J.* 2021, 408, 127275.
47. L. Su, X. Cui, T. He, L. Zeng, et al., *Chem. Sci.* 2019, 10, 2019.
48. S. W. Jang, S. Dutta, A. Kumar, Y.-R. Hong, et al., *ACS Nano* 2020, 14, 10578.
49. Q. Fu, X. Wang, J. Han, J. Zhong, et al. Song, *Angew. Chem. Int. Ed.* 2021, 60, 259.
50. S. Zhang, Y. Li, H. Zhu, S. Lu, et al., *ACS Appl. Mater. Interfaces* 2020, 12, 6250.
51. R. Kumar, A. Gaur, T. Maruyama, C. Bera, et al., *ACS Appl. Mater. Interfaces* 2020, 12, 57898.
52. Y. Yao, X.-K. Gu, D. He, Z. Li, et al., *J. Am. Chem. Soc.* 2019, 141, 19964.
53. L.-P. Han, R.-Q. Yao, W.-B. Wan, H. Shi, et al., *J. Mater. Chem. A* 2019, 7, 25925.

54. L.-W. Chen, X. Guo, R.-Y. Shao, Q.-Q. Yan, et al., *Nano Energy* 2021, 81, 105636.
55. S. Li, L. Shi, Y. Guo, J. Wang, et al., *Chem. Sci.* 2024, 15, 11188.
56. J. K. Nørskov, J. Rossmeisl, A. Logadottir, L. Lindqvist, et al., *J. Phys. Chem. B* 2004, 108, 17886.
57. S. Guo, S. Zhang, and S. Sun, *Angew. Chem. Int. Ed.* 2013, 52, 8526.
58. T. Yu, D. Y. Kim, H. Zhang, and Y. Xia, *Angew. Chem. Int. Ed.* 2011, 50, 2773.
59. J. Wu, L. Qi, H. You, A. Gross, et al., *J. Am. Chem. Soc.* 2012, 134, 11880.
60. V. R. Stamenkovic, B. Fowler, B. S. Mun, G. Wang, et al., *Science* 2007, 315, 493.
61. Y. Liu, A. Ishihara, S. Mitsushima, N. Kamiya, et al., *J. Electrochem. Soc.* 2007, 154, B664.
62. Y. Gorlin and T. F. Jaramillo, *J. Am. Chem. Soc.* 2010, 132, 13612.
63. T. Li, B. Xue, B. Wang, G. Guo, et al., *J. Am. Chem. Soc.* 2017, 139, 12133.
64. Q. Wu, L. Jiang, L. Qi, L. Yuan, et al., *Electrochim. Acta* 2014, 123, 167.
65. D. D. Vaughn Ii, J. Araujo, P. Meduri, J. F. Callejas, et al., *Chem. Mater.* 2014, 26, 6226.
66. Y. Liang, Y. Li, H. Wang, J. Zhou, et al., *Nat. Mater.* 2011, 10, 780.
67. X. Ge, Y. Liu, F. W. T. Goh, T. S. A. Hor, et al., *ACS Appl. Mater. Interfaces* 2014, 6, 12684.
68. H. Yin, C. Zhang, F. Liu, and Y. Hou, *Adv. Funct. Mater.* 2014, 24, 2930.
69. H. Wu and W. Chen, *J. Am. Chem. Soc.* 2011, 133, 15236.
70. S. Isogai, R. Ohnishi, M. Katayama, J. Kubota, et al., *Chem. - Asian J.* 2012, 7, 286.
71. R. Ohnishi, M. Katayama, K. Takanabe, J. Kubota, et al., *Electrochim. Acta* 2010, 55, 5393.
72. A. Takagaki, Y. Takahashi, F. Yin, K. Takanabe, et al., *J. Electrochem. Soc.* 2009, 156, B811.
73. F. Yin, K. Takanabe, J. Kubota, and K. Domen, *J. Electrochem. Soc.* 2010, 157, B240.
74. Z. Wen, S. Ci, F. Zhang, X. Feng, et al., *Adv. Mater.* 2012, 24, 1399.
75. Z.-Y. Wu, X.-X. Xu, B.-C. Hu, H.-W. Liang, et al., *Angew. Chem., Int. Ed.* 2015, 54, 8179.
76. J. Diao, Y. Qiu, S. Liu, W. Wang, et al., *Adv. Mater.* 2020, 32, 1905679.

77. H. Imai, M. Matsumoto, T. Miyazaki, S. Fujieda, et al., *Appl. Phys. Lett.* 2010, 96, 191905.

78. M. Shao, Q. Chang, J.-P. Dodelet, and R. Chenitz, *Chem. Rev.* 2016, 116, 3594.

79. R. Yoo, E. Pranada, D. Johnson, Z. et al., *J. Electrochem. Soc.* 2022, 169, 063513.

80. G. Wu and P. Zelenay, *Acc. Chem. Res.* 2013, 46, 1878.

81. J. Liang, Y. Ling, X.-w. Wu, H. A. Acciari, et al., *J. Appl. Electrochem.* 2019, 49, 1211.

82. A. Zhu, P. Tan, L. Qiao, Y. Liu, et al., *J. Colloid Interface Sci.* 2018, 509, 254.

83. Y. Hori, K. Kikuchi, and S. Suzuki, *Chem. Lett.* 1985, 14, 1695.

84. Y. Hori, K. Kikuchi, A. Murata, and S. Suzuki, *Chem. Lett.* 1986, 15, 897.

85. J.-P. Jones, G. K. S. Prakash, and G. A. Olah, *Isr. J. Chem.* 2014, 54, 1451.

86. J. Li, S. U. Abbas, H. Wang, Z. Zhang, et al., *Nano-Micro Lett.* 2021, 13, 216.

87. L. Wang, W. Chen, D. Zhang, Y. Du, et al., *Chem. Soc. Rev.* 2019, 48, 5310.

88. S. Munir, A. R. Varzeghani, and S. Kaya, *Sustainable Energy Fuels* 2018, 2, 2532.

89. A. Loiudice, P. Lobaccaro, E. A. Kamali, T. Thao, et al., *Angew. Chem. Int. Ed.* 2016, 55, 5789.

90. G. L. De Gregorio, T. Burdyny, A. Loiudice, P. Iyengar, et al., *ACS Catal.* 2020, 10, 4854.

91. X. Feng, K. Jiang, S. Fan, and M. W. Kanan, *ACS Cent. Sci.* 2016, 2, 169.

92. Vayenas, C., White, Ralph, and Gamboa-Aldeco, Maria. *Modern Aspects of Electrochemistry.*

93. G. M. Tomboc, S. Choi, T. Kwon, Y. J. Hwang, et al., *Adv. Mater.* 2020, 32, 1908398.

94. U. Kaiser and E. Heitz, *Phys. Chem.* 1973, 77, 818.

95. M.-G. Kim, Y. Choi, E. Park, C.-H. Cheon, et al., *ACS Appl. Energy Mater.* 2020, 3, 11516.

96. Y. Chen, Z. Fan, J. Wang, C. Ling, et al., *J. Am. Chem. Soc.* 2020, 142, 12760.

97. J. Tian, R. Wang, M. Shen, X. Ma, et al., *ChemSusChem* 2021, 14, 2247.

98. K. Ye, Z. Zhou, J. Shao, L. Lin, et al., *Angew. Chem. Int. Ed.* 2020, 59, 4814.

99. C. J. M. van der Ham, M. T. M. Koper, and D. G. H. Hetterscheid, *Chem. Soc. Rev.* 2014, 43, 5183.
100. L. Du, L. Xing, G. Zhang, X. Liu, et al., *SusMat.* 2021, 1, 150.
101. M. Nazemi, S. R. Panikkanvalappil, and M. A. El-Sayed, *Nano Energy* 2018, 49, 316.
102. J. Wang, L. Yu, L. Hu, G. Chen, et al., *Nat. Commun.* 2018, 9, 1795.
103. H. Wang, Y. Li, C. Li, K. Deng, et al., *J. Mater. Chem. A* 2019, 7, 801.
104. W. Guo, K. Zhang, Z. Liang, R. Zou, et al., *Chem. Soc. Rev.* 2019, 48, 5658.
105. Y. Wang, M.-m. Shi, D. Bao, F.-l. Meng, et al., *Angew. Chem. Int. Ed.* 2019, 58, 9464.
106. L. Li, C. Tang, B. Xia, H. Jin, et al., *ACS Catal.* 2019, 9, 2902.
107. Y.-C. Hao, Y. Guo, L.-W. Chen, M. Shu, et al., *Nat. Catal.* 2019, 2, 448.
108. B. Cui, J. Zhang, S. Liu, X. Liu, et al., *Green Chem.* 2017, 19, 298.
109. Á. B. Höskuldsson, Y. Abghoui, A. B. Gunnarsdóttir, and E. Skúlason, *ACS Sustainable Chem. Eng.* 2017, 5, 10327.
110. Á. B. Höskuldsson, Y. Abghoui, A. B. Gunnarsdóttir, and E. Skúlason, *ACS Sustainable Chem. Eng.* 2021, 9, 8012.
111. N. Gruber and J. N. Galloway, *Nature* 2008, 451, 293.
112. C. Zhang, S. Liu, T. Chen, Z. Li, et al., *Chem. Commun.* 2019, 55, 7370.
113. Y. Tong, H. Guo, D. Liu, X. Yan, et al., *Angew. Chem. Int. Ed.* 2020, 59, 7356.
114. Y. Zhao, F. Li, W. Li, Y. Li, et al., *Angew. Chem. Int. Ed.* 2021, 60, 20331.
115. R. Zhang, Y. Zhang, X. Ren, G. Cui, et al., *ACS Sustain. Chem. Eng.* 2018, 6, 9545.
116. B. Hu, M. Hu, L. Seefeldt, and T. L. Liu, *ACS Energy Lett.* 2019, 4, 1053.
117. L. Zhang, X. Ji, X. Ren, Y. Luo, et al., *ACS Sustainable Chem. Eng.* 2018, 6, 9550.
118. Z. Ma, J. Chen, D. Luo, T. Thersleff, et al., *Nanoscale* 2020, 12, 19276.
119. C. Lv, Y. Qian, C. Yan, Y. Ding, et al., *Angew. Chem. Int. Ed.* 2018, 57, 10246.
120. W. Qiu, X.-Y. Xie, J. Qiu, W.-H. Fang, et al., *Nat. Commun.* 2018, 9, 3485.
121. W. Guo, Z. Liang, J. Zhao, B. Zhu, et al., *Small Methods* 2018, 2, 1800204.
122. Y. Ying, K. Fan, X. Luo, and H. Huang, *J. Mater. Chem. A* 2019, 7, 11444.
123. L. Zhang, X. Ji, X. Ren, Y. Ma, et al., *Adv. Mater.* 2018, 30, 1800191.
124. Y. Luo, G.-F. Chen, L. Ding, X. Chen, et al., *Joule* 2019, 3, 279.

125. X. Ren, J. Zhao, Q. Wei, Y. Ma, H. Guo, Q. Liu, Y. Wang, G. Cui, A. M. Asiri, B. Li, B. Tang, and X. Sun, *ACS Cent. Sci.* 2019, 5, 116
126. H. Wang, X. Mu, Q. Mao, K. Deng, et al., *Chem. Commun.* 2023, 59, 6552.
127. H. R. Inta, D. Dhanabal, S. S. Markandaraj, and S. Shanmugam, *EES Catal.* 2023, 1, 645.
128. Y. Zang, Q. Wu, S. Wang, B. Huang, et al., *J. Phys. Chem. Lett.* 2022, 13, 527.
129. H. R. Inta, D. Dhanabal, S. S. Markandaraj, and S. Shanmugam, *EES Catal.* 2023, 1, 645.
130. H. Zhang, Y. Li, C. Cheng, J. Zhou, et al., *Angew. Chem. Int. Ed.* 2023, 62, e202213351.
131. J. Shi, C. Wang, R. Yang, F. Chen, et al., *Sci. China Chem.* 2021, 64, 1493.
132. J. Li, G. Zhan, J. Yang, F. Quan, et al., *J. Am. Chem. Soc.* 2020, 142, 7036.
133. H. Xu, Y. Ma, J. Chen, W.-x. Zhang, et al., *Chem. Soc. Rev.* 2022, 51, 2710.
134. M. Jiang, A. Tao, Y. Hu, L. Wang, et al., *ACS Appl. Mater. Interfaces* 2022, 14, 17470.
135. D. Kim, D. Shin, J. Heo, H. Lim, et al., *ACS Energy Lett.* 2020, 5, 3647.
136. J. Long, S. Chen, Y. Zhang, C. Guo, et al., *Angew. Chem. Int. Ed.* 2020, 59, 9711.
137. L. Chen, W. Sun, Z. Xu, M. Hao, et al., *Ceram. Int.* 2022, 48, 21151.
138. S. Sethuram Markandaraj, T. Muthusamy, and S. Shanmugam, *Adv. Sci.* 2022, 9, 2201410.
139. T. Muthusamy, S. Sethuram Markandaraj, and S. Shanmugam, *J. Mater. Chem. A* 2022, 10, 6470.
140. D. Wang, Z.-W. Chen, K. Gu, C. Chen, et al., *J. Am. Chem. Soc.* 2023, 145, 6899.
141. Z. Li, Z. Ma, J. Liang, Y. Ren, et al., *Mater. Today Phys.* 2022, 22, 100586.
142. A. Wu, J. Lv, X. Xuan, J. Zhang, et al., *Adv. Energy Mater.* 2023, 13, 2204231.
143. C. Bai, S. Fan, X. Li, Z. Niu, et al., *Adv. Funct. Mater.* 2022, 32, 2205569.
144. H. Zhao, Y. Duan, X. Cheng, C. Fan, et al., *Dalton Trans.* 2024, 53, 15674.
145. T. Mou, J. Liang, Z. Ma, L. Zhang, et al., *J. Mater. Chem. A* 2021, 9, 24268.
146. G. Zhang, Y. Wan, H. Zhao, Y. Guo, et al., *Dalton Trans.* 2023, 52, 6248.
147. T. Wei, H. Bao, X. Wang, S. Zhang, et al., *ChemCatChem* 2023, 15, e202201411.

148. J. Liang, P. Liu, Q. Li, T. Li, et al., *Angew. Chem., Int. Ed.* 2022, 61, e202202087.

149. S. Giddey, S. P. S. Badwal, and A. Kulkarni, *Int. J. Hydrog. Energy* 2013, 38, 14576.

150. M. Jouny, J.-J. Lv, T. Cheng, B. H. Ko, et al., *Nat. Chem.* 2019, 11, 846.

151. S. Jia, L. Wu, H. Liu, R. Wang, et al., *Angew. Chem. Int. Ed.* 2024, 63, e202400033.

Chapter 8

Detection of Electroactive Species through in situ and ex situ Studies

Toufik Ansari and Arindam Indra
Department of Chemistry, Indian Institute of Technology BHU, Varanasi, Uttar Pradesh 221005, India
arindam.chy@iitbhu.ac.in

As discussed in the previous chapters, the electrocatalysts undergo a series of changes on the electrode surface to form the active catalysts (ACs). Especially, 3d-transition metal (TM)-based catalysts face structural modification during the anodic oxidation of the substrates in an alkaline or neutral medium to form metal(oxy)hydroxide. This serves as the AC for anodic oxygen evolution reaction (OER), oxidation of small molecules like ammonia, hydrazine, urea, etc., or organic and biomass oxidation. In addition, an understanding of the structure of different reaction intermediates (IMs) during the electrochemical reaction is crucial to realize the reaction mechanism and for further improvement of the performance of the catalyst. Although ex situ

Materials Design for Electrocatalysis
Edited by Sagarika Bhattacharya and Biswarup Chakraborty

ISBN 978-981-5352-32-0 (Hardcover), 978-1-042-02558-9 (eBook)
www.jennystanford.com

DOI: 10.1201/9781042025589-8

spectroscopic, analytical, and microscopic techniques have been followed to know the structure of the materials after the catalytic process, the information can be achieved only for the relaxed state structure, which could be completely different from the active structure. It is also impossible to detect the reversible changes of the catalyst during the electrocatalytic process by ex situ techniques. Further, tracking the formation of reaction IMs and understanding the reaction mechanism are not possible from the ex situ techniques. In this context, the in situ methods are extremely important. In this chapter, the in situ and ex situ techniques, including different spectroscopies, powder X-ray diffraction (PXRD), and microscopic studies, have been discussed to find out the structure of the AC and the details of the reaction mechanism.

8.1 Introduction

In electrocatalysis, the primary role of a catalyst is to accelerate electron transfer and promote specific reaction pathways, increasing the process's overall efficiency and selectivity.[1,2] Electrocatalysis is vital in various applications, including chemical synthesis, environmental remediation (e.g., electrochemical water treatment), and energy conversion and storage technologies, including fuel cells, electrolyzers, and metal-air batteries.[3,4]

The progress in the field of electrocatalysis can be achieved with a detailed understanding of the reaction mechanism and the AC structure.[5,6] This requires detailed characterization of the electroactive materials using techniques such as powder X-ray diffraction (PXRD), X-ray photoelectron spectroscopy (XPS), X-ray absorption spectroscopy (XAS), infrared spectroscopy (IR), Raman spectroscopy (RS), and different microscopic techniques.[7] All the above-mentioned methods can be operated in situ or ex situ to get an insight into the atomic and electronic structures (ESs) of the catalyst before and after the electrocatalytic reactions.[7,8] It is well understood that the detection of the active sites evolve during the electrocatalytic reactions (e.g., OER, AOR = anodic organic reaction, HER = hydrogen evolution reaction, N2RR = nitrogen reduction reaction, CO2RR = carbon dioxide reduction reaction), is the key to understand the reaction mechanism. Therefore, the development of in situ and operando characterization techniques is crucial for providing real-time information about the electrocatalyst during the reaction.[9]

Recent advances in in situ techniques enable simultaneous monitoring of the reaction process at low pressure and low temperature, allowing for continuous tracking of the catalyst structure and the IMs.[10] Further, the operando studies help to model a more realistic depiction of the catalyst structure compared to in situ approaches at low pressure and temperature. The primary goal of combining in situ, operando, and ex situ approaches is to deepen our understanding of the structure–activity relationship of PCs and ACs.[9,11,12]

8.2 Electrocatalytic Anodic Reconstruction

To correlate the electronic properties, local and long-range bond orders, and particle morphologies to catalytic performance, it is essential to establish structure–activity relationships. However, under applied anodic conditions, TM-based catalysts are known to undergo significant changes on the surface and subsurface, and the as-synthesized material is often referred to as "precatalyst".[13]

In a strongly alkaline medium (pH > 13), TM-based PCs undergo reconstruction, leading to phase transformation and changes in both morphology and composition under oxidizing anodic potentials (Fig. 8.1). This process results in the formation of $M(O)_x(OH)_y$ on the particle surfaces, or in some cases the entire material changes into the AC.[14] The surface overlayer of the AC is often found to be amorphous. In contrast, ultra-small (<10 nm) nanoparticles typically undergo bulk amorphization. The enhanced catalytic performance is attributed to the formation of poorly crystalline or even amorphous $M(O)_x(OH)_y$ phase, which contains numerous defects or vacancies. These vacancies lead to the generation of micropores, and their random distribution can improve bulk porosity, further enhancing the electrochemical properties of the ACs. Among defect types, three strategies, such as anion vacancy, cation vacancy, and multi-vacancy defects are particularly effective in enhancing OER performance.[15]

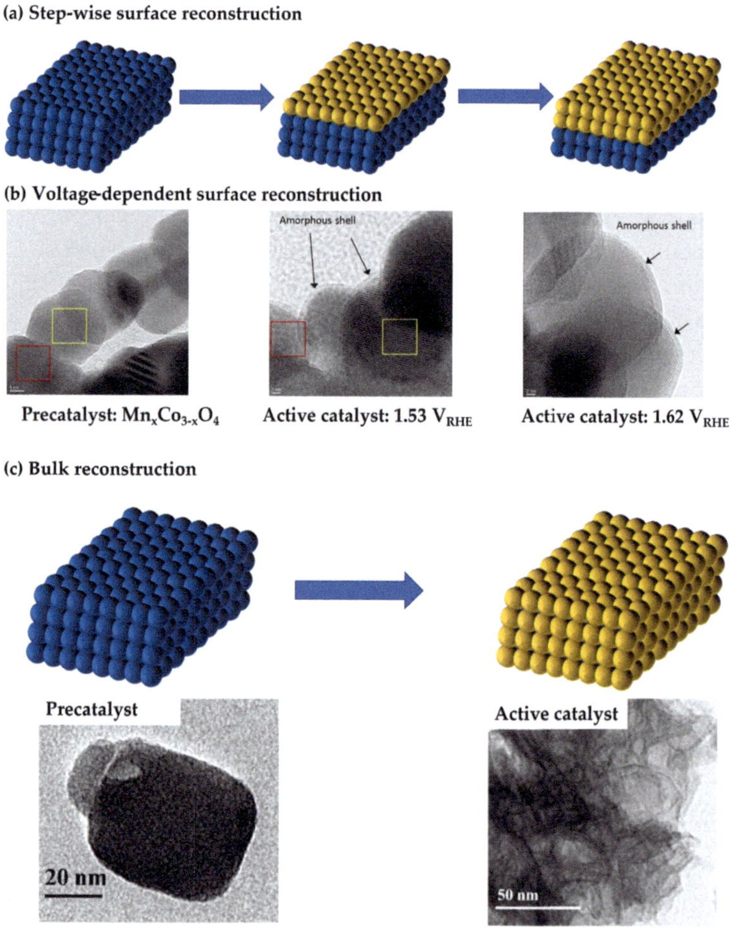

Figure 8.1 (a) Schematic illustration showing the stepwise surface reconstruction of precatalyst into active catalyst M(O)OH at the anode. (b) Voltage-dependent surface reconstruction of $Co_xMn_{3-x}O_4$. (c) Bulk anodic reconstruction of the precatalyst to active catalyst (PBA to Fe-Co(O)OH). The TEM images were reproduced from [3, 13] with permission from Wiley and RSC, respectively.

In the electrochemical reconstruction process, cation vacancies are formed through the leaching of weakly bonded metal species. For example, in Fe-incorporated MOFs, Fe leaching during alkaline OER generates cationic vacancies. These vacancies improve surface adsorption properties by exposing active sites and modifying the ES. The engineering of cation vacancies shifts the catalyst's d-band center, altering the adsorption/desorption dynamics and IM energy levels, thereby impacting OER efficiency.[15,16]

The reconstruction process can be either reversible or irreversible, depending on the structure and the composition of the PC, its morphology, phase, and reaction and operating conditions.[17,18] The reversible reconstructions represent the true catalytic process, where the PC regains its initial phase, morphology, and structure when the applied potential is removed. The reversible activation of the PC has been identified using in situ techniques, such as in situ RS, XPS, and XAS.[18,19] This process cannot be detected by ex situ studies after the removal of the actual reaction conditions. However, for most of the PCs, the activation process is irreversible, leading to a significant enhancement in water oxidation activity. Therefore, the study of the reconstruction chemistry during OER is crucial for a deeper understanding of the actual electrocatalytic process and mechanism, as well as for developing advanced OER electrocatalysts.

In contrast to alkaline media, there are very few reports on the reconstruction process in neutral media. A key challenge is that only a limited number of catalysts can achieve moderate OER performance in neutral conditions.[20] Similar to alkaline environments, surface reconstruction in neutral media (buffer or salt solutions) leads to the formation of high-valence M(O)OH phase, which is regarded as the true catalytic species. This has been confirmed through in situ RS and in situ XAS characterizations.

In addition to strong alkaline and neutral media, electrochemical surface reconstruction has also been studied in acidic conditions. However, a major challenge in acidic media is the severe degradation of electrocatalysts. Surface reconstruction in acidic environments differs significantly from that in neutral and basic conditions, as electrocatalysts are exposed to a highly corrosive and oxidative atmosphere. This results in the selective dissolution of less noble or less stable metal species, which in turn promotes the formation of high-valent species.[21,22]

Recent advancements in characterization techniques including spectroscopy, PXRD, microscopy, and isotope labeling technology have significantly enhanced the efficiency to realize real-time, dynamically evolving information of the catalysts under operating conditions, including insights into their chemical composition, morphology, atomic and ES.

8.3 Techniques

The reconstruction of the catalysts on the electrode surface can be detected using both ex situ and in situ techniques. In this section, we will provide a brief overview on key analytical techniques employed for the detection of the electroactive species.

8.3.1 PXRD

PXRD is a powerful tool for obtaining precise crystallographic information, including crystal phase, available facets, preferred orientation, and lattice parameters. Further, PXRD can offer information about lattice strain, doping effects, and phase separation. However, PXRD cannot be used to find out structural changes for amorphous phases due to the lack of long-range atomic order.[19,23]

Selected area electron diffraction (SAED) patterns also provide key insights into phase transitions, lattice expansion/contraction, and structural disorder.[24]

8.3.2 Raman spectroscopy

Raman spectroscopy is a scattering technique that reveals information about molecular vibrations and rotational energy levels.[8] RS can offer important information about the M-O stretching and bending vibrations as they are highly sensitive to the local ES, oxidation state of the metal ions, and electron donating/withdrawing nature of the ligands.[23,25] Therefore, RS can be effectively utilized to identify the structural changes in metal oxides, metal hydroxides, etc., even when they are in the amorphous phase. The formation of M(O)OH as the active catalysts (ACs) can also be detected by RS. Even the α-, β-, or γ-phase of M(O)OH, formed at the electrode surface, can be detected by RS as they have distinct structural features.[25] In addition, the intensity ratio of the in-plane and out-plane vibrations can be utilized to predict the thickness of the 2D materials. In addition, RS can be utilized to detect the short-lived species.

8.3.3 UV-Vis Spectroscopy

UV-Vis spectroscopy is a powerful tool to monitor the changes in the ES of the molecule/materials. It has been widely used to understand the structural changes of MO_x, $M(OH)_2$, and M(O)OH, where the differences in the λ_{max} for M^{2+}, M^{3+}, and M^{4+} oxidation states are clearly distinguishable.[26,27]

8.3.4 XPS

XPS can detect the elemental composition of the materials, and valence states of the elements present on the surface of both PC and AC.[8] Additionally, XPS is widely used to understand the ES of catalysts, including electron transport and orbital interactions. By analyzing the binding energies of core electrons, XPS provides insights into the catalyst's ES.[28]

8.3.5 XAS

XAS offers comprehensive information on the ESs and coordination (CN) environments of the catalysts. It can be used to detect reaction IMs, track catalyst structural evolution, and identify active components.[29] The XAS can be divided into two categories: the extended X-ray absorption fine structure (EXAFS) and the X-ray absorption near edge structure (XANES).[20] The primary absorption edge in the XANES region is caused by the shift from 1s to 4p states, whereas the pre-edge features are associated with the transition from 1s to 3d states. Because of the close relationship between the pre-edge energy and the formal oxidation state of bulk materials, XANES can be used to identify the valence state of atoms from edge positions. EXAFS, on the other hand, examines the local geometry and bond lengths of the atom by analyzing photoelectron scattering. This technique provides insights into the number of surrounding atoms, their distances, and any structural disorder.[30]

The application of anodic potential raises the oxidation state, leading to a contraction of M–O bonds. This shift in oxidation state, along with changes in bond length or bond order, can be clearly observed using EXAFS. Additionally, changes in the oxidation state affect the spin state of metal ions, which can be studied from the L-edge XANES.[29]

8.3.6 Electron Paramagnetic Resonance Spectroscopy

Electron paramagnetic resonance spectroscopy (EPR) provides information about unpaired electrons in the system.[31] The EPR signal intensity can also be correlated to the number of vacancies in electrocatalysts. Ex situ measurements, where the catalyst is in a relaxed state, can indicate the structural evolution. However, ex situ electrochemical EPR studies in the aqueous medium after freeze-quenching can be directly measured as ice does not absorb microwaves.[32]

8.3.7 Microscopy

The formation of crystalline and amorphous phases can be directly analyzed through microscopic techniques.[33,34] HRTEM offers a detailed understanding of structural changes at the catalyst surface. A more precise information on the structural changes can also be deduced from STEM. The variation in the lattice distances by structural changes can also offer insights into inter-layer expansion, vacancy formation, structural disorder, etc.[35]

In recent years, various in situ characterization techniques have also been developed, including in situ RS, in situ photoluminescence, in situ liquid-phase transmission electron microscopy, and in situ electrochemical scanning tunneling microscopy. Among these techniques, only in situ RS and in situ XAS have been extensively studied.[36]

8.4 Irreversible Anodic Reconstruction: Detection of OER-Active Species

In this section, our primary focus is to understand the structural changes of electrocatalysts during OER by using various ex situ characterization methods. For example, the structural changes of $Ni(OH)_2$ during OER were extensively studied by PXRD, microscopic, and spectroscopic methods (Fig. 8.2).[37] The PXRD of the PC confirmed the presence of the brucite β-$Ni(OH)_2$ phase. TEM and SAED revealed that β-$Ni(OH)_2$ formed single-crystal hexagonal platelets. After electrochemical OER, PXRD of the relaxed state catalyst detected additional diffraction corresponding to the $Ni_{1-x}O$ phase. The $Ni_{1-x}O$ diffraction peaks were weak and broad, indicating small crystal size or low crystallinity, consistent with the nanoporous morphology. TEM of the activated β-$Ni(OH)_2/Ni_{1-x}O$ showed an intriguing ~16 nm surface layer forming after 1000 cycles. This layer feature was absent in the pristine samples, indicating the surface structural reconstruction during OER.[37]

The presence of Ni^{3+} in the $Ni_{1-x}O$ surface layer was further detected by electron energy loss spectroscopy (EELS) (Fig. 8.2). The O–K edge of the $Ni_{1-x}O$ surface layer displayed a pre-peak at ~529 eV, attributed to O-2p states hybridized with the 3d states of Ni^{3+}, this feature was absent in pristine β-$Ni(OH)_2$. Additionally, the Ni–L edges from the NiO surface layer showed a small shoulder on the higher-energy side of the characteristic Ni^{2+} edges, indicating the presence of Ni^{3+}. This clearly demonstrates that the $Ni_{1-x}O$ surface layer contains Ni^{3+} due to Ni deficiency caused by the preferential dissolution of Ni during OER. Additionally, XPS detected characteristic $Ni^{2+}$and Ni^{3+} peaks in the AC (Fig. 8.2).

After OER, a slight blue shift of the Ni K-edge XANES was detected because of an increase in the positive charge density on Ni-ions (Fig. 8.2). The XANES of the OER-catalyst was fitted as the combination of β-$Ni(OH)_2/Ni_{1-x}O$ and $Ni_{1-x}O$. Additionally, Fourier-transformed EXAFS reveals a negative shift in the Ni–Ni bond to a shorter distance after OER along with a decrease in the Ni–Ni coordination number (Fig. 8.2).

The electrochemical activation (800 CV cycles between 1.10 and 1.55 V_{RHE}) of NiO to Ni(O)OH was studied by ex situ RS. Two dominant peaks (E_g and A_{1g}) confirmed the formation of Ni(O)OH. The electrochemical anodic activation of NiGe led to the formation of γ-Ni(O)OH.[38] The Ni 2p XPS detected the peaks for Ni^{3+} in the AC, and a complete conversion of NiGe to γ-Ni(O)OH was confirmed as the peaks for Ni^0 were not detected. Further, the complete oxidation of Ge^0 to GeO_2 was also observed.[38] The RS of the AC revealed two peaks in the range of 450 to 600 cm^{-1}, corresponding to the Ni–O bond. Additionally, a broad band appeared between 850 cm^{-1} and 1200 cm^{-1}, attributed to the ν(O–O) of an active oxygen species ($Ni\text{-}OO^-$) in the oxyhydroxide structure.

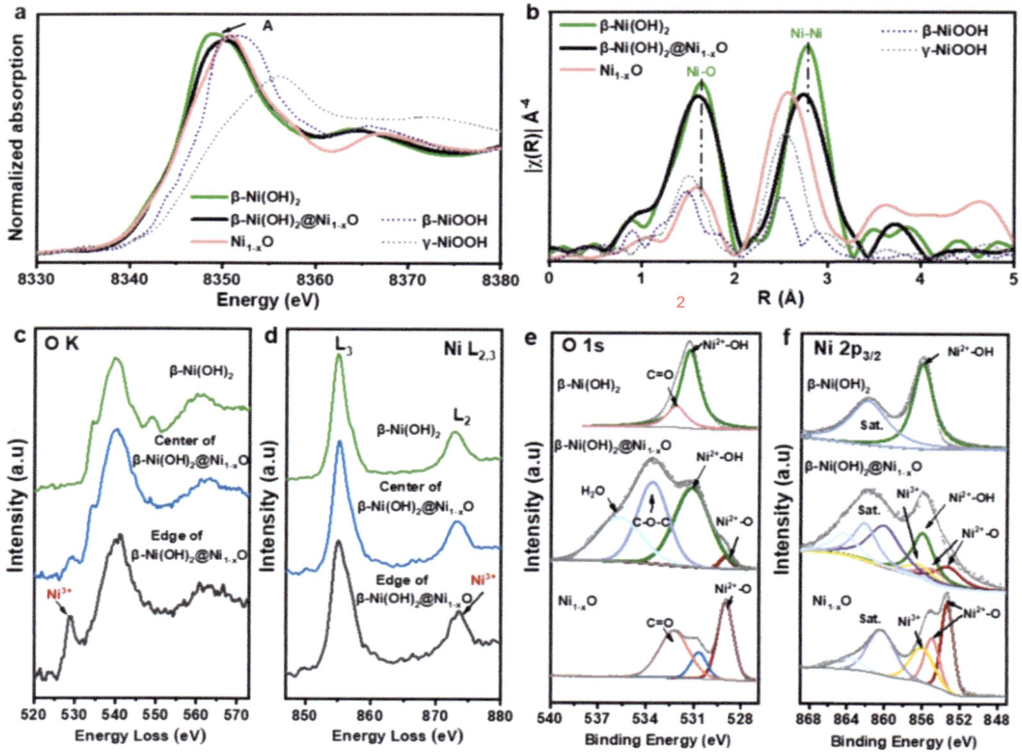

Figure 8.2 (a) Ni K-edge XANES and (b) FT EXAFS for pristine β-$Ni(OH)_2$, activated β-$Ni(OH)_2$@$Ni_{1-x}O$, and $Ni_{1-x}O$. (c) O K-edge EEL spectra of β-$Ni(OH)_2$@$Ni_{1-x}O$; (d) Ni-L edge EEL spectra β-$Ni(OH)_2$@$Ni_{1-x}O$ compared with β-$Ni(OH)_2$. (e) O 1s XPS of pristine β-$Ni(OH)_2$, activated β-$Ni(OH)_2$@$Ni_{1-x}O$, and Ni-deficient $Ni_{1-x}O$; and (f) Ni-$2p_{3/2}$ XPS of pristine β-$Ni(OH)_2$, activated β-$Ni(OH)_2$@$Ni_{1-x}O$, and Ni-deficient $Ni_{1-x}O$. Sat. stands for satellite peaks. The figures were reproduced from [37] with permission from ACS.

For a heterojunction Ni-LDH/α-Fe(O)OH, notable changes were observed in the RS with the variation in the oxidation potential because of the formation of Ni(O)OH/Ni-LDH/α-Fe(O)OH.[39] With the increasing oxidation potential, the intensity of the peaks associated with LDH decreased and new peaks for Ni/Fe(O)OH emerged. At a high potential, the higher intensity of E_{1g} bending and A_{1g} stretching vibration of Ni^{3+}–O was observed compared to that of Fe(O)OH to show a significant electronic interaction between Ni and Fe ions during the oxidation process.

The Fe 2p and Ni 2p XPS of AC revealed that the oxidation states of both Fe and Ni increased until the surface reaction was completed and fully oxidized. Fe withdrew the electrons from Ni to maintain the charge balance. The lattice oxygen (OL) content increased along with Ni^{3+}, enhancing the hybridization between metal 3d and oxygen 2p orbitals. The structural reconstruction induced by α-Fe(O)OH significantly improved the catalyst's electrochemical performance.

The Ni L-edge XAS of Ni(O)OH/(LDH/α-Fe(O)OH) showed a more intense peak than LDH/α-Fe(O)OH and LDH, indicating a higher valence state for Ni. Similarly, the Fe L-edge XAS for Ni(O)OH/(LDH/α-Fe(O)OH) showed an increased peak intensity, reflecting a higher oxidation state and reduced electron density on Fe.

In Ni(O)OH/(LDH/α-Fe(O)OH), the intensity of the M–O peak decreased slightly, while the $-OO^-$ peak appeared at a lower energy, reflecting the enhanced covalency of the M–O bonds and oxidation of OL, generating more active $-OO^-$ species. In the Ni K-edge XANES, the edge was positively shifted for Ni(O)OH/(LDH/α-Fe(O)OH) compared to LDH/α-Fe(O)OH, and the peak at 8350 eV showed higher intensity for Ni(O)OH/(LDH/α-Fe(O)OH) than that of LDH/α-Fe(O)OH. For the Fe K-edge of Ni(O)OH/(LDH/α-Fe(O)OH), the edge peak shifted to a higher energy, and the peak intensity increased significantly compared to LDH/α-Fe(O)OH. The XANES confirmed the increase in the valence state of the metal ions during electro-oxidation. In Ni(O)OH/(LDH/α-Fe(O)OH), the intensities of the Fe–O and Fe–M peaks were decreased, while those of Ni–O and Ni–M were increased, indicating higher local structural disorder due to electrochemical oxidation. Additionally, the Ni–O and Ni–M bond distances in Ni(O)OH/(LDH/α-Fe(O)OH) were contracted compared to LDH/α-Fe(O)OH. This structural change was further supported by Ni K-edge EXAFS oscillation data showing reduced amplitude oscillations and a lower CN number for Ni atoms to reveal a higher defect level and more amorphous features in Ni(O)OH/(LDH/α-Fe(O)OH).

Therefore, Fe offers a crucial role in the formation and stabilization of Ni(O)OH. The charge density around the Ni sites in LDH is likely reduced due to the strong electron-withdrawing and hydrogen-absorbing properties of neighboring α-Fe(O)OH. As a result, the active Ni(O)OH species was successfully generated through anodic activation process.

The importance of structure of the PC and the process of reconstruction was studied by different groups. For example, the anodic reconstruction of CoFe-Prussian blue analog (PBA) and CoFe-nitroprusside (NP) produces the same $M(O)_x(OH)_y$ phase as the AC but with different local ESs. The variation in the applied potential, activation through CV, CA, or CP, etc., have a significant effect in tuning the local atomic structure of the AC and hence the water oxidation activity. The reaction of $Co(OH)_2$@CC (CC = Carbon cloth) with $K_3[Fe(CN)_6]$ and $Na_2[Fe(CN)_5(NO)]$ produced to PC-1 and PC-2, respectively.[3] The strong π-accepting –NO group in PC-2 made a significant difference in the ESs of the catalyst generating high positive charge density on the metal centers. As a result, the activation of PC-2 to AC-2 was facile compared to that of PC-1 to AC-1 (Fig. 8.3).[3]

The PXRD of AC-1 and AC-2 showed the presence of mixed-phase α-$Co(OH)_2$ and β-Co(O)OH. XPS of Co 2p and Fe 2p confirmed a higher positive charge density on the metal ions in AC-2 compared to AC-1 (Fig. 8.3). In the Co K-edge XANES of AC-2, a strong main absorption peak and a weaker pre-edge peak were observed. The reduced intensity of the pre-edge peak suggests a possible presence of LS octahedral Co^{3+}, indicating a more centrosymmetric local structure around Co. Additionally, the shift to higher energy in the Co K-edge XANES for AC-2 compared to AC-1 is consistent with LS Co^{3+}. In contrast, the Co K-edge XANES for AC-1 indicates a mixed-valence state, with both high-spin (HS) Co^{2+} and LS Co^{3+} (Fig. 8.3).

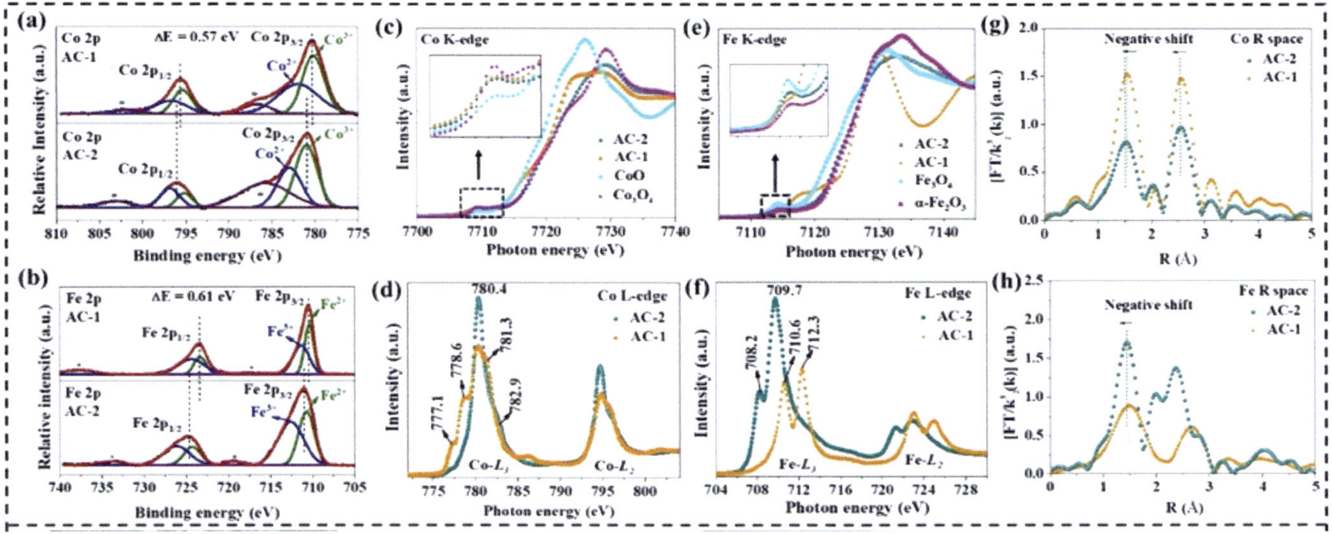

Figure 8.3 (a) Co 2p XPS of AC-1 and AC-2. (b) The Fe 2p XPS of AC-1 and AC-2. (c) Co K-edge XANES of AC-1 and AC-2 compared to reference materials CoO and Co_3O_4. (d) Co L-edge spectra of AC-1 and AC-2 compared to Fe_3O_4 and α-Fe_2O_3. (f) Fe L-edge spectra of AC-1 and AC-2. (g) FT-EXAFS of Co R-space for AC-1 and AC-2 demonstrates a contraction of Co–O and Co–Co bonds in AC-2. (h) The FT-EXAFS analysis of Fe R-space for AC-1 and AC-2 confirms the shortening of Fe–O bond in AC-2. The figures are reproduced from [3] with permission from Wiley.

The Fe K-edge XANES of AC-2 also exhibits a pre-edge characteristic, suggesting primarily octahedrally coordinated Fe^{3+} ions. On the other hand, AC-1 exhibits a clear pre-edge feature that points to a combination of Fe^{2+} and Fe^{3+} oxidation states. Furthermore, a larger concentration of Fe^{3+} species is indicated by the primary absorption peaks in the Fe K-edge XANES of AC-2. The mixed valency of Fe in AC-1 was further confirmed by the Fe L-edge spectrum. The higher intensity in AC-2 indicates a higher concentration of Fe^{3+}.

Similarly, the surface reconstruction of different chalcogenides during OER was studied. The anodic activation of $NiSe_2$ led to the formation of Ni(O)OH through $Ni(OH)_2$.[40] To further study the structure of active $Ni(OH)_2$, ex situ XAS measurements were done on both activated $Ni(OH)_2$ and marketable β-$Ni(OH)_2$. The comparison of the Ni K-edge XANES of activated $Ni(OH)_2$ and marketable β-$Ni(OH)_2$ showed comparable oxidation states and coordination environments around the Ni atoms. However, the edge peak of activated $Ni(OH)_2$ was observed lower than that of commercial β-$Ni(OH)_2$. Furthermore, EXAFS analysis revealed that activated $Ni(OH)_2$ has shorter Ni–O and Ni–Ni bond lengths compared to commercial β-$Ni(OH)_2$. These findings suggest that the IM nickel hydroxide was not β-$Ni(OH)_2$.

From the above examples, it is evident that the AC for OER is M(O)OH. Similarly, in electrochemical organic oxidation, M(O)OH serves as the AC. To determine if the catalyst structure remains consistent after OER and organic oxidation, we have examined the changes in the oxidation state and ES of the catalyst after benzyl alcohol (Bz) oxidation.

The PXRD of the catalyst after Bz oxidation revealed a phase comparable to that of the OER-AC.[42] However, RS revealed a significant difference in the AC structure after OER and Bz oxidation. In Fe–Co(O)OH after OER, a characteristic Co–O vibration appeared, while Bz oxidation led to additional peaks corresponding to Bz and –COOH groups.[41] Furthermore, the peak for M–OCH_2Ph was also observed. The XPS also revealed the variations in the ESs. The Co 2p XPS peak in Fe–Co(O)OH-Bz was shifted to a lower binding energy compared to Fe–Co(O)OH-OER. Additionally, the Fe 2p XPS showed a partial reduction of Fe^{3+} to Fe^{2+} in Fe–Co(O)OH-Bz.

XAS study was conducted to further examine the atomic and ES changes in the catalyst. The pre-edge feature in the Co K-edge XANES, resembling CoO, confirmed the octahedral coordination of Co^{2+}. Additionally, the high absorption peak of Fe–Co(O)OH-Bz was shifted to a lower energy than that of OER catalyst. The Fe K-edge XANES of Fe–Co(O)OH-Bz revealed the presence of mixed-valent Fe^{2+}/Fe^{3+}. The negative shift of the main absorption peak was also observed in Fe–Co(O)OH-Bz relative to Fe–Co(O)OH-OER. Additionally, FT-EXAFS analysis showed a shift in the Co–O and Co–Co bond distances in Fe–Co(O)OH-Bz compared to Fe–Co(O)OH-OER.

Therefore, the XPS and XAS analyses confirm that the Co^{3+} and Fe^{3+} species in Fe–Co(O)OH-OER were largely reduced to Co^{2+} and Fe^{2+} during Bz oxidation. This reduction led to significant changes in the coordination and electronic properties of Fe and Co, including an increment in Co/Fe–O bond length.

EPR was also utilized to understand the anodic reconstruction of $NiFe_{0.5}Sn$ to AC.[42] Compared to $NiFe_{0.5}$-A, $NiFe_{0.5}Sn$-A shows a stronger signal for oxygen vacancies, suggesting that selective etching of Sn promotes oxygen vacancy formation. Under alkaline conditions (pH > 13), the water oxidation mechanism involving $Co^{2+/3+}$ spinels and perovskites has been extensively studied by EPR. In these oxides, surface Co^{4+} sites are formed as the essential IMs during the O_2 evolution process.

The EPR study of Co–P_i showed that an increase in the deposition voltage leads to a higher concentration of Co^{4+} and a decrease in the concentration of Co^{2+}.[32] Long-term water oxidation catalysis can also contribute to changes in the redox characteristics. These findings provide spectroscopic evidence for the formation of Co^{4+} species during water oxidation catalysis.

In conclusion, ex situ techniques can only reveal the properties of electrocatalysts after or before the reactions but cannot detect any dynamic changes in the transition phase of the reactions. Therefore, it could be misleading to simply correlate the OER activity based on the ex situ characterizations.

8.5 In situ and in operando Studies of Anodic Reconstruction

As mentioned in the introduction part, the reversible anodic reconstructions are indicative of the true catalytic process, where the catalyst regains its initial structure once the applied potential is removed. The structural modification in electrocatalysts before and after the OER process can reveal various physical and chemical properties through a range of ex situ characterization techniques. However, it remains challenging to extract essential information about transient IMs that exist only temporarily during the dynamic evolution process. These short-lived IMs, which are undetectable by ex situ experiments, complicate the identification of genuine elementary reactions and limit a deeper understanding of the mechanisms and underlying structural evolution. In this context, in situ and in operando techniques can be applied for a better understanding of the electrocatalytic process.[43]

8.5.1 In situ PXRD

In situ PXRD has been utilized to monitor and study the crystal structural transformations and phase growth of catalysts in real time under OER conditions. However, in situ XRD study faces challenges, particularly in detecting the low crystallinity of the constructed active layers and weak scattering from the solid–liquid interfaces.[44]

The anodic reconstruction of cobalt oxide by the gradual change of electrochemical potentials was studied by in situ PXRD.[45] An increase in the structural coherence length was observed after the immersion of the catalyst film in the electrolyte solution at 1.0 V_{RHE}, and it continued to increase with the increasing anodic potentials. The maximum structural coherence length was recorded at 1.55 V_{RHE}. When the applied potential was increased to 1.62 V_{RHE}, the structural coherence length of the cobalt oxide domains decreased and at a lower potential than 1.2 V_{RHE}, the structural coherence length returned to its initial phase.

This result indicates that cobalt oxide initially exhibited irreversible growth up to an electrode potential of 1.2 V_{RHE}, followed by a reversible structural change. This structural alteration was attributed to the formation of CoO in the crystalline cobalt oxide

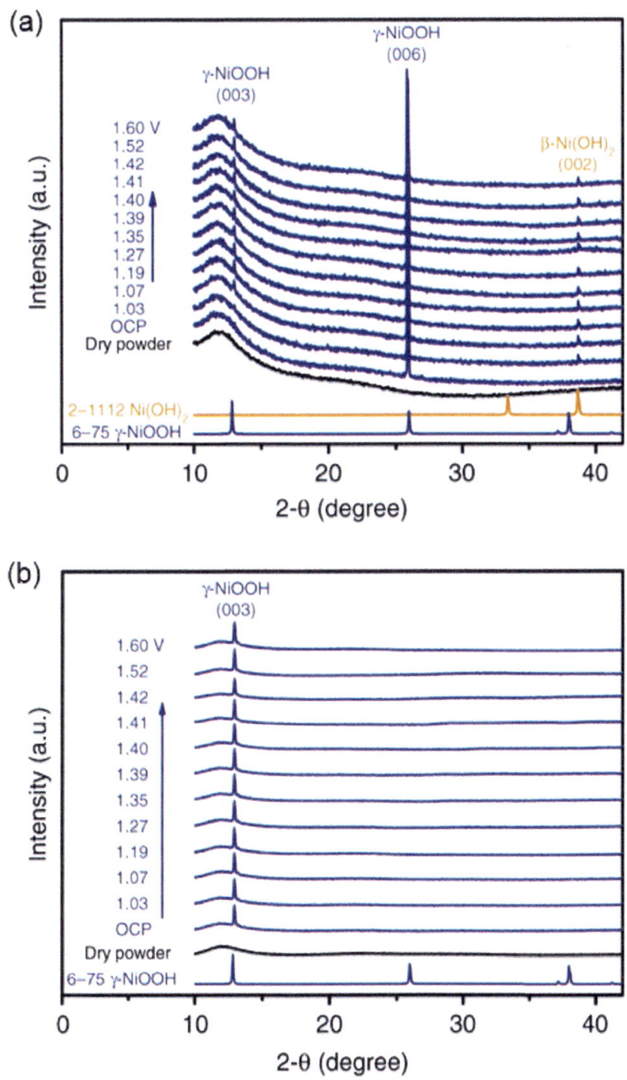

Figure 8.4 Operando XRD of (a) $Ni(OH)_2$/NF and (b) $NiCeO_xH_y$/NF at various applied potentials. The figures are reproduced from [45] with permission from Springer nature.

domains in contact with the electrolyte. Due to its sub-nanometer size, this CoO phase was amorphous and did not generate any additional reflections. Upon reverting to the cathodic scan after exiting the potential range, the amorphous shell recrystallized to Co_3O_4.[44] The PXRD investigations on $Ni(OH)_2$ and $NiCeO_xH_y$ revealed an increase in the crystallinity after the immersion of the catalysts in the alkaline electrolyte solution.[45] In operando conditions, with the applied potential, PXRD detected the transformation of $Ni(OH)_2$ into a mixed-phase structure containing β-$Ni(OH)_2$ and γ-Ni(O)OH (Fig. 8.4). In contrast, the $NiCeO_xH_y$ formed a single γ-Ni(O)OH phase after increasing the electrode potential. The incorporation of Ce into $Ni(OH)_2$ alters the ESs of the Ni species and promotes the structural reconstruction to γ-Ni(O)OH, both of which enhance the kinetics OER.

Other than this, a plenty of PCs like MOFs, phosphides, chalcogenides, carbides, etc., have been studied by in situ PXRD and the formation of M(O)OH as the active phase was detected.[46] Although in situ XRD could be effective for tracking amorphization and significant phase changes, it is less suited for analyzing reconstructed electrocatalysts when only a thin amorphous layer is formed as the active phase.

8.5.2 In situ Raman Study

As a nondestructive analytical technique, in situ RS can provide important information about the chemical structure of the materials. Even, the changes in the amorphous phase can be detected by potential-dependent operando RS. The key advantages of RS associated with the detection of the amorphous phases, not detectable by PXRD, can be studied effectively.

For example, in situ RS was employed to study the gradual phase transition from $NiCeO_xH_y$ to γ-Ni(O)OH (Fig. 8.5).[45] The addition of Ce to the catalyst lowers the energy barrier, enabling the phase transition to occur at a significantly reduced potential of 1.27 V_{RHE} compared to $Ni(OH)_2$ (Fig. 8.5).

Notably, increased OER activity is often associated with a lower potential for the phase shift to the OER-active phase γ-Ni(O)OH. The distinctive peaks between 450 and 600 cm^{-1} are attributed to the γ-Ni(O)OH phase. However, due to the relatively low abundance of Fe(O)OH species on the catalyst surface and the overlay of its representative peaks with other RS signals, the RS studies struggle detecting the formation of Fe(O)OH species.

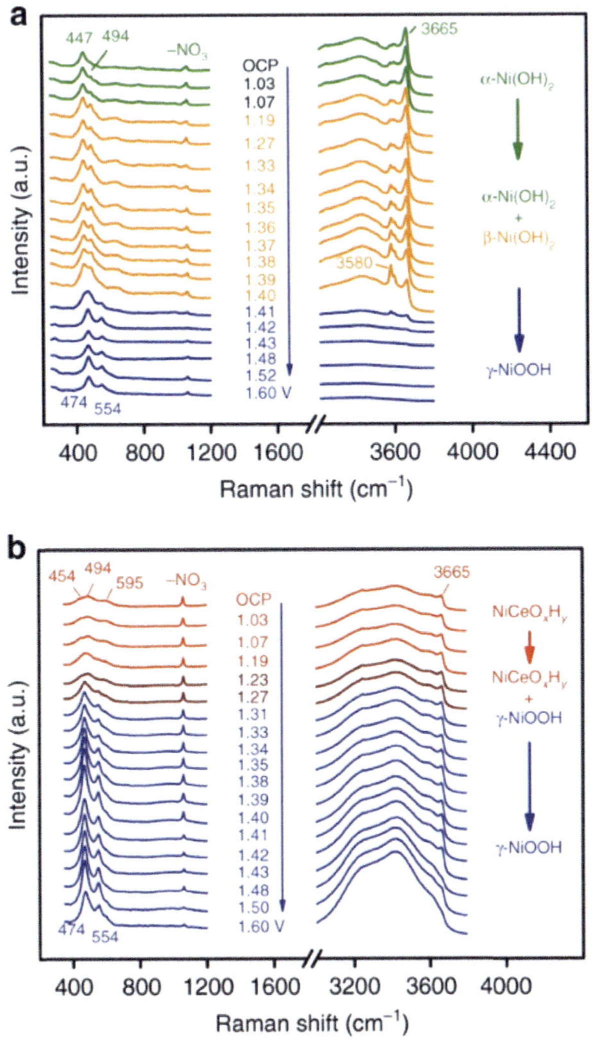

Figure 8.5 In situ Raman spectra of (a) $Ni(OH)_2$ and (b) $NiCeO_xH_y$ from OCP to 1.6 V_{RHE} in 1.0 M KOH solution. The figures are reproduced from [45] with permission from Springer nature.

In situ RS of Ni_3S_2 and $Fe(O)OH/Ni_3S_2$ in the range of 1.04–1.64 V_{RHE} during OER showed peaks between 100 and 400 cm^{-1} corresponding to Ni_3S_2 under OCV and no new peak was detected until the voltage reached to 1.24 V_{RHE}.[47] As the applied potential increased, a pair of new vibrations for Eg Ni–O bending and A_{1g} Ni–O stretching of γ-Ni(O)OH was detected. The surface reconstruction of Fe(O)OH/Ni_3S_2 was started at significantly lower electrochemical potential than that of Ni_3S_2/NF. Further, the shortening of Ni–O bonds was observed with the oxidation of Ni^{2+} to Ni^{3+} under applied anodic potential.[47]

The operando RS of the electrochemical reconstruction of $Ni(OH)_2$ and NiFe-LDH revealed the existence of α-$Ni(OH)_2$ and disordered NiFe-LDH at lower potential (1.1 V_{RHE}).[48] As the potential increased, the new peaks for γ-Ni(O)OH and β-Ni(O)OH appeared, indicating the oxidation of Ni^{2+} to Ni^{3+}. At > 1.6 V, the formation of NiO_x was observed because of the oxidation of Ni^{3+} to Ni^{4+}. The Fe^{3+} species in NiFe-LDH played a crucial role to facilitate the oxidation of Ni^{3+} to Ni^{4+} at a significantly lower potential, thereby enhancing the kinetics of the OER. The introduction of Ce into NiFe-LDH further promotes the oxidation of Ni^{3+} to Ni^{4+} and reduces the OER onset potential.[48] High-valent Ce and Fe function as Lewis acid, increasing the acidity of OH groups bound to Ni and lowering the oxidation potential of Ni^{3+} to Ni^{4+}.

When α-$Co(OH)_2$ was exposed to air, operando RS revealed two distinct peaks at 455 and 518 cm^{-1}, corresponding to Co–O vibrations.[49] The peak at 518 cm^{-1} nearly disappeared at 1.03 V_{RHE} by a broad peak, indicating a transition to less-crystalline hydrous α-$Co(OH)_2$. As the voltage increased, additional peaks appeared at 463 and 572 cm^{-1}, suggesting the formation of γ-Co(O)OH, which subsequently converted to γ-Co(O)OH.

8.5.3 In situ XPS

Exploring the catalyst surface during OER is essential for the understanding of fundamental structure–activity relationships. However, traditional XPS methods require strict ultrahigh vacuum conditions. Ambient-pressure XPS, utilizing synchrotron radiation and incorporating differential-pumping stages, overcomes this limitation and allows for the characterization of electrocatalyst surfaces under more realistic conditions.

The change in the oxidation state of Co in CoO was investigated under operando electrochemical conditions.[50] A $Co(OH)_2$ overlayer was observed to form on Co_3O_4 in both vacuum and hydrated environments. This $Co(OH)_2$ overlayer underwent partial oxidative conversion to Co(O)(OH) after structural stabilization via cyclic voltammetry. However, complete oxidative conversion of the hydroxide layer to oxyhydroxide occurred when the potential was maintained above the Co^{3+}/Co^{4+} transition. Importantly, this full oxidative conversion took place within the so-called precatalytic range, which refers to the range of electrochemical potentials below the operational parameters of the OER.

Furthermore, maximal OER activity on Co oxides is observed only when Co(O)(OH) is present on the surface. Therefore, kinetically promoting the formation of this phase may be crucial for reducing the OER overpotential. Quasi-operando XPS was employed to realize the surface reconstruction process. As the applied potential increased, the oxidation state of Co also increased, with a significant peak shift observed at 1.45 V_{RHE}, indicating the electrochemical activation. Meanwhile, the binding energies in the Ni 2p XPS consistently increased alongside the rising valence state of Ni. In contrast, the Mn 2p XPS revealed that the binding energy increased from the OCV to 1.45 V_{RHE}, then sharply decreased at that voltage before rising again.[51]

To elucidate the mechanism of OER and identify surface active species, it is crucial to understand the near-surface structures in their most stable catalytic state. The Mn 2p XPS spectra reveal that the oxidation state of surface Mn increases at 1.25 V_{RHE} compared to OCV. However, when the applied potential was further increased to 1.55 V_{RHE}, no significant change was observed in the Mn 2p spectra, suggesting that the oxidation state and local environment of the Mn ions remain stable and similar to those at 1.25 V vs. RHE. This indicates that the Mn ions are structurally and electrically comparable at both potentials. Additionally, the peak for oxygen vacancies diminished in the O 1s XPS, while the O_L peak shifted to higher energy. These findings suggest that the OER process occurs after the oxidation of OL and the filling of oxygen vacancies.

Furthermore, beyond the oxides, transition-metal tellurides undergo significant in situ electrochemical tuning in strong alkaline condition, making them highly promising OER catalysts for water electrolysis. The Ni 2p XPS of $NiTe_2$ and $CoTe_2$ revealed Ni^{2+} and Co^{2+} species, but after applying anodic potential, the $NiTe_2$ and $CoTe_2$ structure partially converted to $Ni(OH)_2$ and $Co(OH)_2$.[52] With an increase in the potential beyond 1.4 V_{RHE}, both Ni and Co hydroxides were converted into OER-active Ni(O)OH and Co(O)OH, respectively. This transformation was confirmed by the appearance of additional peaks in the Ni 2p and Co 2p spectra. The Te 3d spectrum further confirmed the surface oxidation of Te, indicated by the formation of TeO_x species.[52]

8.5.4 In situ UV

The redox reactions of transition series metal catalysts can be easily monitored using UV-vis spectroscopy, which is also effective for identifying various electronic transitions. The in situ UV–vis absorption spectrum provides important insights into the reaction IMs and allows the detection of distinctive adsorption peaks that may change as the electronic configuration evolves during the OER process.

The in situ UV–vis spectroscopy studies on Fe(O)OH@NiFe-LDH during OER revealed no significant changes in the catalyst structure till 1.43 V vs. RHE.[34] However, increased absorption was detected after 1.45 V vs. RHE, indicating Ni oxidation and the formation of β-Ni(O)OH. Throughout the CV, an increase in absorption peak for Fe(O)OH@NiFe-LDH was observed, coinciding with the onset potential for OER. As the applied potential increased, the absorption intensified, while a decrease in potential resulted in reduced absorption, indicating reversible reconstruction.

The potential-dependent absorption spectra of Cu_2S/CoFe-LDH were studied from OCV to 1.6 V_{RHE}. The absorption spectra showed no significant changes between 0.6 and 0.9 V_{RHE}, indicating that the Co valence state remained stable. However, when the applied potential was increased from 1.1 V vs. RHE, the new absorption peaks emerged between 450 and 600 cm^{-1}, respectively, indicating the oxidation of Co^{2+} to Co^{3+} and Co^{3+} to Co^{4+}.[53]

For a reported cobalt catalyst, the changes in the oxidation state with applied anodic potentials were studied by in situ UV-vis spectroscopy.[54] At lower potentials, an absorption peak appeared at 405 nm and a broad shoulder was observed as the applied potential was increased. At pH 7, the Co-cat undergoes two distinct redox transitions: Co^{2+}-Co^{3+} oxide and subsequently to a Co^{3+}-Co^{4+} oxide, with midpoint potentials around 1.0 V_{NHE} and 1.2 V_{NHE}, respectively.

XAS study confirmed the simultaneous presence of Co^{3+} with either Co^{2+} (at low potentials and pH) or Co^{4+} (at high potentials and pH). This study demonstrates that in situ UV-Vis spectroscopy can qualitatively track the overall oxidation state changes of cobalt.[54]

The absolute values of Co oxidation states cannot be directly determined from the UV-Vis absorption spectrum. However, this method is highly effective for monitoring changes in oxidation states during time-resolved experiments.

8.5.5 In situ XAS

In situ XAS study offers a detailed understanding of the change in the ES, local CN, and oxidation state during anodic activation of the catalyst. The OER active species of Ni-Bi was detected by recording the Ni K-edge XANES of both activated and unactivated Ni-Bi films at two potentials: 0.4 and 1.0 V_{NHE}. At 0.4 V_{NHE}, the oxidation state of Ni in both films remained at +2, with no OER activity detected in either the activated or non-activated films.[55] However, at 1.0 V_{NHE}, the Ni K-edge XANES of the precatalyst showed an edge energy peak, which was 1.3 eV lower than that of γ-Ni(O)OH and 0.8 eV higher than β-Ni(O)OH. This clearly indicates that the formation of high-valent Ni(O)OH occurs beyond 0.4 V_{NHE}, enhancing the OER activity. Coulometric data revealed that the average Ni valence states at 1.0 V_{NHE} were +3.60 for the activated film and +3.16 for the non-activated film.[55]

The FT EXAFS fitting results for the activated Ni-Bi films indicate that the Ni-O and Ni-Ni distances align with the crystallographic distances of γ-Ni(O)OH. The FT EXAFS fitting results for the non-activated Ni-Bi films reveal a Jahn–Teller distorted coordination pattern around the Ni centers.

After electrochemical reconstruction of the Ni-Bi films, both the changes in the Ni valence state and the coordination geometry suggest a phase transition from β-Ni(O)OH to γ-Ni(O)OH, accompanied by an increase in Ni^{4+} species. The higher electrochemical OER activity observed in the activated films compared to the non-activated film suggests that the γ-Ni(O)OH phase was more active for OER.

In electrodeposited Ni–Fe oxides, significant changes were observed for oxidation states of the metal ions and metal-oxygen bond lengths in response to the applied potential.[56] During the electrochemical reconstruction of $Ni(OH)_2$, the phase transition to γ-Ni(O)OH was confirmed by shorter Ni–O bond lengths and an increased Ni oxidation state that includes both Ni^{3+} and Ni^{4+}. Further, the Fe–O bond length was changed though the oxidation state of Fe remained unchanged.

By combining operando XAS data with DFT calculations, the authors proposed that Fe^{3+} cations are the active sites for the OER in Ni–Fe-based oxides. This finding was further verified by the Mukerjee group, which studied Ni–Co–Fe-mixed oxides using operando XAS. Their findings demonstrated that the presence of Co effectively reduces the local geometry of Ni and Fe, resulting in an optimized bond strength for Fe–OH/(O)OH. Additionally, this study revealed that while Fe tends to stabilize Ni in the 2+ oxidation state, Co promotes the oxidation of Ni to the 3+ state.

XAS demonstrated that the applied voltage increased the oxidation states of both Ni and Fe in Ni–Fe oxide catalysts. Semi-in situ XAS studies revealed that the Fe concentration had a significant effect on the oxidation state of Ni throughout the OER potential. Specifically, Ni was found to be in 4+ oxidation state when the Fe concentration was below 4%, while higher Fe content led to a lower oxidation state of Ni. However no changes in the oxidation state of Fe were observed at different applied potentials.

The gradual changes of NiFe-MOF with anodic potential were detected by using FTIR spectroscopy, combining with XAS.[57] The FTIR, supported by isotope labeling, identified a characteristic vibration peak for surface OOH* under an applied voltage of 1.6 V_{RHE}. When the potential was increased to 1.6 V_{RHE}, operando XAS detected a peak of Ni^{4+} species. Upon reducing the potential back to 1.2 V_{RHE}, the Ni^{4+} species disappeared.

8.5.6 In situ Microscopic Study

The morphological evolution of Co_3O_4 nanoparticles during CA OER was studied by in situ STEM.[58] Over the time, the particle agglomeration increases and gets stabilized. The bright regions in STEM-HAADF images indicate that the particles' amorphization into a lower-density solid was incomplete.

In situ annular dark-field scanning TEM study of OER on Mn_2O_3 revealed that the pristine Mn_2O_3 nanocatalysts had a quasi-spherical shape with a high-quality crystal lattice extending to the surface.[59] Following the electrochemical OER, a surface layer of 0.25 nm was formed. The surface layer exhibited oscillatory growth, indicating a reversible restructuring while the overall size of the nanocatalysts remained unchanged.

8.6 Cathodic Reactions

Cathodic electrocatalytic reactions, such as CO_2 and N_2 reduction and H_2 evolution, play a crucial role in converting electrical energy into chemical energy.

8.6.1 Detection of Active HER Species

HER by electrochemical water splitting can be carried out in different environments, such as acidic, neutral, and alkaline, and catalytic HER in different environments follows completely different mechanisms. Using in situ techniques, the evolution of the catalyst and the reaction IMs have been tracked to understand the process at the molecular levels. Compared to acidic HER, alkaline HER is a multi-step process with the involvement of several IMs as it involves the direct participation and the dissociation of water molecules.[60,61]

In situ RS revealed the influence of interfacial water on alkaline HER with Ru-based electrocatalysts.[62] Their theoretical calculations indicated that the vibration frequencies of the *H–M adsorption bond at the active sites of Ru^{4+}, $Ru^{(4-\delta)+}$, and Ru^{2+} were 1987, 1991, and 1947 cm^{-1}, respectively (Fig. 8.6). Additionally, the RS bands observed at 1825 and 1960 cm^{-1} were attributed to two distinct *H species adsorbed on the Ru sites, each exhibiting different valence states, consistent with the experimental results (Fig. 8.6).[62]

The real catalytic centers of the activated catalysts can be identified by using XAS to provide insight into the specific chemical bonds of the target electrocatalyst under in situ electrochemical activation conditions. For example, the amplitude of the Ni–S bond, measured through the XANES of the activated Ni-BDT sample (at a cathodic current density of 50 mA cm^{-2}), noticeably decreases after 0.5 hours of activation and stabilizes after 1.5 hours.[63] This observation indicates a change in the coordination environment of Ni-BDT during the electrochemical activation process. The fully activated Ni-BDT, referred to as Ni-BDT-A, consists of Ni-Ni and Ni–S bonds, as evidenced by the peak shift from Ni–S to Ni–Ni bonds in the EXAFS.

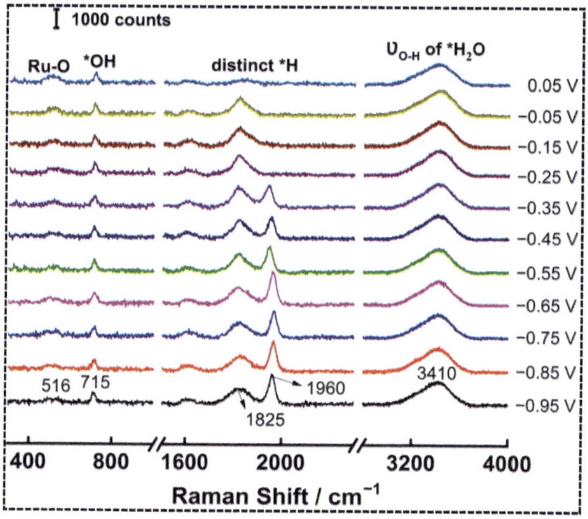

Figure 8.6 In situ RS of the alkaline HER process on Ru catalyst. The figure is reproduced from [62] with permission from Springer Nature.

Furthermore, XAS effectively detected the phase transition in 2H-MoS_2 for HER while also measuring the Ni-Ni and Ni–O bonds. The Mo–Mo and Mo–S bond lengths at OCP were found to be 3.14 Å and 2.40 Å, respectively, confirming the presence of 2H-MoS_2. In the HER channel, 1T-MoS_2 was formed with a bond length of 2.76 Å.[63]

8.6.2 Detection of Active Species of CO_2RR

Electrochemical CO_2RR is highly desirable for achieving carbon neutrality and facilitating intermittent energy storage.[65] For the production of valuable multi-carbon products, it has been reported that the dimerization of *CO and hydrogenation of *CO are the primary C–C coupling pathways (Fig. 8.8). The adsorption configuration and binding strength of these IMs largely determine the final product distribution. Consequently, spectroscopic observations of these key IMs during the CO_2RR process provide valuable insights into the fundamental mechanisms and guidelines for catalyst design.[64,65]

8.6.2.1 In situ Raman spectroscopy

Using potential-dependent in situ RS, IMs in CO_2RR were identified on various Cu (h,k,l) model surfaces. For the Cu (110) surface, C–O(H) stretching vibrations in *COOH species contributed to a shoulder peak at approximately 1046 cm^{-1} (Fig. 8.7).[66] Additionally, experiments involving ^{13}C and D-isotope replacement revealed a distinct redshift in the RS peaks. These results, in conjunction with theoretical simulations of adsorbed species, suggest the presence of key IMs such as *OCCO and *CH_2CHO. In contrast, the RS of the Cu (111) surface showed no evidence of C–C coupling IMs, indicating only the presence of *COOH and *CO species (Fig. 8.7). Overall, these findings demonstrate that Cu (111) favors the formation of C_1 products, while Cu (110) promotes C–C interactions, facilitating the evolution of C^{2+} products (Fig. 8.7).[66]

The structural changes in Cu foil during electrochemical CO_2RR were studied by in situ RS.[67] Initially, the peaks observed at 400 cm^{-1}, 528 cm^{-1}, and 620 cm^{-1} were indicative of a CuO_x phase. At cathodic potential, CuO_x was reduced to metallic Cu on the surface. Additionally, two new peaks (1,390 cm^{-1} and 1,610 cm^{-1}) emerged for the symmetric and asymmetric stretching vibrations of carboxylate groups. Moreover, the peaks in the range of 490–530 cm^{-1} at −0.76 V_{RHE} were attributed to either Cu–C stretching or Cu–O stretching of oxygen or hydroxide species on Cu. Interestingly, as CO surface coverage increases, the peaks at 1,390 cm^{-1} and 1,610 cm^{-1} diminished, while two new peaks emerged for *OCCO IM.

The key conclusions from the in situ RS indicate that CO is the initial IM species, regardless of whether the catalytic phase produces C_2 or C_1 products (Fig. 8). In the CO_2 reduction process, different phases of copper display distinct selectivity, with Cu (110) particularly favorable for C–C coupling and the formation of C_2 products (Figs. 8.7 and 8.8).

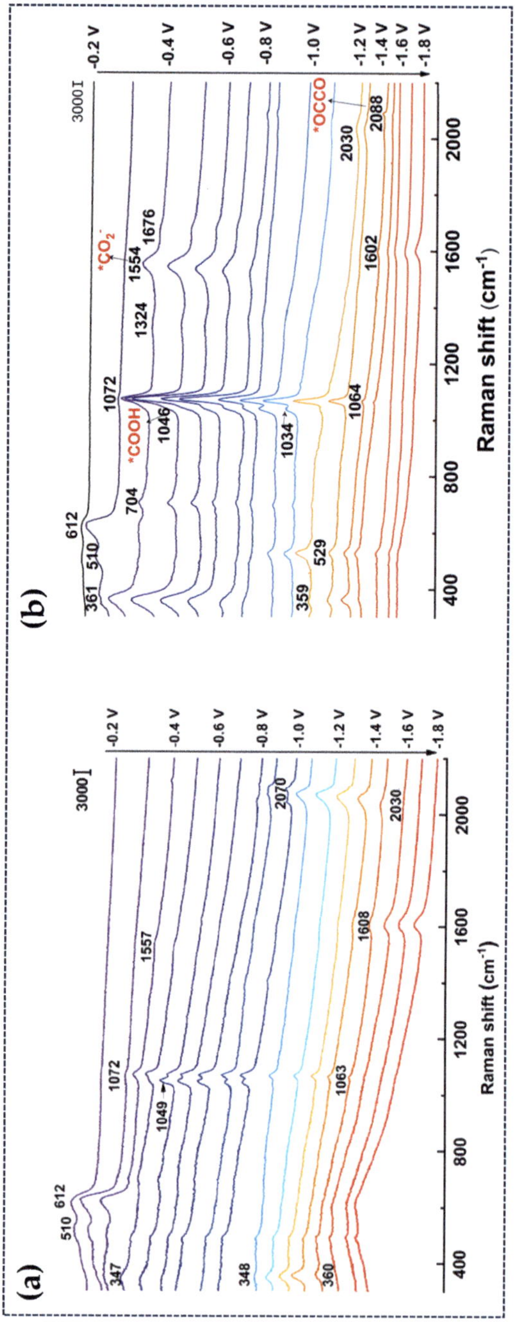

Figure 8.7 In situ Raman spectra during CO_2 reduction on (a) Cu (111) and (b) Cu(110) surfaces in a CO_2 saturated $KHCO_3$ solution. The figures are reproduced from [66] with permission from ACS.

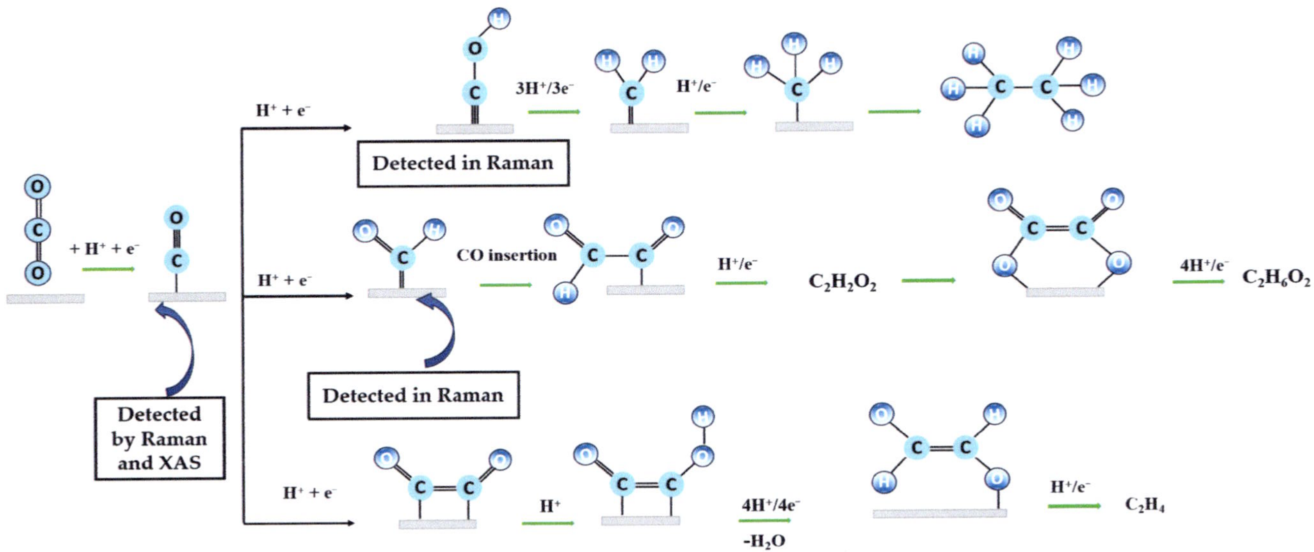

Figure 8.8 Possible reaction pathways for C–C coupling in electrochemical CO_2 reduction to form C_{2+} products on the Cu surface.

8.6.2.2 In situ XAS

Operando XAS has enabled the monitoring of the dynamic reversible conversion of IM species, enhancing our understanding of the catalytically active sites. Yang and co-worker synthesized bicentric copper complexes, such as Cu-1, Cu-2, and Cu-3, and studied their oxidation states and coordination environments during in situ CO_2RR.[68] XANES showed that copper is in +2 state in all three synthesized complexes. FT-EXAFS detected that the average length of the first shell Cu–O/N bonding in Cu-1 was slightly shorter than that in Cu-2 and Cu-3. This is due to the strongly coordinated Cu atoms to N atoms while only weakly interacting with the shared oxygen. In contrast, the contribution of oxygen to the single Cu–N/O coordination in Cu-2 was more pronounced, indicating a stronger ligand field. These findings suggest that both Cu-3 and Cu-1 experience a greater intramolecular strain, which may compromise their electrochemical and thermodynamic stability.[68]

To monitor the Cu evolution, the XANES spectra were recorded with Cu-1. As the applied cathodic potential was increased, the characteristic signals for Cu^{2+} were decreased and the signals of Cu^0 were observed even at low overpotentials. The corresponding FT-EXAFS at all applied potentials showed significant Cu–Cu bonding. During the entire testing period, the Cu–N/O coordination at R = 1.5 Å remained intact, indicating that not all Cu centers were removed from the complex. It is important to note that the presence of Cu^0 state and Cu–Cu bonds in the post-electrolytic sample indicates that the reduction of Cu in Cu-1 was irreversible. Additionally, the electrochemical stability of Cu-2 was evidenced by the FT-EXAFS and in situ XANES. The in situ XAS of Cu-3 also signify the reduction of Cu^{2+} to Cu^0 under cathodic potential. The XANES and FT-EXAFS features at negative potential closely resemble those to copper foil, showing no detectable Cu–N signals.

A similar structural evolution of Cu-1, characterized by the emergence of Cu clusters during CO_2RR, was observed using time-lapse EXAFS at a constant potential of –1.2 VRHE. The presence of Cu^0 and Cu–N/O species in the XANES and FT-EXAFS suggests that the Cu centers in the complex were largely reduced within 8 minutes.

8.6.3 Detection of Active Species of N_2 Reduction

Similar to the CO_2 reduction reaction, in situ RS and XAS studies are being utilized to investigate the actual active sites of the catalysts.

8.6.3.1 In situ Raman spectroscopy

To monitor the changes in surface chemical bonds of the MoO_{3-x}/MXene catalyst, Guo and co-workers used in situ RS (Fig. 8.9).[69] During the N_2RR process, the appearance of a peak around 835 cm^{-1}, attributed to Mo–N stretching vibrations, suggests N_2 chemisorption and activation on the oxygen vacancy-induced unsaturated Mo sites MoO_{3-x} and MoO_{3-x}/MXene. This peak gradually intensified with increasing reaction time (Fig. 8.9). However, when the reaction time was extended, OV-poor MoO_3 and MoO_3/MXene showed no or a faint Mo-N stretching peak, indicating poor N_2 activation (Fig. 8.9).[69]

Shen et al. showed an in situ RS using a Rh_1/MnO_2 catalyst in DE (dilute electrolyte) and salt electrolyte (SE).[70] In the time-dependent in situ RS, the chemisorption of N species on Rh sites produced a distinct Rh–N vibrational band at approximately 480 cm^{-1}. As the reaction time increased, the peak intensities of the Rh–N vibrational band also increased, indicating effective N_2 activation and hydrogenation, which led to the formation of more stable -NH_x IMs on the Rh sites. This process enhances the kinetics of the N_2RR and improves its performance SE. In contrast, only a weak Rh–N vibrational band was observed in DE, indicating that this environment is not favorable for N_2 activation and hydrogenation on Rh sites.

This limitation leads to the formation of restricted -NH_x IMs involved in the NRR process, resulting in low N_2RR activity in DE. The enhanced N_2RR performance of Rh sites in water-in-salt electrolytes (WISE) was further supported by the absence of any Rh–N vibrational band in the RS under Ar conditions. These findings from operando/in situ spectroscopy demonstrate that WISE facilitates improved N_2 activation and hydrogenation on active Rh sites in Rh/MnO_2, leading to higher NRR activity and selectivity.

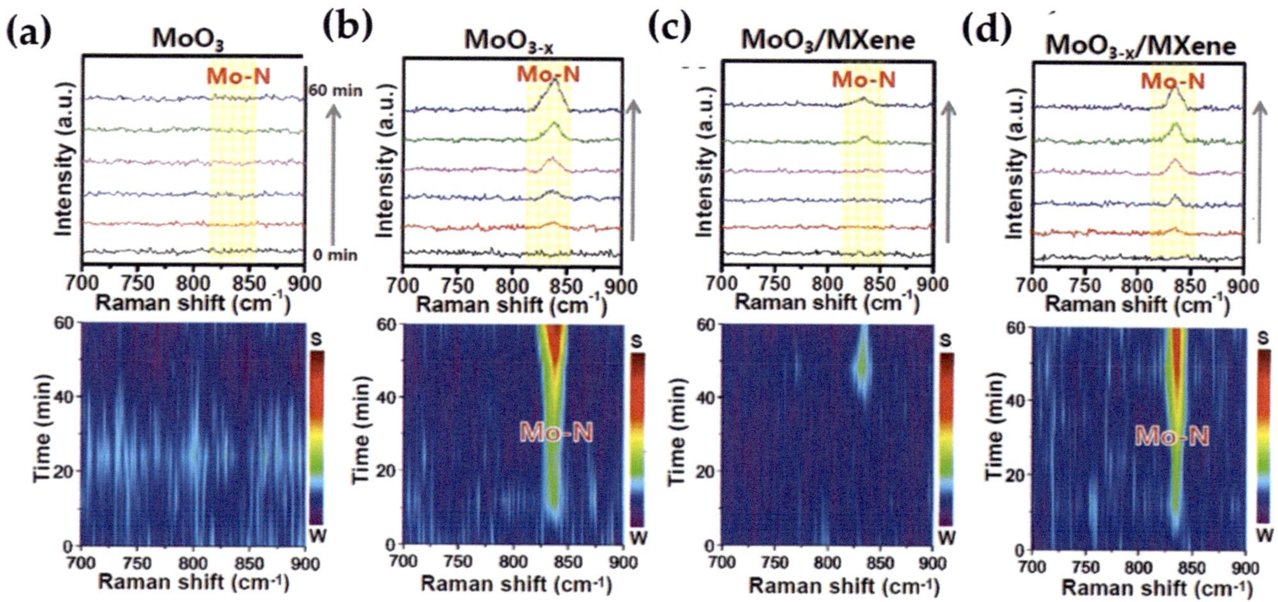

Figure 8.9 Raman spectra (a–d) of the catalysts during the N_2RR at −0.4 V_{RHE}. The figures are reproduced from [70] with permission from ACS.

8.6.3.2 In situ XAS

To gain a deeper understanding of the distinct NRR behaviors of Rh_1/MnO_2 in DE and WISE, operando XAS was conducted under the working NRR conditions.[70] Initially, the electronic and structural evolution of the Rh active centers during the NRR process was monitored. As the potential was increased, the Rh K-edge XANES in WISE showed a shift of the absorption edge toward higher energy, accompanied by an increase in the intensity of the white line. This indicates a rise in the valence state of the Rh atoms due to electron transfer from Rh to N_2.

This indicates that single-atom Rh sites can effectively transfer electrons to the adsorbed N_2 through π back-donation, and enhancing N_2 activation and dissociation. As the potentials increase, the operando Rh K-edge EXAFS in WISE reveal that the Rh-O bond gradually elongates and the intensity of the Rh-O peak increases.

The corresponding WT profiles further illustrate this change, showing a significant high-k shift of the Rh-O intensity, indicating the extended Rh-O bond. This elongation of the Rh-O bond was attributed to the chemisorption of N_2 and chemical IMs on the Rh sites, which distorts the Rh-O conformation.[71] The increased intensity of the Rh-O peak is attributed to the contribution of an emerging Rh–N bond formed by the binding of N species to Rh, further confirming the successful chemisorption of N species on Rh sites. The slow kinetics of the NRR reaction in DE, which leads to its low NRR activity, is reflected in the minimal changes observed in the operando Rh K-edge XANES, EXAFS, and WT profiles under the same NRR conditions. These results starkly contrast with the operando findings in WISE.

8.7 Challenges

(i) Despite significant advancements in optimizing electrocatalysts for the OER, HER, N_2RR, CO_2RR, and AOR, several critical challenges remain unresolved. Currently, multiple mechanistic models can explain OER, HER, N_2RR, CO_2RR, and AOR, yet the transformation process and the formation of different reaction IMs remain unclear.

(ii) Although the catalytic performance of transition metal (TM)-based catalysts has improved significantly, commercialization of these catalysts is still distant because of their poor durability. Therefore, a comprehensive and practical evaluation of these catalysts is essential to bridge the gap between laboratory and real-world applications in energy production. For industrial applications, long-term stability is crucial; however, even highly active catalysts designed for the lab scale studies can only sustain for a few hundred hours under industry conditions. Therefore, it is important to identify the major causes of catalyst degradation. It is essential to understand the relationship between the material's microstructure and the macroscopic environment, including temperature, electrolyte type, and cell type, under operating conditions.

(iii) Precise control over the reaction process and the rational design of catalysts can be achieved by the understanding of interfacial reactions and the catalytic mechanisms. As the complexity of reaction substrates increases, the reaction process becomes more intricate, requiring additional in situ/operando characterizations to detect the catalytic active sites and the mechanism of its formation. To thoroughly examine the IMs and final products formed during the reaction, advanced theoretical simulations should be combined with experimental studies. This approach provides effective and scientifically grounded guidance for the development of innovative catalysts.

(iv) During electrochemical reactions, the electrolyte, applied potential, and reaction products immediately alter the catalyst surface, leading to structural and compositional changes that directly influence catalytic performance.

(v) Additionally, costs related to raw material transportation, catalyst synthesis, and other relevant factors must be considered. Traditional methods of producing nanomaterials are often energy, manpower, and cost-intensive. Other important factors include the availability of raw materials and renewable energy sources, as well as industrial and commercial demands, and site-specific variations in climate and geography.

8.8 Conclusions

In this chapter, we present advanced techniques, particularly in situ XAS and RS, for identifying the actual active catalytic species responsible for the OER, HER, N_2RR, CO_2RR, and AOR. Specifically, the efficiency of energy conversion relies on electrocatalytic reactions involving small molecules, such as HER, OER, and AOR. However, investigating the structural changes that occur during electrocatalysis has proven to be challenging. In situ XAS provides compelling insights into the detailed mechanisms of electrocatalytic water splitting, enabling the identification of active species, monitoring the dynamic structural evolution of electrocatalysts, and observing steady-state reaction IMs. This chapter provides a comprehensive overview of in situ and ex situ characterization techniques for OER, HER, N_2RR, CO_2RR, and AOR electrocatalysts. It emphasizes how these methods enhance our understanding of potential phase transitions, the identification of active sites, and the underlying mechanisms of reactions. Theoretical studies that analyze complex reaction processes using simplified models often struggle to accurately simulate the actual structures and chemical reactions of catalysts. Therefore, in situ and operando techniques are extremely important to understand the structural features of the electrocatalysts in the atomic level and to design more efficient catalyst systems in the future.

References

1. Jiao, Y., Zheng, Y., Jaroniec, M., and Qiao, S. Z. (2015). Design of Electrocatalysts for Oxygen- and Hydrogen-Involving Energy Conversion Reactions. *Chem. Soc. Rev.*, **44**, 2060–2086.
2. Kale, M. B., Borse, R. A., Mohamed, A. G. A., and Wang, Y. (2021). Electrocatalysts by Electrodeposition: Recent Advances, Synthesis Methods, and Applications in Energy Conversion. *Adv. Func. Mater.*, **31**, 2101313.
3. Singh, B., Mannu, P., Huang, Y. C., Prakash, R., et al. (2022). Deciphering Ligand Controlled Structural Evolution of Prussian Blue Analogues and Their Electrochemical Activation during Alkaline Water Oxidation. *Angew. Chem. Int. Ed.*, **61**, 2211585.

4. Wang, J., Wei, J., An, C., Tang, H., et al. Electrocatalyst Design for the Conversion of Energy Molecules: Electronic State Modulation and Mass Transport Regulation. *Chem. Comm.,* **58**, 10907–10924.

5. Zhang, L., Shi, W., and Zhang, B. (2017). A Review of Electrocatalyst Characterization by Transmission Electron Microscopy. *J. Energy Chem.,* **26**, 1117–1135.

6. Lai, F., Shang, H., Jiao, Y., Chen, X., et al. (2020). Recent Progress and Perspective on Electrocatalysis in Neutral Media: Mechanisms, Materials, and Advanced Characterizations. *Interdisciplinary Mater.*, **3**, 492–529.

7. Van S. W. (2023). The Necessity for Multiscale In Situ Characterization of Tailored Electrocatalyst Nanoparticle Stability. *Chem. Mater.*, **35**, 386–394.

8. Chen, S., Ma, L., Huang, Z., Liang, G., et al. (2022). In Situ/Operando Analysis of Surface Reconstruction of Transition Metal-Based Oxygen Evolution Electrocatalysts. *Cell Rep. Phys. Sci.*, **3** .100729.

9. Malkani, A. S., Anibal, J., Chang, X., and Xu, B. (2020). Bridging the Gap in the Mechanistic Understanding of Electrocatalysis via In Situ Characterizations. *IScience*, **23**, 101776.

10. Zhao, Y., Saseendaran. D. P. A., Huang, C., Triana, C. A., et al. (2015). Oxygen Evolution/Reduction Reaction Catalysts: From In Situ Monitoring and Reaction Mechanisms to Rational Design. *Chem. Rev.,* **123**, 6257–6358.

11. Zhang, W., Yang, Y., Tang, Y., and Gao, Q. (2022). In-Situ Reconstruction of Catalysts in Cathodic Electrocatalysis: New Insights into Active-Site Structures and Working Mechanisms. *J. Energy Chem.*, **70**, pp 414–436.

12. Yao, Y., Zhao, G., Guo, X., Xiong, P., et al. (2024). Facet-Dependent Surface Restructuring on Nickel (Oxy)Hydroxides: A Self-Activation Process for Enhanced Oxygen Evolution Reaction. *J. Am. Chem. Soc.,* **146**, 15219–15229.

13. Menezes, P. W., Indra, A., Gutkin, V., and Driess, M. (2017). Boosting Electrochemical Water Oxidation Through Replacement of O_h Co Sites in Cobalt Oxide Spinel with Manganese. *Chem. Commun.*, **53**, 8018–8021.

14. Meng, H., Chen, Z., Zhu, J., You, B., et al. (2024). In Situ Amorphization of Electrocatalysts. *Adv. Funct. Mater.*, **34**, 2405270.

15. Tang, J., Liu, X., Wang, X., Chi, J., et al. (2024). Anionic Vacancy Engineering in the Development of Electrocatalysts for Hydrogen Evolution Reaction: A Review. *Inorg. Chem. Front.*, **11**, 5384–5413.

16. Ma, P., Jia, C., Zhang, Z., and Bao, J. (2024). Structural Reconstruction Chemistry for Alkaline Oxygen Evolution Catalysts. *ChemElectroChem*, **11**, 202400076.

17. Hales, N., Schmidt, T. J., and Fabbri, E. (2023). Reversible and Irreversible Transformations of Ni-Based Electrocatalysts during the Oxygen Evolution Reaction. *Curr. Opin. Electrochem.*, **38**, 101231.

18. Kou, Z., Li, X., Zhang, L., Zang, W., et al. (2021). Dynamic Surface Chemistry of Catalysts in Oxygen Evolution Reaction. *Small Sci.*, **1**, 2100011.

19. Wang, J., Hsu, C. S., Wu, T. S., Chan, T. S., et al. (2023). In Situ X-Ray Spectroscopies beyond Conventional X-Ray Absorption Spectroscopy on Deciphering Dynamic Configuration of Electrocatalysts. *Nat. Commun.*, **14**, 42370–42378.

20. Vitale-Sullivan, M. E., Chang, A., Chou, K.-H., Feng, Z., et al. (2023). Surface Transformations of Electrocatalysts during the Oxygen Evolution Reaction. *Chem. Phys. Rev.*, **4**, 021309.

21. Li, J., Tian, W., Li, Q., and Zhao, S. (2024). Acidic Oxygen Evolution Reaction: Fundamental Understanding and Electrocatalysts Design. *ChemSusChem*, **17**, e202400239.

22. Li, J., Zeng, J., Zhao, F., Sun, X., et al. (2024). A Review on Highly Efficient Ru-Based Electrocatalysts for Acidic Oxygen Evolution Reaction. *Energy and Fuels*, **38**, 11521–11540.

23. Wu, X., Yan, Q., Wang, H., Wu, D., et al. (2024). Heterostructured Catalytic Materials as Advanced Electrocatalysts: Classification, Synthesis, Characterization, and Application. *Adv. Funct. Mater.*, **34**, 2404535.

24. Zhang, L., Shi, W., and Zhang, B. (2017). A Review of Electrocatalyst Characterization by Transmission Electron Microscopy. *J. Energy Chem.*, **26**, 1117–1135.

25. Deng, Y. and Yeo, B. S. (2017). Characterization of Electrocatalytic Water Splitting and CO_2 Reduction Reactions Using In Situ/Operando Raman Spectroscopy. *ACS Catal.*, **7**, 7873–7889.

26. Li, Y., Wu, Y., Yuan, M., Hao, H., et al. (2022). Operando Spectroscopies Unveil Interfacial FeOOH Induced Highly Reactive β-Ni(Fe)OOH for Efficient Oxygen Evolution. *Appl. Catal. B*, **318**. 121825.

27. Long, C., Han, J., Guo, J., Yang, C., et al. (2021). Operando Toolbox for Heterogeneous Interface in Electrocatalysis. *Chem. Catal.*, **1**, 509–522.

28. Corcoran, C. J., Tavassol, H., Rigsby, M. A., Bagus, P. S., et al. (2010). Application of XPS to Study Electrocatalysts for Fuel Cells. *J. Power Sources*, **195**, pp 7856–7879.

29. Cheng, W., Fan, P., and Jin, W. (2024). Visualizing the Structure and Dynamics of Transition Metal-Based Electrocatalysts Using Synchrotron X-ray Absorption Spectroscopy. *ChemSusChem*. e202401306. https://doi.org/10.1002/cssc.202401306.
30. Xie, X., Du, L., Yan, L., Park, S., et al. (2022). Oxygen Evolution Reaction in Alkaline Environment: Material Challenges and Solutions. *Adv. Funct. Mater.*, **32**, 2110036.
31. Bau, J. A., Emwas, A.-H., and Rueping, M. (2024). An Overview of Solid-State Electron Paramagnetic Resonance Spectroscopy for Artificial Fuel Reactions. *iScience*, **25**, 105360.
32. McAlpin, J. G., Surendranath, Y., Dincã, M., Stich, T. A., et al. (2010). EPR Evidence for $Co^{(IV)}$ Species Produced during Water Oxidation at Neutral PH. *J. Am. Chem. Soc.*, **132**, 6882–6883.
33. Zhang, L., Shi, W., and Zhang, B. (2017). A Review of Electrocatalyst Characterization by Transmission Electron Microscopy. *J. Energy Chem.*, **26**, 1117–1135.
34. Li, Y., Wu, Y., Yuan, M., Hao, H., et al. (2022). Operando Spectroscopies Unveil Interfacial FeOOH Induced Highly Reactive β-Ni(Fe)OOH for Efficient Oxygen Evolution. *Appl. Catal. B*, **318**. 121825.
35. Qu, J., Sui, M., and Li, R. (2012). Review Recent Advances in In-Situ Transmission Electron Microscopy Techniques for Heterogeneous Catalysis. *iScience.*, **26**, 107072.
36. Varsha, M. V. and Nageswaran, G. (2020). Operando X-Ray Spectroscopic Techniques: A Focus on Hydrogen and Oxygen Evolution Reactions. *Front. Chem.*, **8**, 00023.
37. Yao, Y., Zhao, G., Guo, X., Xiong, P., et al. (2024). Facet-Dependent Surface Restructuring on Nickel (Oxy)Hydroxides: A Self-Activation Process for Enhanced Oxygen Evolution Reaction. *J. Am. Chem. Soc.*, **146**, 15219–15229.
38. Menezes, P. W., Yao, S., Beltrán, S. R., Hausmann, J. N., et al. (2021). Facile Access to an Active γ-NiOOH Electrocatalyst for Durable Water Oxidation Derived from an Intermetallic Nickel Germanide Precursor. *Angew. Chem. Int. Ed.*, **60**, 4640–4647.
39. Cai, M., Zhu, Q., Wang, X., Shao, Z., et al. (2022) Formation and Stabilization of NiOOH by Introducing α-FeOOH in LDH: Composite Electrocatalyst for Oxygen Evolution and Urea Oxidation Reactions. *Adv. Mater.*, **35**, 2209338.
40. Wang, L., Wang, X., Xi, S., Du, Y., et al. (2019). α-$Ni(OH)_2$ Originated from Electro-Oxidation of $NiSe_2$ Supported by Carbon Nanoarray on Carbon Cloth for Efficient Water Oxidation. *Small*, **15**, 1902222.

41. Singh, B., Ansari, T., Verma, N., Huang, Y. C., et al. (2024). Evaluating the Impact of Anodic Oxidation Reactions on Water Splitting Using Prussian Blue Analog-Derived Metal-(Oxy)Hydroxides. *J. Mater. Chem. A,* **12**, 19321–19330.

42. Chen, M., Lu, S., Fu, X. Z., and Luo, J. L. (2020) Core–Shell Structured NiFeSn@NiFe (Oxy)Hydroxide Nanospheres from an Electrochemical Strategy for Electrocatalytic Oxygen Evolution Reaction. *Adv. Sci.*, **7**. 1903777.

43. Wang, L., Meng, Q., Xiao, M., Liu, C., et al. (2024). Insights into the Dynamic Surface Reconstruction of Electrocatalysts in Oxygen Evolution Reaction. *Renewables*, **2**. 272–296.

44. Bergmann, A., Martinez-Moreno, E., Teschner, D., Chernev, P., et al. (2015). Reversible Amorphization and the Catalytically Active State of Crystalline Co_3O_4 during Oxygen Evolution. *Nat. Commun.*, **6**. 8625.

45. Yan, Z., Sun, H., Chen, X., Liu, H., et al. (2018). Anion Insertion Enhanced Electrodeposition of Robust Metal Hydroxide/Oxide Electrodes for Oxygen Evolution. *Nat. Commun.*, **9**, 2373.

46. Wygant, B. R., Kawashima, K., and Mullins, C. B. (2018). Catalyst or Precatalyst? The Effect of Oxidation on Transition Metal Carbide, Pnictide, and Chalcogenide Oxygen Evolution Catalysts. *ACS Energy Lett.*, **3**, 2956–2966.

47. Chen, M., Zhang, Y., Chen, J., Wang, R., et al. (2024). In Situ Raman Study of Surface Reconstruction of $FeOOH/Ni_3S_2$ Oxygen Evolution Reaction Electrocatalysts. *Small*, **20**, 2309371.

48. Liu, M., Min, K. A., Han, B., Lee, L., et al. (2021). Interfacing or Doping? Role of Ce in Highly Promoted Water Oxidation of NiFe-Layered Double Hydroxide. *Adv. Energy Mater.*, **11**, 2101281.

49. Jing, C., Yuan, T., Li, L., Li, J., et al. (2022). Electrocatalyst with Dynamic Formation of the Dual-Active Site from the Dual Pathway Observed by in situ Raman Spectroscopy. *ACS Catal.*, **12**, 10276–10284.

50. Favaro, M., Yang, J., Nappini, S., Magnano, E., et al. (2017). Understanding the Oxygen Evolution Reaction Mechanism on CoO_x Using Operando Ambient-Pressure X-Ray Photoelectron Spectroscopy. *J. Am. Chem. Soc.*, **139**, 8960–8970.

51. Zhang, N., Feng, X., Rao, D., Deng, X., et al. (2020). Lattice Oxygen Activation Enabled by High-Valence Metal Sites for Enhanced Water Oxidation. *Nat. Commun.*, **11**, 4066.

52. Sivanantham, A., Ganesan, P., and Vinu, A. (2020). Shanmugam, S. Surface Activation and Reconstruction of Non-Oxide-Based Catalysts

through in situ Electrochemical Tuning for Oxygen Evolution Reactions in Alkaline Media. *ACS Catal.*, **10**, pp 463–493.

53. Wang, C., Zhai, P., Xia, M., Liu, W., et al. (2023). Identification of the Origin for Reconstructed Active Sites on Oxyhydroxide for Oxygen Evolution Reaction. *Adv. Mater.*, **35**, 2209307.

54. Risch, M., Ringleb, F., Kohlhoff, M., Bogdanoff, P., et al. (2015). Water Oxidation by Amorphous Cobalt-Based Oxides: In situ Tracking of Redox Transitions and Mode of Catalysis. *Energy Environ. Sci.*, **8**, 661–674.

55. Bediako, D. K., Lassalle, K. B., Surendranath, Y., Yano, J., et al. (2012). Structure–Activity Correlations in a Nickel–Borate Oxygen Evolution Catalyst. *J. Am. Chem. Soc.*, **134**, 6801–6809.

56. Fabbri, E., Abbott, D. F., Nachtegaal, M., and Schmidt, T. J. (2017). Operando X-Ray Absorption Spectroscopy: A Powerful Tool toward Water Splitting Catalyst Development. *Curr. Opin. Electrochem.*, **5**, 20–26.

57. Wang, B., Chu, S., Zheng, L., Li, X., et al. (2021). Application of X-Ray Absorption Spectroscopy in Electrocatalytic Water Splitting and CO_2 Reduction. *Small Sci.*, **1**, 2100023.

58. Ortiz Peña, N., Ihiawakrim, D., Han, M., Lassalle, K. B., et al. (2019). Morphological and Structural Evolution of Co_3O_4 Nanoparticles Revealed by in situ Electrochemical Transmission Electron Microscopy during Electrocatalytic Water Oxidation. *ACS Nano,* **13**, 11372–11381.

59. Zhao, G., Yao, Y., Lu, W., Liu, G., et al. (2021). Direct Observation of Oxygen Evolution and Surface Restructuring on Mn_2O_3 Nanocatalysts Using in situ and ex situ Transmission Electron Microscopy. *Nano Lett.,* **21**, 7012–7020.

60. Pataniya, P. M., Patel, V., Sahatiya, P., Late, D. J., et al. (2022). Hydrogen Evolution Reaction in Acidic and Basic Medium on Robust Cobalt Sulphide Electrocatalyst. *Surf. Interface*, **34**. 102319.

61. Li, J., Tian, W., Li, Q., and Zhao, S. (2024). Acidic Oxygen Evolution Reaction: Fundamental Understanding and Electrocatalysts Design. *ChemSusChem*, **17**, e202400239.

62. Chen, X., Wang, X. T., Le, J. B., Li, S. M., et al. (2023). Revealing the Role of Interfacial Water and Key Intermediates at Ruthenium Surfaces in the Alkaline Hydrogen Evolution Reaction. *Nat. Commun.*, **14**, 5289.

63. Zhu, E., Hu, W., Dong, J., Chen, J., et al. (2024). Advanced Development of in situ Characterization Technique for Electrocatalytic Hydrogen Evolution Reaction. *Adv. Sustain. Syst.*, **8**, 2400224.

64. Birdja, Y. Y., Pérez, G. E., Figueiredo, M. C., Göttle, A. J., et al. (2019). Advances and Challenges in Understanding the Electrocatalytic Conversion of Carbon Dioxide to Fuels. *Nat. Energy*, **4**, 732–745.

65. Yang, C., Wang, Y., Qian, L., Al, E. A. M., et al. (2021). Heterogeneous Electrocatalysts for CO_2 Reduction. *ACS Appl. Energy Mater.*, **4**, 1034–1044.

66. Liu, Z., Zhang, N., and Xiong, Y. (2024). In Situ Raman Characterizations for Enhanced Understandings on Electrocatalysis. *J. Phys. Chem.C*, **128**, 13651–13665.

67. Zhan, C., Dattila, F., Rettenmaier, C., Herzog, A., et al. (2024). Key Intermediates and Cu Active Sites for CO_2 Electroreduction to Ethylene and Ethanol. *Nat. Energy*. https://doi.org/10.1038/s41560-024-01633-4.

68. Yang, B., Chen, L., Xue, S., Sun, H., et al. (2022). Electrocatalytic CO_2 Reduction to Alcohols by Modulating the Molecular Geometry and Cu Coordination in Bicentric Copper Complexes. *Nat. Commun.*, **13**, 5122.

69. Wu, J., Wang, S., Ji, R., Kai, D., et al. (2024). In situ Characterization Techniques for Electrochemical Nitrogen Reduction Reaction. *ACS Nano*, **18**, 20934–20956.

70. Shen, P., Li, X., Luo, Y., Zhang, N., et al. (2022) Ultra-Efficient N_2 Electroreduction Achieved over a Rhodium Single-Atom Catalyst (Rh_1/MnO_2) in Water-in-Salt Electrolyte. *Appl. Catal. B*, **316**, 121651.

Chapter 9

Future Perspectives of Electrocatalysis

Anubha Rajput and Biswarup Chakraborty
Department of Chemistry, Indian Institute of Technology Delhi, Hauz Khas, 11016, New Delhi, India
cbiswarup@chemistry.iitd.ac.in

9.1 Overview

The electrolysis of various substances has emerged as one of the cleanest ways to generate water electrolysis-driven H_2, carbonaceous fuel by the reduction of CO_2 or CO, and to produce ammonia through nitrogen or nitrate/nitrite reduction. Biomass-derived organic compounds electro-oxidation/reduction becomes also a clean route to avoid chemical oxidants. Nevertheless, C-C coupling, C-H activation, and other tricky organic conversions can also be performed with profound selectivity under mild electrochemical conditions, making organic electrochemistry an attractive area of research for the organic chemist. On the other hand, the electrochemical ammonia oxidation reaction (AOR) is an important half-cell reaction of the

Materials Design for Electrocatalysis
Edited by Sagarika Bhattacharya and Biswarup Chakraborty

ISBN 978-981-5352-32-0 (Hardcover), 978-1-042-02558-9 (eBook)
www.jennystanford.com

DOI: 10.1201/9781042025589-9

ammonia fuel cell that has been studied by several material or molecular electrochemists. Furthermore, ammonia decontamination from wastewater through electrolysis can be a potential way of wastewater treatment. In this context, the removal of non-biodegradable organic contaminants or microplastic through electro-coagulation gains significant interest to environmental chemists of civil engineers. Among the mentioned electro-conversions, water electrolysis to hydrogen or CO_2 reduction (CO_2R) plays a big role toward the carbon neutrality target accompanied by a reduced cost of energy generation or conservation. However, replacement of thermodynamically unfavorable water oxidation at the anode with the small molecule oxidation (SMO), like urea, hydrazine, glucose, alcohols, aldehydes, sulfion, etc., have also been demonstrated as a non-conventional albeit efficient, oxidant-free, and sustainable way of producing target substances followed by production of hydrogen through paired electrolysis. The replacement of the OER at the anode with a certain SMO paired electrolysis, cell potential and FE of the electrolyzer to produce H_2 could be improved.[1] The electroreduction of another air-based pollutant SO_x is especially desirable toward the environmental remediation and is the need of the hour.

9.2 Anodic Oxidation Reactions

9.2.1 Water Oxidation Reaction

The water oxidation reaction, or OER, is important during water splitting. However, it is limited by the high-energy $4e^-/4H^+$-transfer process and, therefore, sluggish kinetics. The major challenge posed in front of the scientists is increasing the energy conversion efficiency and lowering the OER overpotential. Despite the remarkable activity of RuO_2 and IrO_2 toward the OER, they suffer a major challenge related to their cost. The majority of synthetic transition and/or non-transition metal-based catalysts are examined toward OER in an alkaline medium due to the low thermodynamic potential and comparatively faster OER electro-kinetics. Transition metal oxide catalysts like perovskites, metal oxide or (oxy)hydroxide, and spinel oxides, as well as non-oxide catalysts like chalcogenides, nitrides, phosphides, high-entropy alloys, and even monometallic catalysts, have been shown (summarized in Chapter 6) to display appreciable OER activity along with low synthesis cost.[2] However, these transition metal-based catalysts are prone to inevitably undergo in situ phase transformation during alkaline electrolysis, some being too labile to transform even in the absence of potential bias. Several methods have been adopted to improve the activity as well as stability of the electrocatalysts in acidic as well as basic media, but prior to it, a comprehensive understanding of the OER mechanism is important. The widely used methods include tuning of particle size, morphology control, oxygen vacancy creation, heteroatom doping, interface engineering, etc. Theoretical studies also helped to model the reactive surface of the catalyst, estimating band energy and mechanistic interpretation. A majority of the reported studies are well correlated by theoretical modeling. Significant progress in the material design for water splitting has made it possible to commercialize the technique to build up stacks for bulk-scale hydrogen production. However, for the stack building or water electrolyzer fabrication, noble metal-based electrode could not be substituted due to their extreme robustness.

9.2.2 Breaking the Scaling Relationship

The activity of an electrode material is often commanded by several reactivity descriptors, such as structural, electronic, and kinetic factors, which directly affect the catalysis. Scaling of these intrinsic structural and electronic properties can describe and/or predict the catalytic performance. The scaling relationship in adsorbate evolution mechanism (AEM) can help in quickly screening the catalyst, but it imposes a limit on OER activity. Switching the OER pathway to (lattice oxygen oxidation mechanism) LOM can break the scaling relationship in conventional AEM to go beyond the activity limitation of the volcano plot. The origin of the high overpotential for the OER is the strong scaling relations among the three oxygen-containing intermediates (*OH, *O, and *OOH) that impose a minimum intrinsic overpotential of about 0.37 V on all OER catalysts. Under this scaling relationship, intensive efforts have been made to propose catalytic descriptors, such as d-band center, bulk O-2p band center (e.g. occupancy, metal–oxygen covalency, electronegativity, coordination number, outer electron numbers, bulk thermochemistry, and so on) aiming to bridge the structures and catalytic behaviors of catalysts and optimize the activity to reach the apex of the volcano plot. On the other hand, some recent advances have been achieved to circumvent the scaling relation limit for further improving the OER activity beyond the state-of-the-art noble metal catalysts. It is essential to have an in-depth understanding of the material–property relationships before designing advanced materials and structures with superior electrocatalytic properties.[3]

Despite being successful at the laboratory scale, the OER catalysts face stability issues in extreme operating conditions, that is, at the industrial scale. Alongside this, the fabrication of water-splitting electrolyzers is also limited by low ionic conductivity, slow $[OH]^-$ transportation, and contamination in the electrolytes. The local acidic conditions of proton exchange membrane (PEM) electrolyzers, which are state-of-the-art, result in lower ohmic loss, faster kinetics, and higher energy efficiency. However, the metal-based OER catalysts undergo severe corrosion in such acidic conditions. Therefore, devices using neutral or mildly basic conditions need to be developed, keeping in mind the associated lower ionic conductivity, and slow reaction kinetics in the neutral media. The electrolyzers bypassing the use of expensive and fragile membranes should be developed for sustainable electrochemical H_2 production.

9.2.3 Ammonia Oxidation Reaction

Ammonia acts as a viable energy source in various fields like agriculture, pharmaceutics, and chemical industries, etc., and is produced on an industrial scale (176 million tons a year) in well-established ammonia production plants. The electrochemical oxidation of NH_3 (AOR), an H_2 carrier gas, is a sustainable method of producing green hydrogen and nitrogen, as well as the generation of electricity via direct ammonia fuel cells (DAFCs), which is a well-established and efficient technology for energy conversion. Along with N_2, $[NO_2]^-$ and $[NO_3]^-$ are readily formed due to the overlap of electrochemical potentials for the parallel pathways of

AOR, which also have significance in the agricultural, chemical, and pharmaceutical industries. Currently, technologies utilizing ammonia as feedstock for DAFC to generate hydrogen find great significance. In the context of H_2 production, the AOR is thermodynamically much more favorable, requires only 0.06 V than the OER, which requires 1.23 V.[4] However, the AOR also suffers sluggish kinetics due to the involvement of multi-electron and proton transfer. Therefore, electrocatalytically active catalysts that can accelerate the rate of reaction are highly desirable. Transition metal-based compounds act as potential catalysts. AOR with ruthenium, iron, and nickel complexes has been explored along with some transition metal-imide, and nitride compounds, which have also been studied recently. However, due to several disadvantages concerning the difficulty in the recovery of the homogeneous catalysts, the focus is shifted more toward the heterogeneous catalyst. Currently, Pt outstands other heterogeneous catalysts for AOR, but the Pt surfaces are highly susceptible to stripping/poisoning, which greatly limits their long-term use as a commercial catalyst. To be cost-effective, Pt can be alloyed with other metals like Ir, Cu, Ni, and Zn, thereby altering the surface electronic structure and altering the surface adsorption energies of reaction intermediates. Likewise, multimetallic catalysts greatly improve the catalyst performance by the synergistic interplay of various constituents. In recent times, Ni-based catalysts are most studied and are active toward ammonia oxidation as well as electrochemical urea oxidation (UO).[5] Owing to the high onset potential of Ni-based electrocatalysts, several NO_x species are often produced as side products, thereby reducing the yield of desired products. The composite catalysts of Ni with other metals like Cu, Co, Mn, etc., have been found to be successful in overcoming the catalyst efficiency and product selectivity issues. The carbon-based, metal-free catalysts are also becoming the center of focus due to low cost, high electrical conductivity, structural tunability, good selectivity, and solid stability in acid/alkaline conditions. Graphene and N-doped graphene aerogel (NGA) are a few electrocatalysts that have been successful in replacing metal-based catalysts.[6]

Alongside the development of more efficient, selective, and stable electrocatalysts for AOR, efforts are required for a comprehensive understanding of the AOR mechanism on the surface of the catalyst.

The formation of reaction intermediates of AOR, and the selectivity of certain products over others, needs a detailed analysis. DFT calculations, machine learning, and operando spectroscopic studies are the primary tools for gaining insight into the detailed mechanism and scaling up the AOR process for industrial applications. The development of catalysts for the AOR in non-aqueous electrolytes also requires attention. The majority of ammonia, a potent fuel, remains unrecovered from industrial wastewater and domestic sewage and goes wasted. Sophisticated infrastructure and techniques need to be developed for the recovery, capture, and storage of the raw materials for ammonia fuel. Infrastructures for the production, storage, and transportation of valuable oxidation products are also necessary. Ammonia oxidation-based technologies like DAFCs also suffer several drawbacks like slow reaction rates when used at low temperatures and catalyst degradation at high-temperature operation. Along with the upgradation in the existing technologies, the development of other devices and techniques based on ammonia oxidation also holds a lot of potential in addressing the environmental challenges.

9.2.4 Biomass Oxidation Reaction

Biomasses are organic wastes broadly constituting crop/vegetation residues and domestic wastes. It is an abundant H_2 carrier that can act as a potential CO_2-neutral substitute for conventional fossil fuels. However, biomass oxidation without proper carbon capture also results in CO_2 emission. Generation of the same amount of H_2 from electrochemical biomass oxidation requires ca. 3 times less energy than the OER, accompanied by the production of several value-added chemicals.[7] Some biomass constituents include glycerol (biodiesel waste), glucose/5-hydroxymethylfurfural (HMF)/methanol/ethanol (biomass-derived), and urea (waste-water derived).

Glycerol oxidation can take place at different carbon atoms in the structure, which may be slower in an acidic medium, while over-oxidation of some products may occur in an alkaline medium. Pt or Pd offer high catalytic activity but suffer deactivation due to overoxidation or CO poisoning. Gold (Au) may be used as an alternative, which offers relatively high stability and higher activity. Combining noble metals with earth-abundant metals (Ni, Fe, Co, and

Cu) can not only increase the activity and stability of the catalyst but also the selectivity toward glycerol oxidation.[8] Certain p-block elements like Bi, Sb, Sn, and Pb can also be alloyed with Pt to improve the catalyst activity and stability. HMF oxidation converges toward 2,5-difurandicarboxylic acid (FDCA) as the final oxidation product, requiring a lower potential than water oxidation. Non-noble metals like Cu show sluggish activity toward water oxidation, thereby selectively oxidizing HMF.[9] Other earth-abundant metals provide a research scope toward the selective oxidation of HMF, which is a promising strategy for improving HMF oxidation in the future. Methanol and ethanol electro-oxidation decreases electricity consumption by up to 50% as compared to water oxidation, where ethanol oxidation is comparatively difficult to achieve due to the presence of a C-C bond. Pt and Pd have been considered to be more active toward C-C bond cleavage but have also been combined with earth-abundant metals to improve the activity and stability of the catalyst. In addition to the aforementioned alcohols, cellulose, and lignin can also be considered as much more abundant and cheaper proton carriers for electrochemical H_2 production at much lower oxidation potential as compared to water oxidation. Electrochemical UO poses a promising method to obtain useful products from urea, which is generated as agricultural, domestic, as well as industrial waste in huge amounts. However, UOR is kinetically sluggish due to a $6e^-$ transfer and complex reaction steps. Pt, Rh, IrO_2, RuO_2, and Ni are the benchmark catalysts reported to be active for UO that reduce the reaction energy barrier.[10] Pt/C still remains the benchmark, and the catalyst efficiency can further be increased by techniques like active site engineering, lattice defects, electrode design for gas bubble detachment, electrolyte diffusions, etc. Further improvements in the sluggish kinetics of biomass oxidation can be done by engineering the catalysts according to the chemical structure and reaction pathway of the biomass substrates. Biomass oxidation is still in the developing stage, with the oxidation technologies confined to the laboratory scale. The alkaline biomass oxidation technologies can be enhanced further to be utilized at the industrial scale, which can be more efficient, sustainable, and serves as a transition toward net-zero CO_2 emission as compared to the conventional H_2 production methods.

9.3 Cathodic Reduction Reactions

9.3.1 CO_2 Reduction Reactions

Enormous research is being done to convert or recycle CO_2 to serve a dual purpose, that is, to circumvent the rising CO_2 concentration in the air and to produce carbon-based value-added products. Electrochemical CO_2R is a clean and sustainable way for CO_2 recycling. However, owing to the thermodynamic stability of CO_2 molecules, the reaction kinetics of CO_2R are severely compromised, and CO_2 activation requires a high potential. Due to the high reduction potential, the process is often accompanied by the formation of several side products and the competitive H_2 formation from the HER. Pure-metal-based electrocatalysts like Pt, Ni, Fe, Al, Ti, and Ga produce CO as the major reduction product, but they suffer from CO surface poisoning. Elements like Au, Ag, and Zn also produce CO as the major reduction product with higher efficiency, while Sn, Pb, In, and Hg produce formate as the major CO_2R product. This suggests that alloying different elements can considerably affect the electronics, surface structure, and binding energetics of the reaction intermediates and help in designing a more efficient electrocatalyst for CO_2R. Moreover, the particle size has also been found to affect the product selectivity in metals such as Cu, Ag, and Au.[11, 12] Besides, metal oxides/chalcogenides/pnictides-based catalysts, and metal-porphyrin-based catalysts have recently been identified as exceptionally active and stable catalysts toward eCO_2R reaction.[13] The metal-free carbon-based CO_2R catalysts are also a hot topic of research. Carbon-only catalysts like graphene or nonmetal doped (e.g., N, S, P, etc.) graphitic catalysts have been shown to exhibit enhanced stability and selectivity.

Most of the reported catalysts focus on the production of single carbon-based products (CO, CH_4, CH_3OH, HCOOH) and are not advanced enough to be commercialized. Besides improvement in the existing technologies, the catalytic processes to produce multi-carbon-based products like alcohols and hydrocarbons can be developed in the future by designing materials that promote intermediate adsorption, C-C bond formation, and offer selective reaction pathways. The lower solubility issue of CO_2 in water can be sorted by introducing ionic liquids, which offer higher CO_2 adsorption on catalysts and reduce the reaction energy. Although ionic liquids may hamper large-scale utilization due to high cost, high catalytic currents and stability still make them the potential electrolytes for future developments. Besides, gas-phase eCO2R is also being investigated using gas-diffusion electrodes (GDEs). The gas-phase eCO_2R shows up to seven times more current density than the liquid-phase electrolysis. However, the research on the GDEs is still in its infancy and holds great catalytic feasibility in the future.

9.3.2 N_2 Reduction Reaction

The electrochemical N_2 reduction reaction (NRR) to ammonia can be extremely challenging due to the exceptional stability of N_2 molecules, owing to the high dissociation energy of triple bonds.[14] The reduction of N_2 under normal conditions is unachievable, even in strongly acidic environments. The involvement of $6e^-$ and $6H^+$ in the N_2 reduction to ammonia further complicates the process energetically.[15] During the reduction, the N_2 molecules are adsorbed on the electrode surfaces and are reduced with the concurrent combination with protons to produce ammonia. However, the current electrocatalytic methods for NRR suffer major drawbacks due to a lack of sufficient activity and selectivity toward the production of ammonia. Another threat to the NRR efficiency is the competitive HER, which can even be the dominant reaction in the majority of systems, thereby severely compromising the Faradaic efficiency of the NRR reaction. Metal surfaces like Ru(0001) and Re(111), single-atom catalysts made by anchoring metal atoms on catalyst supports, and metal nanoclusters have been found to show moderate activity toward NRR. Transition-metal nitrides, carbides, and oxides have also been recognized as cost-effective and efficient catalysts. Volcano plot predicts IrO_2 to be the most promising catalyst for NRR, but it is likely to be poisoned by proton adsorption. Au as single atoms and nanoclusters has also shown great potential toward NRR. Due to the high cost of catalytically active noble metal catalysts, metal-free electrocatalysts can be used as an alternative, which, along with being cost-effective, also provide higher catalytic activity. Carbon-based nanomaterials have a high surface area and high porosity, which facilitate mass transfer at a higher rate. Heteroatom doping (e.g. N, B) in the carbon-based electrocatalysts further improves the catalyst activity by altering the electronic structure, surface structure, and introducing defects, etc. However, these catalysts may suffer stability issues. Carbon cloth treated at high temperatures has also been seen to be active toward NRR to ammonia.[16]

The overall N_2RR electrocatalysis is in the primitive stage of development. Certain attempts have been made to increase the Faradaic efficiencies, but the N_2RR process still produces ammonia at a slow rate and efficiency, which is several times lower than that required for commercialization purposes. Concerning the catalyst structure and the reaction setup, the catalyst may undergo surface poisoning and certain cell components like cell membrane may undergo decomposition during the catalysis. Therefore, the long-term stability of the catalyst and the reaction setup needs major improvement by altering the choice of electrolytes (aqueous/non-aqueous), cell setup (liquid/gas phase), and temperature/pressure optimization. Furthermore, theoretical studies and operando techniques can be utilized to fully elaborate the poorly understood N2 reduction pathway on the catalyst surface. This would help in designing the catalyst and optimizing the catalyst surface to offer more selectivity of product formation and increase the Faradaic efficiency of the desired product.

9.3.3 NO_x Reduction Reaction

Nitric oxide (NO), nitrogen dioxide (NO_2), nitrous oxide (N_2O), $[NO_2]^-$, and $[NO_3]^-$ can commonly be described as NO_x that are harmful to the environment. For the electrochemical reduction of NO_x, N_2 and NH_3 are two desired valuable products. The lower N-O/N=O bond dissociation energy in NO_x compared to N_2 dissociation energy makes it a more energy-efficient method of producing NH_3 and H_2.[17] However, the reduction of NO_x is a complex process, involving multiple reaction pathways and reaction products like N_2 and N_2H_4, along with NO_2, and N_2O. This impedes the overall Faradaic efficiency and selectivity of the reaction. Moreover, catalyst deactivation and electrode/membrane degradation are also several challenges. The single-atom catalysts designed for NO_x reduction purposes include Al, Mn, Fe, Cu, and Nb with B, N co-doped CNTs. Other single atoms like Bi and Au may suppress the competitive HER. The metal alloys can further act as better catalysts due to the combination of unique properties of individual metals. Besides, metal oxides have also been examined for electrochemical reduction of NO_x due to their stability in harsh electrochemical conditions. However, metal oxides suffer major product selectivity issues. Transition metal chalcogenides preferably form NH_3 due to strong adsorption of NO intermediate at S/Se vacancy sites, which also suppresses the competing HER. Carbon-based electrocatalysts are also active due to their electrical conductivity, stability, and corrosion resistance. CoS nanosheets are among very few metal sulfide-based catalysts reported for electrocatalytic NOR.[18] Metal sulfides/nitrides/borides electrocatalysts remain relatively unexplored toward electrochemical reduction of NO_x and need further attention.

Apart from the electrocatalyst design, the poor solubility of NO_x gases in aqueous electrolytes greatly limits the NO_x reduction efficiency. EDTA/Fe^{2+} metal complex as an electrolyte additive is shown to capture NO_x rapidly and increases the efficiency of the reduction reaction. Electrolytes like H_2SO_4, $HClO_4$, and Na_2SO_4 can efficiently solve the solubility and mass transfer issues without the use of additives.[19] The electrochemical devices also affect the reaction efficiency and product yield. The H-type cell used for NO_x reduction effectively hinders the flow of $[NH_4]^+$ to the anodic chamber, therefore inhibiting its oxidation to N_2, N_2O, or NO. However, $[NH_4]^+$ interacts with the nafion membrane and affects its stability. The flow cells, on the other hand, simultaneously remove the formed NH_3, thereby preventing further reactions from occurring. Furthermore, GDEs-based flow cells can further enhance the efficiency of the gas-phase reactions.

Besides electrochemical methods, other techniques like photocatalysis, photo-electrocatalysis, and thermal catalysis may be considered for NO_x reduction in the future. Solar NO_x to NH_3 conversion also holds great potential. Electrocatalysts with higher stability, activity, and selectivity need to be researched. Other than the catalyst, the reaction medium also needs optimization. The use of ionic liquids may increase the NO_x solubility and increase the conductivity, thereby increasing the overall efficiency of the reaction. Moreover, a stable and more efficient catalyst, as well as reaction setup, needs to be developed in the future for the commercialization of electrocatalytic NO_x reduction to yield valuable products.

References

1. Chen, Z., Han, N., Zheng, R., Ren, Z., et al. Design of earth-abundant amorphous transition metal-based catalysts for electrooxidation of small molecules: Advances and perspectives. *SusMat.* 2023, 3 (3), 290–319.
2. Song, F., Bai, L., Moysiadou, A., Lee, S., et al. Transition metal oxides as electrocatalysts for the oxygen evolution reaction in alkaline solutions: An application-inspired renaissance. *J. Am. Chem. Soc.* 2018, 140 (25), 7748–7759.
3. Hammer, B. and Nørskov, J. K. Theoretical surface science and catalysis—calculations and concepts. In *Advances in Catalysis*, Vol. 45; Academic Press, 2000; pp. 71–129.
4. Yang, Y., Kim, J., Jo, H., Seong, A., et al. A rigorous electrochemical ammonia electrolysis protocol with in operando quantitative analysis. *J. Mater. Chem. A* 2021, 9 (19), 11571–11579.
5. Rothery, R. A. and Weiner, J. H. Shifting the metallocentric molybdoenzyme paradigm: The importance of pyranopterin coordination. *J. Biol. Inorg. Chem.* 2015, 20 (2), 349–372.
6. Zhou, Y., Zhang, G., Yu, M., Wang, X., et al. Free-standing 3D porous N-doped graphene aerogel supported platinum nanocluster for efficient hydrogen production from ammonia electrolysis. *ACS Sustainable Chem. Eng.* 2018, 6 (7), 8437–8446.
7. Bambagioni, V., Bevilacqua, M., Bianchini, C., Filippi, J., et al. Self-sustainable production of hydrogen, chemicals, and energy from renewable alcohols by electrocatalysis. *ChemSusChem* 2010, 3 (7), 851–855.
8. Song, T., Gao, F., Zhang, Y., Chen, C., et al. Efficient polyalcohol oxidation electrocatalysts enabled by PtM (M = Fe, Co, and Ni) nanocubes surrounded by (200) crystal facets. *Nanoscale* 2020, 12 (17), 9842–9848.
9. Nam, D.-H., Taitt, B. J., and Choi, K.-S. Copper-based catalytic anodes to produce 2,5-furandicarboxylic acid, a biomass-derived alternative to terephthalic acid. *ACS Catal.* 2018, 8 (2), 1197–1206.
10. Boggs, B. K., King, R. L., and Botte, G. G. Urea electrolysis: Direct hydrogen production from urine. *Chem. Commun.* 2009, (32), 4859–4861.
11. Reske, R., Mistry, H., Behafarid, F., Roldan Cuenya, B., et al. Particle size effects in the catalytic electroreduction of CO_2 on Cu nanoparticles. *J. Am. Chem. Soc.* 2014, 136 (19), 6978–6986.

12. Back, S., Yeom, M. S., and Jung, Y. Active sites of Au and Ag nanoparticle catalysts for CO_2 electroreduction to CO. *ACS Catal.* 2015, 5 (9), 5089–5096.

13. Costentin, C., Drouet, S., Robert, M., and Savéant, J.-M. A local proton source enhances CO_2 electroreduction to CO by a molecular Fe catalyst. *Science* 2012, 338 (6103), 90–94.

14. Cui, X., Tang, C., and Zhang, Q. A review of electrocatalytic reduction of dinitrogen to ammonia under ambient conditions. *Adv. Energy Mater.* 2018, 8 (22), 1800369.

15. van der Ham, C. J. M., Koper, M. T. M., and Hetterscheid, D. G. H. Challenges in reduction of dinitrogen by proton and electron transfer. *Chem. Soc. Rev.* 2014, 43 (15), 5183–5191.

16. Li, W., Wu, T., Zhang, S., Liu, Y., et al. Nitrogen-free commercial carbon cloth with rich defects for electrocatalytic ammonia synthesis under ambient conditions. *Chem. Commun.* 2018, 54 (79), 11188–11191.

17. Wan, H., Bagger, A., and Rossmeisl, J. Electrochemical nitric oxide reduction on metal surfaces. *Angew Chem. Int. Ed. Engl.* 2021, 60 (40), 21966–21972.

18. Zhang, L., Zhou, Q., Liang, J., Yue, L., et al. Enhancing electrocatalytic NO reduction to NH_3 by the CoS nanosheet with sulfur vacancies. *Inorg. Chem.* 2022, 61 (20), 8096–8102.

19. Krzywda, P. M., Paradelo Rodríguez, A., Benes, N. E., Mei, B. T., et al. Effect of electrolyte and electrode configuration on Cu-catalyzed nitric oxide reduction to ammonia. *ChemElectroChem* 2022, 9 (5), 1–7.

Index